Introduction to Safety Science

Transportation Human Factors: Aerospace, Aviation, Maritime, Rail, and Road Series

Series Editor

Professor Neville A. Stanton

University of Southampton, UK

Automobile Automation
Distributed Cognition on the Road
Victoria A. Banks and Neville A. Stanton

Eco-Driving
From Strategies to Interfaces
Rich C. McIlroy and Neville A. Stanton

Driver Reactions to Automated Vehicles
A Practical Guide for Design and Evaluation
Alexander Eriksson and Neville A. Stanton

Systems Thinking in Practice
Applications of the Event Analysis of Systemic Teamwork Method
Paul Salmon, Neville A. Stanton, and Guy Walker

Individual Latent Error Detection (I-LED)
Making Systems Safer
Justin R.E. Saward and Neville A. Stanton

Driver Distraction
A Sociotechnical Systems Approach
Kate J. Parnell, Neville A. Stanton, and Katherine L. Plant

Designing Interaction and Interfaces for Automated Vehicles
User-Centred Ecological Design and Testing
Neville Stanton, Kirsten M.A. Revell, and Patrick Langdon

Human-Automation Interaction Design
Developing a Vehicle Automation Assistant
Jediah R. Clark, Neville A. Stanton, and Kirsten Revell

Assisted Eco-Driving
A Practical Guide to the Design and Testing of an Eco-Driving Assistance System (EDAS)
Craig K. Allison, James M. Fleming, Xingd A. Yan, Roberto Lot, and Neville A. Stanton

Introduction to Safety Science
People, Organisations, and Systems
David O'Hare

For more information about this series, please visit: https://www.crcpress.com/Transportation-Human-Factors/book-series/CRCTRNHUMFACAER

Introduction to Safety Science

People, Organisations, and Systems

David O'Hare

CRC Press is an imprint of the
Taylor & Francis Group, an **informa** business

First edition published 2022
by CRC Press
6000 Broken Sound Parkway NW, Suite 300, Boca Raton, FL 33487-2742

and by CRC Press
4 Park Square, Milton Park, Abingdon, Oxon, OX14 4RN

CRC Press is an imprint of Taylor & Francis Group, LLC

ISBN: 9780367462826 (hbk)
ISBN: 9781032231969 (pbk)
ISBN: 9781003038443 (ebk)

DOI: 10.1201/9781003038443

Typeset in Times
by Deanta Global Publishing Services, Chennai, India

This Book is Dedicated to the Memory of
Captain Alwyn ('Gordon') Vette and
Professor Stanley N. Roscoe

Contents

SECTION 1 Individuals

SECTION 2 Organisations

SECTION 3 Systems

Preface

Underlying almost all human history is the dynamic interplay between risk and opportunity. Early hunters had to balance the potential risks of confronting large wild beasts with the opportunity to obtain valuable nutrition. For thousands of years since the first Polynesians set sail across the Pacific, humans have ventured out into the unknown in pursuit of opportunities and potential rewards but at some degree of risk to themselves. In the less prosperous parts of the world, individuals take on significant risks to their own health and welfare to engage in hazardous work to provide services such as construction and raw products (e.g. the minerals cobalt and lithium) required by the more prosperous countries. In contrast to the ever-present dangers facing workers in much of the world, people in more prosperous societies expect to work and travel in almost total freedom from risk. Every technological disaster from the sinking of the *Titanic* to the Chernobyl nuclear disaster or to the Boeing 737 Max crashes is viewed with much horror and concern and accompanied by demands that these systems should be instantly 'fixed' to prevent such events from happening again.

Highly publicised disasters such as those just mentioned are of course just the 'tip of the iceberg' of industrial safety problems. Significant expenditures of time and effort are now spent in technologically advanced nations on 'health and safety' with high degrees of regulation on all hazardous industries and pursuits. Nevertheless, significant numbers of workers and travellers will experience harm in any given year. 'Fixing' these problems first requires that we understand the genesis of harm and then apply this knowledge to remedial measures. This book is about the various attempts to understand the problems of safety in modern society. These range from an exclusive preoccupation with individual failings such as 'human error' to a similarly myopic focus on organizational structures or processes as the sole source of everything that goes wrong.

For the past few decades, terms such as 'systems failure' or 'systemic failure' have become familiar as the official response to virtually everything from major technical disasters to everyday road traffic crashes. Although commonly used, these terms are actually not well understood even by people working in the fields of health and safety, whether they be psychologists, medical personnel, engineers or managers. Certainly, the scientific advances that underlie this perspective are very little known and generally not mentioned in the literature designed to aid the process of managing safety in the modern world. This book was written to guide the interested reader through the three main paradigms that have guided safety science – the individual, the organisational and the systems-based.

The book is not a practical guide or 'how to' manual for safety practitioners but is intended for those with an interest in developing a deeper understanding of how we view safety and risk in modern life. Having said that, I believe there is much here to guide the safety practitioner towards a deeper and more useful understanding of their craft, as much of what passes as purely practical knowledge is necessarily based on a foundation of more conceptual or theoretical knowledge. For example, the practical endeavours of aircraft designers and motor mechanics draw heavily, whether explicitly or by informed intuition, on their knowledge of aerodynamics and internal combustion, respectively. So it is with the theory and practice of safety.

David O'Hare,
Dunedin, August 2021.

Acknowledgements

This book could not have been written without the support of a great many people. The University of Otago has been my academic home for nearly 40 years, and the continuing support of the University and of the Department of Psychology is greatly appreciated. Professor Mike Colombo and Professor Jamin Halberstadt were particularly significant in supporting my later career development. The University of Otago has been generous in supporting my scholarly activities and providing ongoing access to essential resources. Many academic colleagues in the field have provided support and friendship, including Professor Mark Wiggins and Dr Ross St George, who have been with me from near and afar over many decades. Dr John Heydon has been especially generous in giving me the benefit of his extensive experience at the front-line of workplace health and especially for his valuable suggestions for Chapter 1. Dr Karl Bridges very kindly agreed to assist with writing the final chapter as well as providing valuable comments on earlier chapters. My thanks to my fellow academics Jackie Hunter and Ted Ruffman in particular, along with the other regular attendees at the morning coffee and quiz group: Du Kangning (Jacqui), Kong Qiuyi (Shirley), Kate Fahey, May Huang and the irrepressible Roger Yan, among them. A special thanks to all my wonderful graduate students over the years, especially Rachel Goh, Keryn Pauley, Douglas Owen and others who have gone on to outstanding careers in human factors and safety. Most importantly, without my family, Dianne, Calum and Kirstie, I wouldn't be here writing this, so my love and gratitude know no bounds.

Author biography

David O'Hare, PhD (University of Exeter), is Emeritus Professor of Psychology at the University of Otago, Dunedin, New Zealand. He is mainly known for his research in pilot cognition and decision making, attracting funding from agencies such as the Federal Aviation Administration (FAA) and NASA. He has also led research projects into general aviation accidents and injuries and the role of case-based learning in aviation safety. He co-founded the New Zealand Ergonomics Society and continues to advise on human factors and safety to a wide range of organisations and regulatory bodies. He has authored books on aviation safety such as *Flightdeck Performance* (1990, with Stanley Roscoe) and *Human Performance in General Aviation* (1999). More information can be found at safetyscience.nz.

1 Introduction to Safety

Throughout recorded history, the great civilisations of Greece, Rome, Egypt, Persia, China, South East Asia and the Indus Valley, South America, Renaissance Europe and others have constructed numerous impressive monuments. Some of these such as the Colosseum, Pyramids of Giza, Great Wall, Angkor Wat, Machu Picchu and Notre Dame remain to this day, while others have vanished into the dust of history. Each project depended on vast amounts of human labour as well as extraordinary engineering and technical skills. Life for the labourers who worked on these projects would have been, in the words of the 17th-century English philosopher **Thomas Hobbes**, largely 'poor, nasty, brutish and short'. Workers on the Great Pyramid are known to have died at a young age (30–35) many with severe bone and skeletal damage.[1] In the 21st century, the conditions of workers engaged in similar enterprises vary dramatically from one part of the world to another. It has been reported, for example, that hundreds of young migrant workers are dying each year on construction projects in Qatar.[2] More than 30 have died working on stadium construction for the FIFA 2022 World Cup alone.

THE TOLL OF TOIL

In the most developed countries of Europe, North America and Australasia, legal protections for the health and safety of workers have grown over the past couple of hundred years. Prior to the Industrial Revolution, interest in worker health and safety was patchy and mainly confined to observed links between certain kinds of work and the occurrence of particular illnesses. Lead and mercury poisoning was noted by the ancient Greeks and by Renaissance physicians who noticed that medieval scribes who were in the habit of dipping their quills in metallic ink and then putting them in their mouths were exhibiting symptoms of what we now know to be lead poisoning.[3] What is considered the very first book on occupational health by **Bernardo Ramazzini** appeared around 1700. He noted that working postures were associated with various health consequences.[4] He was probably the first person to advise workers to vary their postures and not to sit or stand for excessive periods. His warnings of the dangers of a sedentary life now seem remarkably prophetic! He also noted the negative effects of frequent repetition of movement – now commonly referred to as *repetitive strain injury* (RSI) *or occupational overuse syndrome* (OOS), which are now seen as having a substantial psychological as well as a physical component.

In the 1760s the world changed dramatically. Historians refer to the period of rapid technological developments in machinery for weaving and textiles, the use of coal for iron smelting and the invention of the steam engine as the *industrial revolution.* In particular, the discovery of how to produce cast iron along with **James Watt**'s steam engine led to the building of the first iron bridge (1779), iron boat (1787) and,

DOI: 10.1201/9781003038443-1

of course, the development of the railway industry. The growing population and need for raw materials as well as finished produce led to rapid market expansion and a new world of technological change. Inevitably, the new machinery and new ways of working led to new ways of becoming sick and injured. Child labour was exploited in factories and elsewhere. Life in the factories was 'bitter and hard'.[5] England introduced legislation to control the use of children as chimney sweeps in 1788 and this began the series of legislative controls over work and safety that have continued until the present day.

Further legislation to control the conditions of child labour appeared initially in England in 1802 and then in the form of the first Factory Act of 1833. The fact that it took legislation to ensure children under 18 did not work more than 69 hours a week provides a small glimpse into the conditions and expectations of work less than two hundred years ago.[6] This act introduced the idea of an inspectorate whose job it would be to visit premises to ensure compliance with the laws and regulations. Successive Factory Acts in the late 19th century introduced further improvements to working conditions.

SINKING OF THE *TITANIC* ON MAIDEN VOYAGE, APRIL 1912

The *Titanic* left Southampton on Wednesday, 10th April, and after calling at Cherbourg, proceeded to Queenstown, from which port she sailed on the afternoon of Thursday, 11th April, following what was, at that time, the accepted outward-bound route for mail steamers from the Fastnet Light, off the southwest coast of Ireland, to the Nantucket Shoal light vessel, off the coast of the United States. This track, usually called the Outward Southern Track, was followed by the *Titanic* on her journey.

An examination of the North Atlantic route chart shows that this track passes about 25 miles south (that is outside) of the edge of the area marked 'field ice between March and July', but from 100 to 300 miles to the northward (that is, inside) of the dotted line on the chart marked, 'Icebergs have been seen within this line in April, May and June'.

The *Titanic* followed the Outward Southern Track until Sunday, the 14th April, in the usual way. From 6 p.m. onwards to the time of the collision, the weather was perfectly clear and fine. At a little before 11.40 p.m., one of the look-outs in the crow's nest struck three blows on the gong, which was the accepted warning for something ahead, following this immediately afterwards by a telephone message to the bridge 'Iceberg right ahead'. At 11.40 p.m. on that day, she struck an iceberg and at 2.20 a.m. on the next day she foundered.[7] An estimated 1,517 of the 2,224 passengers and crew lost their lives.

The sinking of the *Titanic* was a significant international event and some immediate safety changes followed shortly thereafter. Chief amongst these were more frequent ice patrols in the North Atlantic, requirements for round-the-clock manning of the ship's radio station and stricter requirements regarding the number of lifeboats and the conduct of lifeboat drills. The most notable and long-lasting impact was the creation of an *International Convention for the Safety of Life at Sea*, subsequently referred to as 'SOLAS' in 1914. This remains one of two major international standards governing the safety of maritime operations with the other being the *International Convention on Standards of Training, Certification and Watchkeeping* (STCW; 1978). Both the SOLAS and STCW protocols are under the responsibility of the International Maritime Organization (IMO), a body established in 1948 by the United Nations.

The regulation of safety at work which began in the late 18th and 19th century in England followed a similar pattern into the 20th century. Legislation was introduced by Act of Parliament to regulate safety in a particular industry such as mining (various Mines and Quarries Acts between 1954 and 1971) or manufacturing (e.g. Factories Act, 1961). The legislation would set out various regulatory requirements and an inspectorate would be established to visit premises and assess compliance with these requirements. Non-compliance could be punished with prosecution leading to fines or other sanctions.[8] Shipping was regulated by separate Acts such as the Merchant Shipping Acts (1894–1906) which applied to the *Titanic*.

A major change in the regulation of workplace safety began in Britain in the 1970s. Although the rate of workplace injuries and fatalities had dropped dramatically from the early part of the 20th century, when it is estimated that the fatality rate for factory workers was 17.5 per 10,000 workers, to a rate in the 1960s of 4.5 per 10,000, there were still over 1,000 work fatalities and half a million injuries per annum.[9] Many people including worker trade unions were questioning whether the existing patchwork of regulations and sporadic enforcement activities was effective in protecting workers and those exposed to workplace activities. In 1970, the UK government set up an independent committee to look at health and safety at work which was chaired by **Lord Alfred Robens**, formerly chairman of the British Coal Board.

The Robens Committee Report in 1972 proposed quite revolutionary changes to the philosophy and practice of safety regulation. The committee concluded that complex regulations propped up by inspections and punishments fostered a climate in the workplace that actively, if unintentionally, worked against the interests of safety. For management, the regulations could be seen as effectively setting the maximum standards which need only be barely achieved, and that the risks of discovery and prosecution from a limited and over-stretched inspectorate were also minimal. For both employees and employers, the existing approach encouraged them to view safety as an external problem looked after by the government rather than an intrinsic issue for themselves. Accepted by the government, the Robens report became the basis of the Health and Safety at Work Act of 1974 which replaced the existing patchwork legislation under one umbrella applying to all workers. The Act also established the UK *Health and Safety Commission* (HSC) now known as the *Health and Safety Executive* (HSE) whose mission is 'to prevent workplace death, injury or ill health ... by working with dutyholders to help them understand the risks they create and how to manage them'.[10]

The philosophy advocated by the Robens Committee and enshrined in much work safety regulation ever since has been described as 'regulated self-regulation' or 'meta-regulation'. The reliance on setting minimum standards and then policing compliance with those standards by an inspectorate was rejected in favour of an approach based on setting out the general health and safety goals to be achieved, and providing support and resources to workplaces to achieve those goals. Progress would be encouraged primarily by incentives and positive reinforcement rather than by fines and punishments, although these would remain as the 'last resort'. From this brief summary, three distinct eras of workplace health and safety can be distinguished as shown in Table 1.1.

The emphasis in the 1974 legislation was that industries and businesses should be guided to take control of their health and safety performance. Measures were to be 'proportionate' and 'sensible' with respect to the risks involved. In other words, the costs of improving safety and health were to be explicitly weighed against the value of the potential benefits.

WORKPLACE SAFETY LEGISLATION AROUND THE WORLD

The United States: Economic historian **Mark Aldrich** of Smith College, a prestigious liberal arts college in Massachusetts, has written extensively about workplace safety in the development of the United States. Little evidence exists prior to the 1880s because no one in the United States much cared about it. The process of opening up and 'developing' a whole new continent, and the consequent wealth that could be achieved, encouraged a zeal for productivity and an indifference to workers: 'what is clear that nowhere was the new work associated with the industrial revolution more dangerous than in America'.[11] Comparisons between reported fatality rates for activities such as mining and railway work showed these to be twice as dangerous for workers in the United States compared with the UK.

TABLE 1.1
Three Broad Eras of Workplace Health and Safety Regulation in the UK and Elsewhere.

	'Make Your Own Rules'	'Command and Control'	'Meta-Regulation"
'Who?'	Firm or industry makes the rules	Government makes the rules and enforces compliance through inspections	Government stimulates self-regulation by requiring safety management systems (SMSs) and audits
'When?'	Industrial Revolution (1760 onwards)	Factory Acts (late 19th century) to various Acts of the 1960s and 1970s	Since UK Health and Safety at Work Act (1974). Pre-1980s, before Robens Report (1972)
'How?'	Exploitative of workers except for rare examples of benevolent owners	Reactive and punitive. Firms aim for compliance with a minimum required by regulation	Proactive. Focused broadly on the identification and management of risks and hazards. Safety auditing

Improvements began in 1893 with the Safety Appliance Act, the first Federal work safety law which mandated the use of improved braking and coupling systems for the railroads. The dangers of American mines were evident in an annual death toll of over 3,000 workers. This grim statistic led to the establishment of the US Bureau of Mines in 1910 although it was not until half a century later that another mining disaster in West Virginia led to an Act of Congress in 1969 and the establishment of the Mine Safety and Health Administration (MSHA) which regulates safety in the mining industry.[12] Although safety at work was much improved over the wild west years of the late 19th century, the annual toll of 14,000 deaths, 2.2 million disabling injuries and hundreds of thousands of known occupational diseases led to increasing pressure on the US Congress to improve workplace safety across the board. The Occupational Safety and Health Act was passed into law in 1970. This created the Occupational Safety and Health Administration (OSHA) within the US Department of Labor (easy to find at www.osha.gov) to develop and enforce safety and health regulations as well as maintaining record-keeping, training and research programmes. At the latest count (2018), the United States recorded 4,493 annual workplace fatalities, around 4.6 million injuries requiring medical attention with an estimated annual cost of US$ 170.8 billion.[13] A breakdown of the US occupational fatalities into sectors is shown in Figure 1.1.

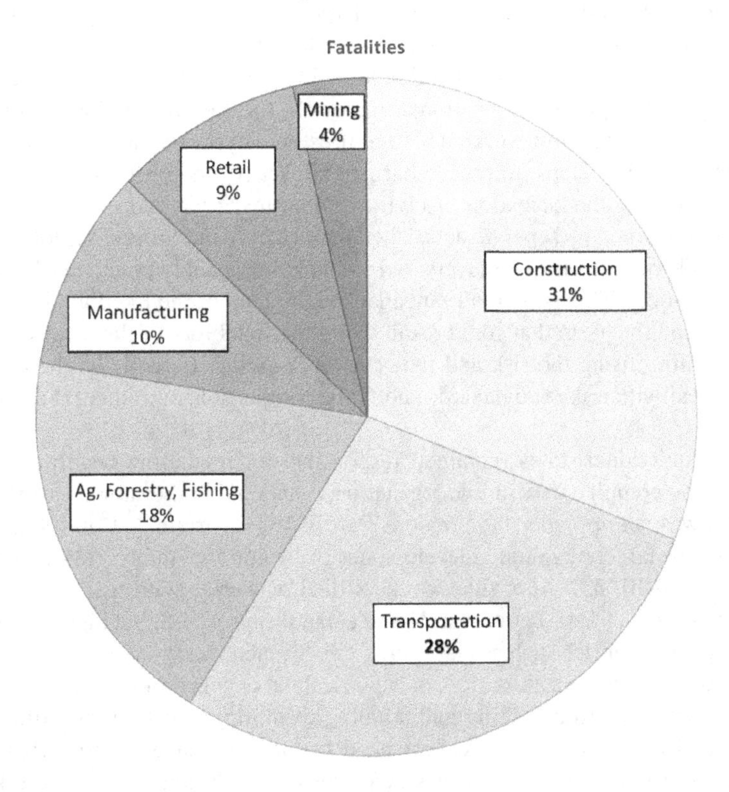

FIGURE 1.1 A breakdown of US occupational fatalities for the year 2017. Source: https://stats.bls.gov/iif/oshwc/cfoi/cftb0313.htm.

The research aspect of occupational health and safety in the United States is largely taken up by the National Institute of Occupational Safety and Health (NIOSH) which was also created by the 1970 Occupational Safety and Health Act. Unlike their counterparts in many other countries, the US approach remains largely in the 'command and control' column of Table 1.1. OSHA inspectors can assess compliance with regulatory requirements and issue 'citations' for violations. The financial penalties are desultory, ranging from a maximum of US$ 1,000 for hazards that result in injury or illness to a maximum of US$ 7,000 for hazards that create a serious risk of death or serious injury. If the employer wilfully permits such violations without taking any action then the very top maximum penalty of US$ 70,000 can in theory be levied. If the probability of being inspected is low (say 1%) then the expected cost of not complying (probability of detection multiplied by potential fine) may be as low as US$ 10! As two law professors pointed out: 'This example may seem extreme but it is not ... because most industries are seldom inspected by the agency'.[14] This stems from the fact that there are only around 3,000 inspectors covering over 140 million workers in the United States. The researchers cautiously recommend a mixture of enlisting cooperation from employers along with backup options of sanctions for non-compliance as the most effective approach to safety regulation.

Australia: Australia is one of the six largest countries on the planet although the number of employed workers (approximately 12 million) is less than that of California or Texas. The Work Health and Safety Act (2011) came into force at the beginning of 2012 providing a national framework for 'the elimination or minimisation of risks arising from work'. This requires employers to 'take positive, proactive, and systematic steps' to reduce harm. The Act covers most business activities in Australia, with the exception of civil aviation, maritime and offshore activities which are governed by separate acts. The Work Health and Safety Act followed the UK principle of requiring employers to do what is 'reasonably practicable' to ensure health and safety. This is set out to mean a weighing up of the likelihood of the risk occurring and the harm that could result with the availability of the means of eliminating or minimising the risk and the costs of so doing. Essentially, the measures taken to deal with risks and hazards should be 'proportionate' rather than 'safety at any cost'.

Whilst individual states remained responsible for regulating health and safety and ensuring compliance with the legislation, a national body known as Safe Work Australia was set up under the previous Safe Work Australia Act of 2008. Its key functions are largely strategic (development of health and safety policy) and educational. In 2010, 337 Australian workers died at work[15] with a fatality rate per 100,000 workers of 1.9 – better than New Zealand (approx. 4.8) but higher than leading European countries such as Germany (1.2), Sweden (1.23) and Denmark (1.53).[16]

New Zealand: European settlers in New Zealand engaged in a variety of hazardous work activities from fishing and whaling to mining and forestry. Attempts to manage and reduce the harms experienced in the workplace led to a Regulation of Mines Act (1874) and a Factories Act (1891). New Zealand has long taken an unusual path to injury compensation for workers with legislation introduced in 1900 to provide automatic weekly compensation for workplace injuries. This approach was enshrined with the Accident Compensation Act (1972) and the creation of the Accident Compensation Commission (ACC) to provide automatic compensation for all workplace and motor vehicle injuries.[17] New Zealand remains covered by a no-fault injury compensation scheme that provides for costs of treatment and rehabilitation as well as coverage of up to 80% of wages or salary whilst off work. The scheme receives just under a quarter of a million work-related injury claims per annum. The high likelihood of receiving some compensation or treatment may thus incentivise the reporting of even relatively minor injuries thus inflating the New Zealand injury figures relative to other comparable countries.

EXPLOSION AT THE PIKE RIVER COAL MINE, 19 NOVEMBER 2010

'The Pike River underground coal mine lies high in the rugged Paparoa Range on the West Coast of the South Island. Access to the mine workings was through a single 2.3 km stone drift, or tunnel, which ran upwards through complex geological faulting to intersect the Brunner coal seam.

On Friday 19 November 2010, at 3.45 pm, the mine exploded. Twenty-nine men underground died immediately, or shortly afterwards, from the blast or from its toxic atmosphere. Two men in the stone drift, some distance from the mine workings, managed to escape. The immediate cause of the first explosion was the ignition of a substantial volume of methane gas. The mine was new and the owner, Pike River Coal Ltd had not completed the systems and infrastructure necessary to safely produce coal. There were numerous warnings of a potential catastrophe at Pike River. One source of these was the reports made by the underground deputies and workers. For months they had reported incidents of excess methane (and many other health and safety problems). In the last 48 days before the explosion there were 21 reports of methane levels reaching explosive volumes. The reports were not heeded'.[18]

The Health and Safety in Employment Act (1992) was based on the principles proposed in the UK by the Robens Committee Report and placed an emphasis on the duties of employers, in consultation with workers and unions, to manage health and safety in the workplace. Prescriptive rules and an emphasis on compliance was replaced by an emphasis on managed self-regulation. As in the UK, the regulator's principal role was to set goals to be achieved and lay out the general principles to be followed, whilst leaving it to individual industries and businesses to develop the specific means to achieve those goals.

As noted above, New Zealand's rates of occupational injury have been higher than those in comparable countries. There have been three separate reviews of work-related fatal injuries. The first, covering the decade between 1975 and 1984, found an overall rate of 7.2 per 100,000 workers. A second study covering the next decade (1985 to 1994) showed a trend towards lower fatality rates with an overall rate of

5.03 per 100,000. Certain industries (agriculture, forestry and fishing) and occupations (agricultural pilot, helicopter pilot) had rates many times higher than the average. This is one reason for the discrepancy in rates between a country such as New Zealand, with a high proportion of workers engaged in primary industry, and other nations more oriented towards manufacturing. The most recent review, covering the period between 2005 and 2014, estimated the overall work-related fatality rate at 4.8 per 100,000 workers, a very slight improvement on the previous survey.[19] However, in contrast to the previous studies, the figures included work-related fatalities occurring in vehicle crashes on public roads. These accounted for nearly a quarter (22.5%) of all the work-related fatalities. This is not dissimilar to the proportion of work-related fatalities in Australia (approx. one-third) that occurred in motor vehicle crashes.[20]

The coal mine explosion at Pike River in 2010 led to a renewed public interest in the issue of workplace safety and led the Royal Commission of Enquiry to advocate for a wholesale overhaul of the country's health and safety legislation. The most immediate response was the (re)establishment of a strengthened mining inspectorate. This had largely dwindled over the previous decades as several different government departments became involved in permitting and overseeing the mining industry. An independent task force was established to review New Zealand's health and safety regulation and in 2013 recommended sweeping changes including the establishment of a new body called Worksafe New Zealand to regulate health and safety at work across the country.

Worksafe was set up as part of the new Health and Safety at Work Act (2015). Its functions are largely in-line with those of agencies such as the UK Health and Safety Executive and Safe Work Australia. All these bodies develop strategies for managing and improving workplace safety; promoting education and training, and providing resources for businesses to assess and manage their risks and hazards. Compliance and enforcement actions are also available as required. Worksafe New Zealand currently employs over 200 health and safety inspectors. There remain separate regulators for civil aviation (CAA) and maritime activities (Maritime NZ).

Rest of the World: A database of the world's health and safety legislation covering Albania to Zimbabwe and most places in between has been compiled by the International Labour Organization – an agency of the United Nations. Neither of the two most populous countries (China and India) has unified health and safety legislation of the kind that has been adopted in most Western nations. China, with three-quarters of a billion workers, has a large number of laws, regulations and standards in different areas administered by the State Council's department for work safety supervision and administration. There is also a National Center for Occupational Health and Safety (SAWS). India, having been colonised by the British, has a set of health and safety laws reminiscent of early post-war Britain, such as a Factories Act (1948) and a Mines Act (1952). The International Labour Organization (ILO) note that India has 'no primary occupational safety and health legislation'. There is a National Safety Council (NSC) whose activities seem to be largely training and promotional.

Accurate data on workplace injuries in China and India are difficult, if not impossible, to obtain. At best, fatal injury rates in these countries appears to be at least

20 times higher than in the UK with the *Times of India* reporting an annual death toll of 48,000 from workplace accidents in that country.[21] It is worth noting that for quite some time, Western nations have been 'outsourcing' much of their hazardous work to countries with much weaker legislative requirements and less organised labour forces. This transfers the burden of risky work so that injury figures are simultaneously decreased in Westernised nations and increased in other countries. For example, almost all of the world's ships end their lives in Bangladesh or India where the hazardous work of dismantling is undertaken by poor workers from as far afield as Nepal. Shipbreaking is an extremely hazardous occupation with immediate threats to safety from toxic fumes, asbestos and fuel as well as longer-term threats to health. As noted at the start of the chapter, these same poor countries may also provide large numbers of foreign 'guest' workers to wealthier countries to work on their more hazardous projects.[22]

RELATIVE BURDENS OF WORKPLACE INJURY AND ILLNESS

Whilst work is necessary to sustain life and provide for human wants and needs, it comes with a cost. The best estimate of the global cost of occupational injury and illness is around US$ 2.8 trillion, equating to 4% of global GDP.[23] The United States has one of the highest overall costs (US$ 250 billion) but lowest proportions of GDP (1.8%). The estimated annual figures for the UK are (£15 billion/2% GDP), Australia (AU$ 57.5 billion/5.9% GDP) and New Zealand (NZ$ 20.9 billion, 3.4% GDP). These figures can vary depending on the method of calculation and what is included as a 'cost' but nevertheless all indicate that the costs of work to society are considerable. Workplace injuries constitute a relatively small proportion of the total (13.6%, or 318,000 annual deaths globally) compared with work-related occupational diseases and ill health (86.4%, or 2,022,000 deaths annually). The relative burden of workplace injury and illness to global workplace harm is illustrated in Figure 1.2.

The likelihood of becoming ill *from* work is considerably greater than the likelihood of becoming injured *at* work. The exact ratio varies considerably from country to country and region to region. In New Zealand and the United States, there are 15–17 times as many estimated deaths from work-related illnesses each year as there are deaths from workplace accidents. For Australia, the ratio is about 32 times higher for illnesses compared to injuries, and for the UK, the ratio is 127 times higher. There is not too much that can be made of such comparisons. The UK and New Zealand report very similar rates of occupational disease but New Zealand's reported worker injury rate is so much higher that the ratios between illness and injury vary considerably.

Needless to say, there are considerable variations in the accuracy of occupational injury reporting between different countries. There are even bigger variations in occupational illness monitoring and reporting. This makes comparisons between countries and regions fraught with difficulties, and even tracking rates over time is subject to influence from these variations in reporting and accuracy of tracing. As previously noted, whether or not those occupational injuries that occur in vehicle crashes on public roads are counted in the occupational injury tallies varies between

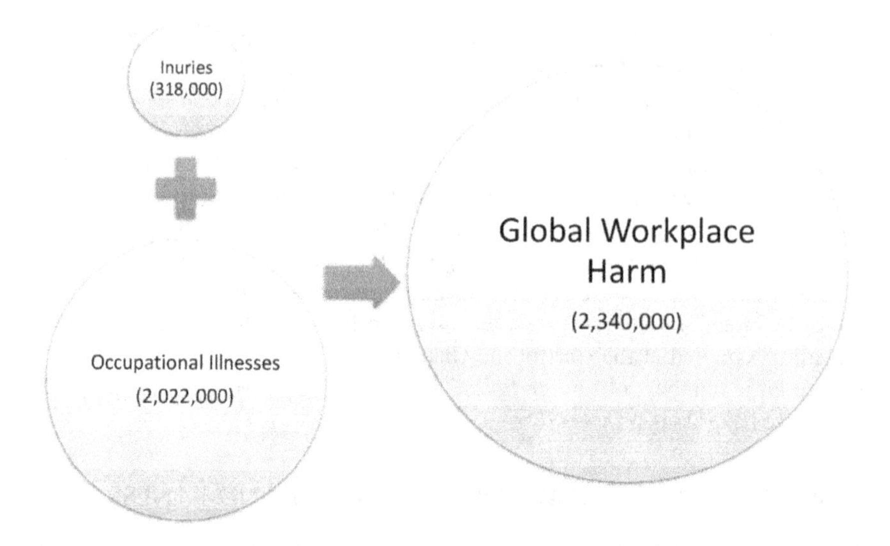

FIGURE 1.2 The relative proportions of the number of people suffering workplace injury and workplace illness to the estimated global numbers harmed annually. Figures are from the International Labour Organization (ILO).

countries and has varied over time, with the relatively recent realisation that such injuries account for a significant proportion of the total of occupational-related injury.

The best-known example of occupational illness is undoubtedly that of exposure to asbestos fibres leading to respiratory diseases such as mesothelioma and lung cancer. Bernardo Ramazzini, mentioned earlier, was the first 'modern' writer to outline some of the hazards (such as chemicals, dust, metals, etc.) faced by workers in various occupations. Later, pioneering occupational physician **Donald Hunter**, a doctor at the London Hospital for nearly forty years, catalogued a range of occupational illnesses such as 'Baker's Asthma'; 'Cotton Weaver's Bottom'; 'Painter's Palsy'; 'Coal Miner's Lung' and 'Stonecutter's Consumption'. These picturesque terms often referred to rather unpleasant and deadly diseases such as silicosis or coal-tar-induced cancer. **Lewis Carroll** based one of his characters in *Alice in Wonderland* on an occupational disease syndrome found in millinery (hat) workers. This was 'Mad Hatter Syndrome' and was caused by exposure to mercury vapours produced in the preparation of fur pelts.

Hunter published his comprehensive guide to occupational health in 1955: *The Diseases of Occupations* has long been the standard reference source in occupational medicine and a tenth edition of Hunter's book was published in 2010.[24] Many of Hunter's examples, including diseases associated with mining, asbestos and other industries using toxic substances such as mercury, lead and phosphorous raised immediate issues about regulation and control. Left unregulated, most employers downplayed or minimised the risks of their operations and the damaging effects manifesting in their workforces, and many still do. As described above, the natural reaction to this state of affairs was the steady introduction of regulations designed to improve and control workplace safety. Since the middle of the 20th century, however, the prevailing philosophy has been to de-emphasise external control and punishments in favour of an emphasis on guided self-regulation. The burning question is whether this approach has been more successful in managing the risks and hazards of work than previous approaches.

DOES 'REGULATED SELF-REGULATION' WORK?

Between the unfettered free rein of the early industrial revolution and the comprehensive state controls of the Soviet command system lies a continuum of regulatory approaches. The terms 'meta-regulation' and 'self-regulation' are used interchangeably to describe approaches whereby the targets of regulation such as airlines or chemical manufacturers are empowered to constrain their own activities in the interests of safety. Under 'meta-regulation' the government agency provides strategic direction and guidance to businesses or industries to set effective strategies to manage their own safety risks. The latter can be referred to as 'self-regulation'. Hence, 'meta-regulation' refers to the regulation of self-regulation. The emphasis in these approaches is to incentivise the targets of regulation to manage their own affairs without heavy-handed government intervention. This may be done by setting safety targets to be achieved rather than prescribing particular methods or approaches. Nevertheless, the possibility of state-controlled sanctions, such as fines or even closure, is retained if the targets should fail to exercise effective self-regulation.[25]

Has the move away from heavy-handed government regulation based on policing and punishment heralded in a new golden dawn of safety? Quite an extensive literature exists to debate the underlying philosophies of different approaches to regulation and safety, but surprisingly little actual evidence exists on the effectiveness of meta-regulation and self-regulation. One problem is conceptual in that it is difficult to precisely define what qualifies as examples as there clearly exists a continuum of possible approaches to safety regulation rather than just two or three discrete alternatives. A second problem is that there are numerous built-in differences between one industry and another and between one time period and another that make it exceedingly difficult to disentangle the exact effects of an ill-defined regulatory approach to an ill-defined set of occupational problems.

There is some evidence to suggest that a greater emphasis on self-regulation or 'regulated self-regulation' is likely to prove more effective with industries that have considerable public exposure and where the costs of low-probability/high-consequence events are particularly visible and likely to greatly impact on the activity itself. The best example would be commercial aviation where air crashes attract considerable public attention and can affect public confidence in the activity to an extent that business suffers considerably. The recent examples of the Boeing 737 Max crashes resulting in considerable loss of business for the manufacturer as well as for customer airlines operating the aircraft would be a case in point. In the aviation industry case, the goals of the regulator are highly coincident with the business goals of the industry participants and so there is a high degree of cooperation and mutual interest in achieving them. The Boeing Max example also illustrates the dangers of too high a degree of overlap between regulator and regulatee as the FAA strove to expedite the Boeing programme in the face of competition from European competitor, Airbus.

On the other hand, industries and activities either that are not in the public eye (e.g. most industrial manufacturing) or that lack the ability to effectively self-regulate (e.g. many small- to medium-size enterprises) may not be so effectively managed in this way and may perform better under approaches that are nearer to the 'command and control' end of the spectrum with more emphasis on inspection and punishment. In many countries, such as New Zealand, the majority of business enterprises are very small (60% have less than 6 employees) and large enterprises with more than 50 employees comprise less than 5% of workplaces.[26] Small workplaces lack the resources and access to specialised expertise that can be provided by large organisations.

University of Pennsylvania Law School professor **Cary Coglianese** has written extensively on the topic of safety regulation. In summing up the pros and cons of popular self-regulation strategies such as setting performance standards to be achieved, he recommends that

> Successful regulation requires that regulators understand their options when choosing among different types of regulation ... regulators need to analyse carefully the likely consequences of each relevant design choice under the conditions anticipated. They also need to review and reevaluate regulations after they are adopted to ensure that they are working as intended.[27]

As discussed above, most Western governments have adopted the Roben philosophy of the benefits of increasing self-regulation over government control as the basis for their approaches to modern safety management. One consequence of this has been the increasing importance of risk assessment in general, and *safety management systems* (SMSs) in particular, as the key elements of the regulated approach to self-regulation in the workplace and elsewhere. These will be discussed in more detail in Chapter 2 but it is worth foreshadowing the lack of convincing evidence for the effectiveness of SMSs. One expert has gone so far as to point out that 'undue faith in management systems can result in complacency or a failure to recognize important hazards and address them'.[28] This was illustrated by the explosion of the Esso natural gas plant in Australia in 1998.

ESSO LONGFORD NATURAL GAS PLANT EXPLOSION AND FIRE, 25 SEPTEMBER 1998

The Longford gas plant in Victoria to the east of Melbourne was the onshore processing facility for natural gas supplies from production platforms in the Bass Strait, which separates mainland Australia from Tasmania. On Friday, 25th September 1998, at about 12.26 a.m., a vessel in one of the three gas plants fractured, releasing hydrocarbon vapours and liquid. Explosions and a fire followed. Two Esso employees were killed. Eight others were injured. The fire was not extinguished until 27th September.[29]

A great deal of the Royal Commission of Inquiry into the Longford gas plant explosion was taken up with discussing the technical matters involved. However, the main emphasis in the report was on the management of safety at the plant. In particular, the (on-paper) very comprehensive safety management system (SMS) known formally as the 'Operations Integrity Management System' (OIMS). OIMS comprised over 140 manuals and documents and covered risk assessment, personnel and training, incident investigation and eight other elements. The Commissioners noted the gap between the system which was held up as 'world class' and 'best practice' and its implementation which seems to have been sadly lacking. In fact:

> OIMS … comprised a complex management system. It was repetitive, circular, and contained unnecessary cross-referencing. Much of its language was impenetrable. These characteristics made the system difficult to comprehend both by management and by operations personnel … Even the best management system is defective if it is not effectively implemented.

The Longford case illustrates an ever-present danger of the meta-regulation/self-regulation approach to safety, which is that resources can be almost exclusively directed towards the paperwork of safety – ensuring comprehensive manuals and documentation, whilst insufficient attention is directed towards personnel 'buy-in' and practical understanding. Even auditing by the external regulator may not necessarily ensure that there is a deep-down proactive response to the real risks and hazards. The Longford plant, for example, had just passed an external audit of their safety systems less than six months before the deadly fire and explosion.

Almost all the research on safety management and regulation has been conducted in Western capitalist societies where businesses strive to maximise profit by increasing revenue and/or reducing costs. Expenditure on safety is inherently vulnerable to cost-cutting as it can appear unproductive and unrelated to the productive goals of the enterprise. In Western countries, it is also not uncommon for health and safety issues to be used as a bargaining chip in industrial negotiations between employers and worker unions. From what little evidence there is – for example, safety in the Soviet nuclear programme (see Chapter 5) – other bureaucratic and organisational pressures can similarly act to downgrade safety in non-capitalist systems. Yet other issues arise in the context of those countries (e.g. India, China) with large labour surpluses and a ready supply of potential replacements for injured workers. In this context, large expenditures on elaborate safeguards and precautionary measures can appear particularly hard to justify.

TRANSPORTATION SAFETY

The first modern production internal combustion engine motor vehicle was developed and built in Germany by Karl Benz in 1896, although it was the American Henry Ford who pioneered mass production of the car in the early 20th century. The automobile has since proved to be one of the deadliest inventions in human history! Currently, motor vehicle crashes lead to 1.35 million worldwide fatalities annually with a further 20 to 50 million injuries – many of them long-lasting. The cost of these automobile crashes is estimated at around 3% of global GDP.[30] At the time of writing, global Covid-related deaths had reached 5.3 million. As noted above, there is some overlap between transportation deaths and injuries and occupational deaths and injuries as many of the latter take place on public roads. The burden of motor vehicle-related injury is unevenly distributed around the world with the majority of deaths and injuries occurring in low- to middle-income countries.

Other transportation modes involve considerably fewer adverse outcomes. In 2019, scheduled commercial air transport, for example, experienced just 6 fatal accidents worldwide involving 239 fatalities.[31] This is close to the average of seven fatal accidents and 292 fatalities per annum recorded over the past 5 (2015–2019) years. In contrast to the carefully kept figures for worldwide airline transport, it is much more difficult to estimate the toll from what is known as 'general aviation'. Basically, this refers to everything else that involves non-military aircraft (both fixed-wing and helicopters). Non-scheduled flights, charter flights, search and rescue, agricultural operations, logging, flight training and pleasure-flying are just some of the activities taking place under this heading. There is no doubt that these activities are several orders of magnitude more risky than the extensively regulated and controlled commercial civil aviation sector. There is no global tally of general aviation deaths and injuries but the country with the greatest amount of general aviation activity, the United States, averaged 1,282 general aviation accidents, including 223 fatal accidents per annum over the most recent decade (2009–2018). The 166 fatal accidents in 2018 involved 277 fatalities. The UK had a total of 72 fatal general aviation accidents between 2010 and 2015 involving 105 fatalities.[32]

A similar situation prevails in the maritime domain where global commercial shipping accidents and fatalities are recorded but no comprehensive information exists about local shipping, domestic ferries and recreational or pleasure craft. On the global commercial fleet, data covering the period between 1995 and 2000 showed that there have been on average 222 ship losses per annum involving 455 fatalities.[33]

Data on railway safety are even more sparse, although the International Union of Railways (UIC) publishes an annual report covering most of the countries of Europe, with additional data from Iran, Russia and South Korea. In total, these countries reported 2,361 'significant accidents' in 2018 involving well over 1,000 fatalities. Very few of these railway accidents involved derailment or trains colliding with the majority involving other individuals being hit by a train. A significant number also involved railway-crossing accidents. Recent, innovative approaches to reducing the frequency of such collisions are discussed in Chapter 4. With one of the biggest rail networks in the world, Indian Railways has over 1.4 million employees and more than 1,000 reported accidents since 2004.[34]

The US National Safety Council (NSC) has taken aim at the number of preventable deaths – both in the workplace and elsewhere. In 2017, of the 169,936 preventable deaths from unintentional injury, the overwhelming majority (40,231) occurred on the road, dwarfing maritime (466), air (385) and railway transportation (439) deaths. In fact, preventable motor vehicle deaths accounted for 93.5% of all transportation fatalities in the United States. Overall, transportation contributed exactly 25% of all the preventable unintended deaths in the United States that year.[35] It is important to note that these statistics are not to be simply equated with instances of 'human error' as most modes of transportation are heavily dependent on infrastructure – the design of roadways; the signalling and maintenance of railway tracks; airport design and air traffic control, to name but a few of the key factors beyond the control of the driver or operator.

These bald statistics disguise the true human costs of unwanted adverse outcomes in modern transportation systems. Modern aviation has been at the forefront of the application of knowledge obtained from scientific studies of individual human behaviour and this has steadily reduced the dangers of air transportation. Most associated with the scientific discipline of human factors and ergonomics, similar approaches have been taken with regard to motor vehicle and railway accidents, and most recently in the maritime domain. Although these are a key part of the safety science story, they by no means exhaust the potential contributions of modern safety science to reducing the harms related to modern transport or the larger area of preventable occupational injury.

THE 'SAFETY JOURNEY'

All biological organisms seek to survive – at least long enough to procreate. An urge to throw oneself over cliffs (as Walt Disney falsely suggested was the case for the lemming (a small rodent) in a 1958 nature documentary) would be counterproductive, to say the least. It is natural in this sense to see safety as the highly desirable state that maximises survivability and minimises the potential for harm. Indeed, most dictionaries define safety as 'freedom from harm or loss'. *Safety Science* is, therefore, the application of scientific methods, theories and principles to the promotion of a state of productive well-being, free from harm or loss, and the avoidance of a state of injury, suffering and possible death. In practice, safety science is oriented towards human activities such as work and travel where harm and loss are the unintended, but nevertheless possible consequences of engaging in the activity. It does not pertain to activities where harm is clearly intended, such as warfare. Ironically, many of the advances in Safety Science have come from military research designed to reduce the adverse effects of weapons on the people who use them!

Unlike physics, philosophy or physiology, Safety Science is not a single unified discipline. Instead, it combines influences from a variety of other separate disciplines. Chief amongst these are psychology, sociology and another multidisciplinary scientific discipline known as systems science. Many other disciplines have a significant contribution to make to the practical business of managing safety in the workplace and elsewhere. Medicine, physiotherapy, occupational therapy, physiology, engineering, ergonomics and human factors, and industrial hygiene are just some of the fields that contribute to the management of safety and harm reduction in the workplace and elsewhere. In practical terms, the management of occupational health and the management of occupational safety are separated by different training backgrounds (largely medical for the former; largely psychological or engineering-related for the latter) and spheres of operation.

This book was written to provide an introduction to the development and current state of the scientific approach to safety, increasingly known as Safety Science. The rest of the book is divided into three sections with each section covering one of the three fundamental approaches to Safety Science (see Figure 1.3), namely the psychological with its focus largely on the individual (Chapters 2–4), the sociological with its focus largely on the organisational (Chapters 5–7) and the systems approach which takes the whole system as its unit of analysis (Chapters 8–11). Each of the chapters addresses an important question about safety. Table 1.2 lists these questions in plain language.

A key ingredient of modern Safety Science is the science! This presents a considerable challenge as the elements of safety – workers and travellers; businesses and organisations; and governments and regulators cannot be taken apart and easily experimented on in the way of classical physics or chemistry. Instead, we must rely on following the principles of the scientific method – openness, accurate description, reproducibility, testability – to provide reliable evidence and credible theories. It is important to note that none of these criteria require the scientist to reduce phenomena

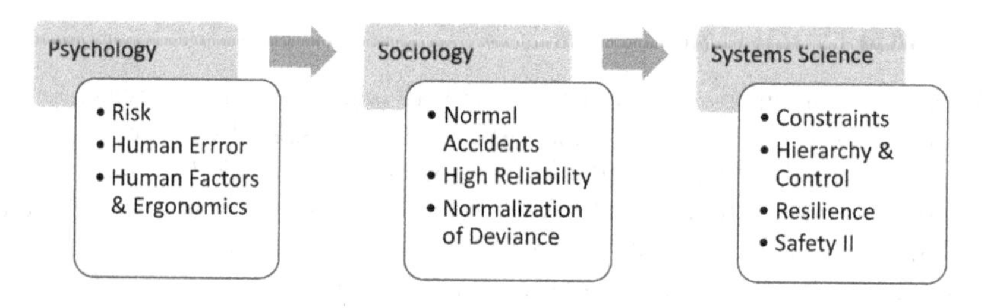

FIGURE 1.3 The 'safety journey' ahead.

TABLE 1.2
Eleven Questions about Safety That Are Addressed in the Chapters of This Book.

Chapter Number	Question
1	Why does safety matter?
2	Are most accidents caused by deliberate risk-taking?
3	Are most accidents due to human error?
4	Are people set up to fail by design flaws?
5	Are some organisations just 'accident-prone?'
6	How can high-risk organisations manage to operate safely?
7	How can organisations become less safe without anyone noticing?
8	How can working safely be achieved?
9	What needs to be controlled to prevent accidents?
10	Is safety the avoidance of failure or the achievement of success?
11	How can we understand the causes of accidents?

to their constituent parts and to subject these components to scrutiny one-by-one. Aptly known as 'reductionism' this approach is synonymous with the great 18th- and 19th- century Western advances in chemistry and physics and was the model largely adopted by late 19th- and 20th-century psychologists. Whilst undeniably productive, this approach has recently fallen out of favour in many areas such as ecology and climatology where it has become increasingly apparent that many phenomena can only be properly understood in the context of other related phenomena. Understanding the complex inter-relationships between parts (e.g. individual workers) and the whole (e.g. the network of employers and regulators) is the goal of systems science. The implications of this approach for Safety Science will be discussed in the third and final section of the book.

It is essential to note that in this context 'theory' is not the opposite of 'practice' but, in fact, good theory leads to good practice. The eminent social psychologist **Kurt Lewin** is famously quoted as saying that 'There's nothing so practical as a good theory'. In fact, an early autobiography of him was entitled *The Practical Theorist.* During the course of our journey, we will encounter many theories. Theories are usually developed to provide a coherent explanation that makes sense of a set of observed phenomena and can generate new predictions and interventions. **Alfred Wallace** and **Charles Darwin**'s Theory of Natural Selection, first proposed in 1858, is an example of a good theory that has withstood the test of time and continues to provide a satisfactory explanation of key biological phenomena.

Nothing in Safety Science (or most other fields for that matter!) approaches the stature of Darwin's theory. Many people profess authority and expertise in matters of safety but not all have approached it from a scientific point of view and there is often an absence of good empirical evidence to support some of the theories and practices that are advanced from time to time in the safety field. This is particularly true in regard to the theoretical side of safety science which has been largely lacking. One of the distinguishing features of the present book is that it is focused on reviewing both empirical evidence and conceptual theory that have been developed scientifically. Over time, most theories are superseded by newer, better theories in a continuous process of development and refinement. Any scientific theory should come with the proviso that 'this is the best account we have at the present moment'. With these caveats in mind, we can map out the journey ahead.

The remainder of this book is designed to take readers on a 'journey' across the three main perspectives on Safety Science culminating in the third leg where we encounter the most modern ideas, some of which challenge long-held notions about human error and organisational design for instance. The chapters follow a rough chronological development from some of the earliest, individualist approaches in Safety Science which began in the late 19th century, through the detailed study of organisations in the mid-20th century, to the recent emphasis on systems as a whole which has emerged strongly in the 21st century. However, this arrow-like developmental path towards the current systems perspective is only partly accurate. As will be discussed in the third section, systems science has actually been around for quite a long time. On a timeline, the earliest developments in systems science in the 1940s predate some of the organisational perspectives covered in the second section, so the development of Safety Science resists any attempt at neat chronological pigeonholing.

Both an unguarded factory saw and an exploding nuclear reactor can affect the lives of one or more individuals in painful and long-lasting ways. Both can be, and often are, reduced to the individual acts and decisions that are most closely related in time and space to the event itself. Whilst psychological approaches often seem to provide the most natural and appropriate explanations of such events, organisational and systemic levels of analysis are just as relevant and potentially more powerful. Only by gaining a comprehensive understanding of these events can we hope to develop the appropriate means of ensuring safe work and safer workplaces.

Many interesting and worthy books have been published over the last decade or so in the field of Safety Science. These range from attempts to provide comprehensive sources of information on safety and safety management from one particular perspective, such as **Ian Glendon**'s series of popular texts reviewing the psychological research and managerial practices in the area, to multiple, rather more polemical works from **Sidney Dekker**, now based in Australia, and **Eric Hollnagel**'s research collections as well as his more popular, briefer volumes presenting his own novel perspectives on modern Safety Science.[36] In contrast, the present introductory volume attempts neither to be a comprehensive source textbook nor to advance any particular theoretical framework or point of view but to bring together a wide range of empirical and conceptual advances in the field into one organised whole.

Each chapter can be read as a stand-alone guide to a particular topic, such as human error. Extensive source referencing and other information are provided by way of numbered references in each chapter, and these are gathered together under the heading of 'Notes' at the end of each chapter. These enable both the more inquisitive to check the sources for themselves and the reader who wishes to delve further into a topic to find additional material to follow up. In the course of our 'journey' we will look at the major 'landmarks' along the way. In a relatively short exposition, the aim is to be illustrative and informative rather than exhaustive, with the aim of equipping the reader with the necessary tools and perspectives to assess for themselves the progress that has been made in developing a new and exciting science of safety.

NOTES

1. https://www.newscientist.com/article/mg14920131-100-pyramids-broke-the-backs-of-workers/
2. https://www.theguardian.com/global-development/2019/oct/02/revealed-hundreds-of-migrant-workers-dying-of-heat-stress-in-qatar-each-year
3. These examples come from the first chapter in a useful text: Friend, M.A., & Kohn, J.P. (2014). *Fundamentals of Occupational Health and Safety, 6th Ed.* Lanham, MD: Bernan Press.
4. Franco, G., & Fusetti, L. (2014). Bernardo Ramazzini's early observations of the link between musculoskeletal disorders and ergonomic factors. *Applied Ergonomics, 35*, 67–70.

5. The information about the industrial revolution comes from former Cambridge Professor of Modern History J.H. Plumb's authoritative work. See: Plumb, J.H. (1950). *England in the Eighteenth Century (1714–1815).* Harmondsworth: Penguin.
6. Thomson, D. (1950). *England in the Nineteenth Century.* Harmondsworth: Penguin.
7. British Board of Trade. (1912). *British Wreck Commissioner's Inquiry Report on the Loss of the "Titanic". (s.s.).* British Board of Trade. A copy of the report can be accessed at: https://www.titanicinquiry.org/BOTInq/BOTReport/botRep01.php
8. Simpson, S. (2013). Principal health and safety acts. In J. Channing (Ed.), *Safety at Work.* Boca Raton, FL: Taylor and Francis.
9. Sirrs, C. (2015). Accidents and apathy: The construction of the 'Robens Philosophy' of occupational safety and health regulation in Britain, 1961–1974. *Social History of Medicine, 29*(1), 66–88.
10. From the HSE website: https://www.hse.gov.uk/enforce/index.htm
11. Aldrich, M. (1997). *Safety First: Technology, Labor, and Business in the Building of American Work Safety 1870–1939.* Baltimore, MD: The Johns Hopkins University Press.
12. This information comes from the second chapter ('Safety Legislation') of Friend, M.A., & Kohn, J.P. (2014). See Note 3 for full reference details.
13. These numbers come from the U.S. National Safety Council: https://injuryfacts.nsc.org/work/work-overview/work-safety-introduction/
14. Shapiro, S.A., & Rabinowitz, R.S. (1997). Punishment versus cooperation in regulatory enforcement: A case study of OSHA. *Administrative Law Review, 49*(4), 713–762.
15. MacDonald, W., Driscoll, T., Stuckey, R., & Oakman, J. (2012). Occupational health and safety in Australia. *Industrial Health, 50*, 172–179.
16. https://injuryfacts.nsc.org/international/work-related-injuries-around-the-world/work-related-deaths-around-the-world/. This is run by the U.S Government National Safety Council.
17. https://www.acc.co.nz/about-us/who-we-are/our-history/
18. Royal Commission on the Pike River Coal Mine Tragedy. (2012). *Volume 1 + Overview.* Wellington, NZ: Royal Commission.
19. The most recent study is: Lilley, R., Maclennan, B., McNoe, B.M., Davie, G., Horsburgh, S., & Driscoll, T. (2020). Decade of fatal injuries in workers in New Zealand: Insights from a comprehensive national study. *Injury Prevention.* doi: 10.1136/injuryprev-2020-043643. The previous study was: Feyer, A.-M., et al, (2001). The work-related fatal injury study: Numbers, rates and trends of work-related fatal injuries in New Zealand 1985-1994. *New Zealand Medical Journal, 114*, 22–28.
20. See reference details in Note 15 above.
21. The ILO has a useful website with information about health and safety regulation around the globe: https://www.ilo.org/dyn/interosh/en/f?p=14100:1:0::NO:::
22. See: https://www.who.int/healthinfo/global_burden_disease/estimates/en/
23. Takala, J., et al. (2014). Global estimates of the burden of injury and illness at work in 2012. *Journal of Occupational and Environmental Hygiene, 11,* 326–337.
24. Baxter, P.J., Aw, T.-C., Cockcroft, A., Durrington, P., & Harrington, J.M. (Eds.). (2010). *Hunter's Diseases of Occupations, 10 Ed.* Boca Raton, FL: CRC Press.
25. Coglianese, C., & Mendelson, E. (2012). Meta-regulation and self-regulation. *Penn Law School Public Law and Legal Theory Research Paper No 12–11.* Downloaded from http://ssrn.com/abstract=2002755
26. The data comes from an annual survey of New Zealand workplaces by the government Ministry of Business, Industry and Enterprise (MBIE): https://www.mbie.govt.nz/assets/8e3c619785/national-survey-of-employers-2016-17-summary-findings.pdf

27. Coglianese, C. (2017). The limits of performance-based regulation. *University of Michigan Journal of Law Reform, 50*(3), 525–564.
28. Gunningham, N. (2011). Investigation of industry self-regulation in workplace health and safety in New Zealand. Hawker, ACT: Gunningham and Associates. Downloaded from: HYPERLINK "https://regnet.anu.edu.au/sites/default/files/publications/attachments/2015-04/NG_investigation-industry-self-regulation-whss-nz_0.pdf" Investigation of Industry Self-regulation in Workplace Health and Safety in New Zealand (anu.edu.au)
29. This brief outline is based on the Royal Commission of Inquiry into the Longford disaster. See: Parliament of Victoria. (1999). *The Esso Longford Gas Plant Accident Report of the Longford Royal Commission.* Melbourne: Government Printer for the State of Victoria. Downloaded from: https://www.parliament.vic.gov.au/papers/govpub/VPARL1998-99No61.pdf
30. World Health Organization. (2018). *Global Health Estimates 2016: Deaths by Cause, Age, Sex, by Country and by Region, 2000–2016.* Geneva: World Health Organization.
31. https://www.icao.int/safety/Pages/Safety-Report.aspx
32. U.S. figures are provided by the annual AOPA Air Safety Institute report known as the Nall Report. The 30th report provides data for the years 2009–2018 (inc). The UK figures are from the latest review of general aviation crashes conducted by the UK Air Accidents Investigation Branch. (2018). *Annual Safety Review 2017.* Air Accidents Investigation Branch. Downloaded from: https://assets.publishing.service.gov.uk/media/5a9ff085ed915d07a3b5dd3a/AAIB_Annual_Safety_Review_2017_Lo_res.pdf. The Nall Reports can be found at: https://www.aopa.org/training-and-safety/air-safety-institute/accident-analysis/joseph-t-nall-report
33. Lloyds Register provides 'wreck returns' for shipping from 1890 onwards. See: https://hec.lrfoundation.org.uk/archive-library/casualty-returns
34. https://www.bbc.com/news/world-south-asia-10680206
35. These data came from the NSC website: https://www.nsc.org/newsroom/accidental-deaths-hit-highest-number-in-recorded-u
36. The third edition of '*Human Safety and Risk Management*' by Ian Glendon and Sharon Clarke was published in 2016 by CRC Press in Boca Raton, FL. Sidney Dekker is well known for books such as '*Just Culture: Balancing Safety and Accountability*' originally published in 2007 with a second edition in 2016 (Boca Raton, FL: CRC Press).

Section 1

Individuals

2 Risk

For most of human history, the outcome of human actions and decisions has been seen as either a matter of fate or the whim of the goods. For many people alive today, this is still the case. The idea that there is, in fact, a pattern and order to human actions, and that we are not simply playthings of the gods is, according to **Peter L. Bernstein**, the 'revolutionary idea that defines the boundary between modern times and the past'.[1] The tools that we now use to think about decision making in the face of unknown possibilities were largely developed during the 17th and 18th centuries during the period of human development referred to as the Renaissance.

Central to these new ideas was the concept of risk. Mathematicians had already begun to work out the basics of probability – the chance of throwing a total of nine when rolling two dice, for example, is twice the chance of throwing a three.[2] The wise gambler will use this knowledge to bet accordingly. The most significant breakthrough was achieved by the French mathematician **Blaise Pascal** who argued that decisions should be based not only on the probability or likelihood of an occurrence but also on their gravity or significance. This is the foundational idea of all risk management and forms the basis of most current definitions of risk.

The International Organization for Standardization (ISO), for example, has recently defined the four components of risk (see Figure 2.1) comprising a *risk source* (often referred to elsewhere as a hazard); an *event* (either something that happens unexpectedly or something that does not happen as expected); *consequences* (the gravity or significance of the outcome as Pascal first noted); and *likelihood* or chance of something happening.[3] Almost every activity involves risk and every organisation must, therefore, continuously manage risk. Whilst mathematics dictates how we should combine estimates of the likelihood of an event with the consequences of that event to produce an overall estimate of risk, it neither helps us identify the sources of risk in the first place nor determines how we perceive the likelihood of occurrence or judge the gravity of an outcome.

Identifying sources of risk and judging the likelihood of untoward events are matters of human judgement. To understand risk management, we need to understand the nature of the perceptions and judgments that underlie these core processes. What makes something appear risky? How do we judge the likelihood of an event? We will then examine whether people necessarily avoid risks or they are sometimes drawn to risk-taking – mountain climbing or investing in crypto-currency to give two examples. We will look at some of the more quantitative approaches to risk assessment before evaluating the usefulness of risk matrices in safety management systems (SMSs).

DOI: 10.1201/9781003038443-3

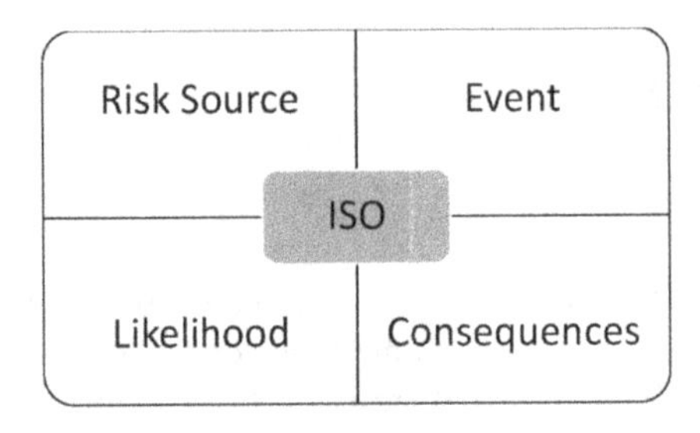

FIGURE 2.1 The four components of risk according to the International Organization for Standardization (ISO).

PERCEIVING RISK

All insurance is based on the idea of attaching a monetary value to perceived risk. Almost any event can be insured – house fire, car theft, personal accident, loss of a ship, etc. Should the event actually take place then the insurer pays out the agreed sum. In return, the insurer receives a fixed amount usually paid as an annual premium. This cost will depend on the insurer's estimate of the likelihood of the event occurring. Insurers have been active since at least the 17th century with the rapid expansion of ocean-based trade. Insurance companies sprang up everywhere by the mid-18th century.[4] To accurately price risk, one must have an accurate perception of the likelihood of these events taking place. For example, to insure against loss of life one needs to know what the likelihood of death is at any given age. This, of course, will vary with occupation, geographical location and the like.

With improved record keeping of births and deaths, it has become possible to build better estimates of life's risks. An interesting example was provided by two physicists, **Bernard Cohen** and **I-Sing Lee**.[5] Their method was to examine various sources of data for different categories of risk – e.g. 'Accidents', 'Disease', 'Tobacco', 'Marital Status', 'Geography', etc., and to express risk in terms of days of life lost. Some examples are shown in Table 2.1.

Most people find some of these data surprising as they do not entirely correspond with our usual perception of the risks of life. For example, very few people would have considered that '... not being married is one of the greatest risks people voluntarily subject themselves to'. Many people would also be sceptical of the supposedly negligible level of added risk posed by the nuclear industry. Of course, it needs to be pointed out that these data were collected in the 1970s or earlier and many things have since changed. It is also clear that marital and socioeconomic status, for example, are broad labels that may represent a constellation of other factors and are not risk sources in the sense defined earlier.

What this does demonstrate is that our perception of risk may not simply mirror the figures relied on by insurers and actuaries to guide the pricing of risk. The seminal studies of perceived risk were conducted by **Paul Slovic**, **Baruch Fischoff** and **Sarah Lichtenstein**.[6] In 1976, Paul Slovic and his colleagues founded a non-profit research organisation ('Decision Research') located in Eugene, Oregon. Since then,

TABLE 2.1
Some Actuarial Data on Life Risks from Cohen and Lee (1979).

Risk Source	Days of Life Expectancy Lost
Unmarried (male)	3,500
Smoker (male)	2,500
Heart disease (male)	2,300
30% overweight	1,300
Smoker (female)	800
Low socioeconomic status	700
All accidents (female)	297
X-rays	6
Diet drinks	2
Routine radiation from the nuclear industry	30 min

Slovic has continued to research and publish in the area of risk research over four decades amassing an extraordinary total of nearly a quarter of a million scholarly citations.

Their initial approach involved asking groups of university students and adult communities to complete questionnaire ratings of 'the likelihood of dying as a result of this activity or technology'. Their results show some matches with the actuarial statistics. For example, people accurately rated smoking as a hazardous activity. In other cases, there were marked differences. For instance, accidents were generally over-rated and diseases, generally, under-rated. Nuclear power generation was viewed with suspicion and rated as highly hazardous. Certainly, the 'psychometric approach' used by Slovic and his colleagues can be criticised as relying on people trying to answer questions that they really have little or no idea about. Lacking personal experience of most of the hazards involved, they have no option but to rely on vague general knowledge and media reports.

Delving a little more deeply, Slovic had other groups rate the characteristics they most closely associated with individual hazards. Examples of these attributes included voluntariness, controllability, familiarity, immediacy of consequences, threat to future generations and the degree to which the risks are equitably distributed, amongst others. Using standard psychometric techniques to bring out the inter-relationships in these data results in graphical representations such as shown in Figure 2.2. This shows a two-dimensional space in which the axes represent the characteristics which were most closely associated with the judgements of the hazard's riskiness.

The most important determinant of the perceived riskiness of a hazard was the degree to which the hazard evoked feelings of dread defined by catastrophic potential, lack of control, and the inequitable distribution of risks and benefits. In the figure, hazards rated highest on this dimension are located to the far right of the x-axis with those lowest on the left. The vertical axis represents ratings of familiarity with new hazards with unknown effects at the top and old, familiar hazards with known

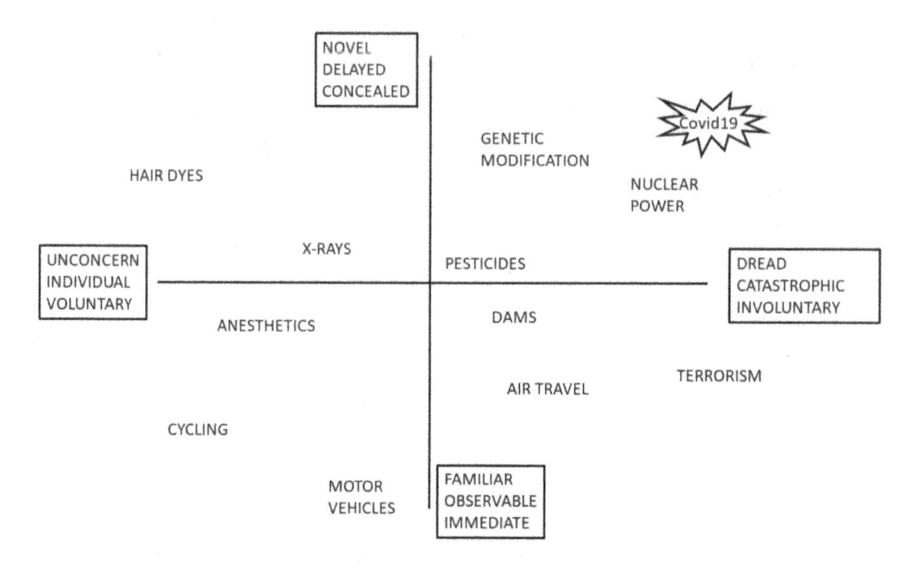

FIGURE 2.2 Dimensions of risk perception as shown in early studies of risk perception. Covid-19 would likely appear in the upper-right quadrant as a novel hazard that has evoked feelings of dread and caused catastrophic harm.

effects at the bottom. The lower-left quadrant contains highly familiar hazards that typically affect only one individual at a time such as bicycles and chainsaws. In contrast, those hazards located in the upper-right quadrant are typically newer with the potential to affect a large number of people at a time.

Armed with this 'psychometric map', we can easily predict the risk perceptions of any new biological, social or technological hazard. At the time of writing, the coronavirus or Covid-19 had appeared and rapidly spread around the globe.[7] It would rate high on dread risk of uncontrollable catastrophic potential. If asked to rate the riskiness of this new hazard, it would undoubtedly be rated 'extremely risky' as predicted by its position in the upper-right-hand quadrant of Figure 2.2. Similar, but more familiar, viral illnesses, such as measles or common influenza, which kill more than half a million people annually, would be in the lower-left quadrant. Slovic has suggested that we are especially sensitive to new threats with as-yet-unknown effects.[8] These events can be said to hold 'signal potential' as a potential warning of future catastrophic potential. It would make a certain amount of sense from an evolutionary point of view to be especially wary of these kinds of risks until their mechanisms and possible effects become better understood.

The ratings of risks used in these studies reflect the overall averages of diverse groups of individuals. Later studies began looking at individual and group differences more closely.[9] A telephone survey of 1,800 US residents provided a large enough sample to compare the risk perceptions of male and female and white and non-white. Risk perceptions of a wide range of environmental (e.g. ozone depletion), technological (e.g. nuclear power, air travel) and lifestyle (e.g. smoking, alcohol) risks were measured along with a range of attitudinal and demographic characteristics.

One group of respondents gave consistently lower ratings to virtually every risk. This was the subgroup of white males. This was true for average risk ratings as well as for the proportion rating a hazard as 'high risk' which was also always lower for white males than any other group. The researchers were able to show that this group was characterised by being more politically conservative with greater trust in institutions and authority.

This phenomenon has since become known as the 'white male effect' and has been further investigated and, more recently, linked to climate change scepticism. More recently, a diverse group of researchers (inevitably including Paul Slovic) linked risk judgements to selective information processing driven by a motivation to conform to a particular kind of social identity. For example, scepticism about the risks of guns (in the United States) might derive from a desire to confirm a particularly white, male identity rooted in gun ownership. The researchers classified their telephone survey respondents into four groups based on 'cultural worldviews'. The first was a concern for individual rather than collective interests, as reflected in questions such as 'The government interferes far too much in our everyday lives'. The second was a preference for egalitarianism over hierarchy reflected in questions such as 'We have gone too far in pushing equal rights in this country'.

The results showed the overall 'white male effect' as previously but also showed that the most extreme risk scepticism was associated with those individuals whose cultural worldview could be described as individualist and hierarchical. Such individuals are extremely hostile to gun control and highly sceptical of environmental risks due to the threat both possess for their strongly held cultural identities. Unsurprisingly, perhaps, the same groups have been shown (again within the United States) to be strongly supportive of climate change denial. As the authors put it: '... individuals are disposed selectively to accept or dismiss risk claims in a manner that expresses their cultural values'. The fragmentation of mass media and speed of dissemination of 'information' (true or false) on social media have made it increasingly easy to exist solely in a warm bath of soothing, identity-supportive ideas.

In conclusion, whilst our perception of a small number of risks such as falling from a height may be hardwired into us, our perception of almost all the risks of life and work are multiply determined by a range of psychological, social and cultural factors. These perceptions may, or as is often the case, may not coincide with the judgements of expert risk managers based on statistical and actuarial data. Drawing attention to the risks of a particular hazard, or, conversely, emphasising the relative benefits of safety precautions, may well not have the desired impact on behaviours if these assessments are incongruent with individual perceptions of these hazards. Furthermore, these perceptions can be specific to particular groups or subgroups defined by age and gender, social status, ethnicity and other social and cultural markers. Advising, communicating and managing risk are, therefore, extremely complex subjects broadly related to how an individual sees themselves in the world. Even if everyone agrees on the risks associated with a given hazard, it does not follow that everyone will choose to behave the same way towards that hazard. This reflects differences in risk tolerance or the amount of risk individuals are willing to bear.

RISK TOLERANCE

One of the most basic pieces of financial advice is that the longer your investment time horizon (e.g. a 25-year-old saving for retirement) the greater should be your tolerance of risk. Conversely, a 64-year-old should have their investments in low-risk areas. Tolerating higher risk usually results in greater volatility of outcomes – i.e. higher highs and lower lows. An individual's tolerance could be determined by both dispositional factors (e.g. a general aversion to risk) and situational factors (as in the time horizon for retirement saving).

Psychologist **Lola Lopes** has extensively investigated human behaviour in economic choices. Economists and psychologists have conducted thousands of studies where participants have to express a preference for one choice (usually expressed in terms of a probability of getting a particular outcome) over another. Although not at all representative of most real-world situations involving risk (e.g. should I continue to the top of the mountain or turn back?), these artificial tasks allow for very careful control and isolation of the relevant variables and are, therefore, valuable in generating theoretical explanations of how people approach risky situations.

Lopes has proposed that there are two key drives in human behaviour.[10] One is a drive towards safety and security. The other is a drive towards potential. The drive towards safety shows up as fear and a focus on the worst outcomes in any situation; the drive towards potential shows up as hope and a corresponding focus on the best possible outcomes in any situation. We can all show a drive towards safety and caution in some circumstances and a drive towards potential in others. Thus, Lopes presents a two-factor theory of risk-taking in which actual behaviour depends on both a dispositional characteristic (some people are generally a lot more cautious; some people are generally a lot more risk-seeking) and also a situational aspect – people can choose to be very cautious by purchasing insurance policies whilst also engaging in mountain climbing or base jumping.

There is considerable evidence that certain dispositional characteristics have clear psychobiological foundations. The area of personality classification has been a fraught one with intense debates over the exact number of personality dimensions. Although there is a current acceptance of what is known as the 'Big Five' dimensions of personality (Extraversion, Neuroticism, Agreeableness, Conscientiousness and Openness to Experience), earlier work has also emphasised characteristics such as anxiety and sensation-seeking.[11] In fact, these two, perhaps, provide the closest bases for Lopes' drives towards safety and potential. Sensation-seeking, in particular, has been extensively studied and linked to a wide range of real-world behaviours.

First described by **Marvin Zuckerberg** in the 1960s, sensation-seeking involves a drive to seek out new and exciting experiences. Sensation-seeking tendencies are measured by a self-report scale, which has been modified over the years. Correlations between scores on the Sensation Seeking Scale (SSS) and a wide variety of behaviours have been demonstrated.[12] These include involvement in extreme sports; alcohol and drug use; gambling; unprotected sex; choice of occupation; travel and certain driving behaviours. Generally, SSS scores decrease with age and are typically higher in males than females. This is certainly consistent with data showing that males are more likely than females to be admitted to emergency rooms following accidents and are much more likely (88% compared to 12%) to feature in the 'Darwin awards' – a term coined to commemorate those receiving fatal injuries whilst pursuing 'idiotic risks'.[13]

It is important to note several qualifications. First, sensation-seeking is not a drive towards risk-taking *per se*. People who undertake extreme sports are typically meticulous in preparing for their activities and regard themselves as carefully controlling as much of the risk as possible. Second, risk-taking is extremely specific to a particular domain or activity. Mountain climbers may well purchase insurance, invest in low-risk term deposits and wear their seat belts all whilst preparing to engage in activities with a relatively high risk of injury. Researchers have repeatedly demonstrated that risk-taking does not generalise across domains.[14] Finally, it is important to note that whilst tendencies to seek out thrilling experiences or avoid anxiety-producing situations may well have strong physiological foundations, biology is not destiny. In other words, behaviours depend on situational demands – including social and cultural factors – as well as dispositional ones.

Just as was noted previously with regard to risk perception, risk tolerance or risk-seeking is not a characteristic of any one group or the product of any one factor. Preference for a seemingly riskier choice in any given situation is multiply determined by a variety of dispositional (e.g. gender, anxiety, sensation-seeking, etc.) and situational factors (e.g., importance of safety versus security, time horizon, etc.). It is certainly the case, however, that the appraisal of risk can be accompanied by feelings or emotions. Both fear and hope can be experienced quite viscerally as well as cognitively.

George Loewenstein and colleagues developed a theoretical framework to explain the often-observed divergence between cognitive appraisals of risk (e.g. 'I know the probability of this flight crashing is extremely low') and the affective or emotional appraisal of the same event (e.g. 'I am terrified of getting on this plane').[15] Paul Slovic's work on the perception of risks described earlier highlighted the frequent discrepancy between public perceptions and statistical evidence with the former driven by affective responses to hazards in the form of feelings of fear and dread. Loewenstein proposed that our reactions to risk are driven largely by *anticipatory* feelings experienced at the time of risk appraisal. Such feelings can be evoked by the way in which risk is presented, for example, in a highly vivid dramatic style. Colourful reporting of the outbreak of a new threat, such as the Coronavirus ('Covid-19') in 2020, can markedly increase fear, heighten the perceived risk and lead to a number of risk-avoiding behaviours. Paul Slovic has subsequently labelled the reliance on 'risk as feelings' as the 'affect heuristic' whereby judgements and decisions about risk are made quickly and easily on the basis of one's emotional reactions rather than slowly and deliberately based on conscious reflection on probabilities and outcomes.[16]

These theoretical explanations have important practical implications for communicating and talking about risks. If behaviours that seem undesirably risky are driven by emotional or affective responses then providing information about probabilities and outcomes will have little or no effect. Adolescent smoking and fast driving would seem to be two examples where this might be the case. Public unease about nuclear power generation was never assuaged by the supply of reassuring statistics about low probabilities of malfunctions or the numbers of lives actually lost in nuclear meltdowns.

The role of affective responses in risk also seems to underlie the often-noted negative correlation in people's minds between the risks and benefits of any given action. In other words, the higher the perceived risk the lower the perceived benefit and vice versa. It follows that if an action is perceived to be highly beneficial then its riskiness automatically seems lower. This might well explain a number of workplace behaviours where the benefits of not following procedures, taking shortcuts or not using personal protective equipment (PPE) may be seen as greatly outweighing their risks. In the next section, we briefly look at whether most accidents are caused by deliberate risk-taking.

RESCUE OF THE *REINDEER*

The *Reindeer* was originally a gun vessel launched in 1883 in the UK. She enjoyed an interesting career including active involvement in anti-slavery operations. At the start of the First World War, she was converted into a salvage ship and was involved in salvaging German submarines as well as the wreck of the *R38* airship, which had crashed into the river Humber in the North West UK. She then moved to Canada, where on 12 March 1932 she and her crew of 30 were caught in a severe gale. Eventually, she began to leak and her Captain called for assistance.

This was provided by the Canadian Pacific liner *Montcalm*, which was heading for the safety of port but on receiving the distress call from the *Reindeer*, and at considerable risk to her own passengers and crew, turned back into the storm to rendezvous with the *Reindeer*. By this time, the seas were so huge that the *Reindeer* could not launch her lifeboats. Eventually, a boat was sent out from the *Montcalm* and with considerable difficulty and at great risk, all 30 men were safely transferred to the *Montcalm*. Nine officers later received Sea Gallantry medals for their bravery.[17]

RISK-TAKING AND ACCIDENTS

Although it seems intuitively reasonable that taking unnecessary risks should be associated with accidents and injuries, the scientific literature does not provide much support for this. Many studies of road traffic accidents have found a link between the tendency to report willingness to commit violations of traffic rules (e.g., tailgating, speeding, etc.) with self-reported accident involvement but not with objective records of crash data. The link between self-reports of risk-taking and accidents might be inflated by a tendency to report similar levels of both to appear consistent. Additionally, self-reported accident data have often been shown to be rather unreliable. On the other hand, some objectively observable risky behaviours, such as insufficient following headway to the vehicle ahead, have been found to be associated with drivers' accident involvement.

Studies of risk-taking in occupational areas have generally failed to show a clear link between questionnaire measures of risk-taking and injuries and accidents at work.[18] As noted above, there is little consistency in risk-taking across domains so a tendency to take more risks in one area (e.g. financial) would be unlikely to predict risk-taking in another (personal or working habits). The question of exactly what constitutes risk-taking is another problematic area for this line of research. Risks, in the sense of a non-zero probability of harm, are inherent in almost all human activity; so there is no risk-free way of engaging in forestry, fishing, flying or anything else.

Willem Wagenaar, Professor of Experimental Psychology at Leiden University in the Netherlands, was a highly distinctive and creative cognitive psychologist interested in paradoxes of rationality such as that involved in gambling, where gamblers know that the odds are (literally) stacked against them but gamble (and lose) nonetheless. His most famous scientific publication was a study of his own memory spread over a 6-year period, but he was, perhaps, most widely known for his role as an expert psychological witness in the trial of John Demjanjuk, a Ukrainian-born US citizen who was accused initially of having served as a guard known as 'Ivan the Terrible' at Treblinka and later as a guard at Sobibor, where he was an accessory to the murder of over 28,000 people brought in to the camp by train from the Netherlands.

Wagenaar has carefully examined the question of whether deliberate risk-taking is a common feature in accidents as is widely believed. For this to be the case, he argues, those involved have made some sort of consideration of the risk associated with different courses of action and must then have specifically chosen one course of action with a greater degree of risk than other possible courses of action. Whilst there is no absolutely definitive way of determining what was going through people's heads at the time of a given event, retrospective analyses of reports of accidents provide little evidence of a deliberate process of evaluating and weighing up risks occurring in the minds of the individuals involved.

Wagenaar's analysis of a large number of accidents at sea reported by the Dutch Shipping Council found that each accident on average had 23 separate 'causes' (some many more than this).[19] In around half the cases examined, information about the potential dangers was not even available or not recognised as problematic. In only one case, were the researchers able to find a deliberate acceptance of risk. In relation to the *Herald of Free Enterprise* capsize (see '*Herald of Free Enterprise*, Zeebrugge, 6 March 1987'), he notes that

> the actions preceding the capsizing were, although extremely dangerous, not the result of a deliberate acceptance of risk. The assistant bosun failed to wake up but not because he thought the risk was acceptable. The captain failed to notice the open bow doors, but not because he thought he could risk it.[20]

None of the actors involved could have foreseen the unlikely combination of events that conspired to lead the *Herald* to disaster. And none of the events would have led to the disaster had it not been for fundamental design flaws with the open-deck roll-on roll-off ferries such as the *Herald*.[21]

This research shows how the people at the centre of these events are typically faced with a bewildering number of circumstances to quickly understand and evaluate. A large number of rapidly changing circumstances and the diffusion of information between the participants would make these events extremely difficult to accurately comprehend. Wagenaar concluded that in real operations, people generally do not take deliberate risks but apply well-practised habitual procedures and routines that have worked in previous circumstances. The problem, as Willem Wagenaar points out, is not what people are doing (which only appears incorrect in hindsight) but with the complex, almost impossible environment with which people are confronted.

The complexity of the circumstances with which people are confronted combine to create events that are so novel that, as Willem Wagenaar succinctly puts it, 'people do not believe that the accident about to occur is at all possible'. Hence the label 'impossible accidents'. Wagenaar is particularly scathing of industrial safety programmes that focus on telling people to act safely or that try to select out 'accident-prone' individuals. These strategies have been repeatedly shown to be ineffective. As we shall see later, safety has to be tackled at the systems level.

As the rescue of the *Reindeer* example shows, risk-taking also has a social and cultural dimension. The captain of the *Montcalm* chose the risk of returning to the dangerously stormy seas rather than continuing to the safety of port because of a sense of duty and responsibility of one seafarer to another. Similar duties and obligations play a part in the actions of other groups such as mountain rescue teams and volunteer firefighters.

HERALD OF FREE ENTERPRISE, ZEEBRUGGE, 6 MARCH 1987

The *Townsend Thoresen* roll-on roll-off ferry left Zeebrugge in Belgium and capsized 24 minutes later as it accelerated to cruising speed for the cross-channel crossing to Dover in the U.K. This caused a bow wave to enter the car deck through the bow loading doors which had been left open. The volume of water entering through the open doors caused the vessel to rapidly lurch to port and then to slowly capsize to port in relatively shallow water. A total of 188 lives were lost. The assistant bosun had gone to sleep in his cabin and failed to close the bow doors. No other crew members noticed this absence. The company had applied direct pressure on crews to depart early. The departure from Zeebrugge had been carried out in a hasty fashion. In accordance with normal operating procedures, the ship's master took the absence of negative reports of problems as confirmation of readiness. Despite requests, there were no TV monitors on the bridge showing the bow door area.[22]

Risk-taking, then, is not some simple characteristic that people have or don't have to some degree or other. There are underlying biopsychological factors (such as sensation-seeking and anxiety) that play a role. At the same time, the type of task and situation – whether it involves financial, social, recreational or working activities plays a major role. Situational pressures or demands such as the time and resources available can influence choices to speed up, cut corners or neglect precautionary measures. This is particularly so if these pressures are recurrent leading to habitual patterns of behaviour. All of these factors can be seen at play in the *Herald of Free Enterprise* disaster. We will return to this case in detail in Chapter 11. Both psychological processes, such as the balance between slower deliberative decision-making processes and more rapid affective responses, significantly determine our perception of risk as do social and cultural factors, such as entrenched work practices as well as a duty of care or sense of mutual obligation. All of these can be highly important determining factors in observed behaviour, so that what appears simple (e.g. risk-taking or neglect of duties) is, in fact, complex and multiply determined.

ADAPTING TO RISK

Individual tolerance or preference for risk plays a major role in a highly controversial theory proposed by **Gerald Wilde**, now a Professor Emeritus at Queen's University, Toronto. Wilde observed that traffic accident rates per head of population tended to remain fairly static over time but tended to decline in terms of distance driven. As both highway engineering and car design have greatly improved over the years, Wilde drew the conclusion that people must, therefore, be accepting some accident rate as 'normal' and trading off the benefits of better roads and vehicles for increased speed and mobility. In other words, Wilde proposed a homeostatic mechanism whereby drivers compare 'perceived risk' with their 'target risk' and adjust their behaviour accordingly.[23] If the level of 'perceived risk' exceeds that of their 'target risk', then the adjustment should be to drive more cautiously; if the level of 'perceived risk' is less than the level of 'target risk', then the adjustment should be towards greater speed, more aggressive cornering and so forth. Figure 2.3 presents a simplified outline of this process. The theory is known as the 'risk homeostasis theory'.

For evidence, Wilde draws on several field studies of driver behaviour in vehicles with and without various safety systems such as anti-lock brakes (ABS). One study cited by Wilde involved a Munich taxi fleet with some vehicles equipped with ABS and some not. Over a three-year period, accident rates did not differ significantly between the ABS-equipped taxis and the others but accelerometer data indicated more extreme deceleration (i.e. harder braking) in the ABS fleet. Observational data also suggested that the ABS cars were driven more aggressively. Similar findings were made in another study of Oslo taxis equipped with ABS. Video recordings of the vehicles were made. The ABS-equipped taxis followed closer to the vehicles in front than the non-ABS-equipped taxis.[24]

A detailed analysis of trends in US traffic data showed that Wilde's original assumptions regarding static accident rates per capita and declining rates per unit of time were incorrect.[25] Many critics have also argued that the basic psychological assumptions of the theory which require people to actively hold some 'target

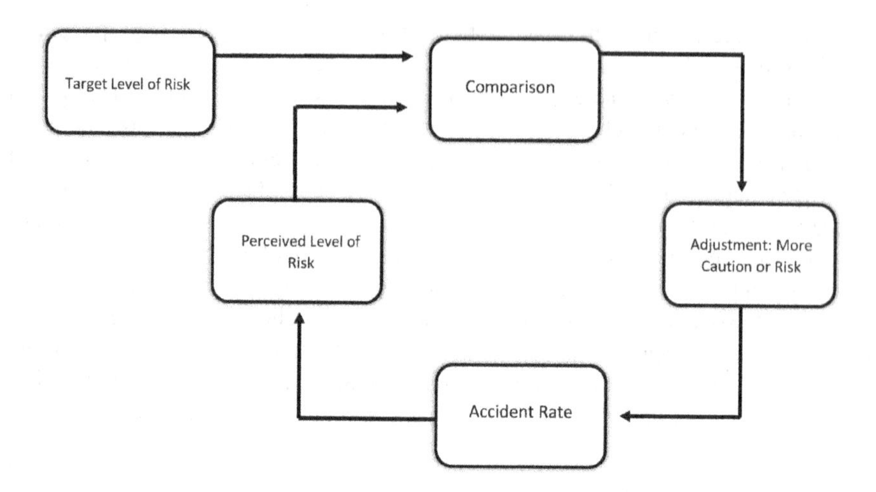

FIGURE 2.3 The homeostatic theory of risk perception. People are predicted to act more or less cautiously depending on how the current level of risk compares with their 'target' level of risk.

risk' level in mind are highly implausible. However, even the fiercest critics have conceded that the observation that people do adjust their behaviour in response to perceived risk is correct. The taxi fleet studies do show that drivers may compensate for the increased control and safety offered by active driving features such as ABS by driving a little faster and more aggressively. In contrast, passive features such as airbags that remain quietly in the background until required, generally have little or no effect on drivers' behaviour.

The notion of behavioural adaption to risk has some interesting implications for the increasingly automated vehicles that are currently on the roads. Whilst we do not yet have fleets of fully autonomous vehicles, we do already have highly autonomous vehicles that can maintain lane position and following distance, brake automatically in response to hazards and so forth. Whilst the intention of the manufacturers is to increase vehicle safety through these additional automated systems, it remains to be seen if drivers will adapt to the benefits offered by these systems by using freed-up attentional resources and cognitive capacity for other pursuits, thus potentially negating the intended benefits.

Extensive research has already been conducted on the implications of varying degrees of automation on driver performance.[26] Examples of behavioural adaptation have already been noted with drivers of automated vehicles increasingly likely to engage in extraneous activities such as using DVD players, eating, grooming, texting or even reading magazines. Almost all the research conducted on behavioural adaptation to risk has been conducted with driving and very little evidence exists of behavioural adaptation in other domains. It seems more than likely, however, that similar effects would accompany safety initiatives in other areas of the workplace. This can lead to perverse effects whereby planned safety improvements do not actually lead to increased safety, as people adjust their behaviour to save time or improve output instead.

QUANTIFYING RISK

As noted earlier, the ISO model of risk consists of four components as shown in Figure 2.1. By far, the simplest approach involves simply enumerating the hazard sources for any given activity. This is the kind of approach often taken by Coroners' Courts or Boards of Inquiry. For example, the Coroners' Report into the tragedy at the Australian Dreamworld amusement park in 2016 listed a series of significant hazards: the spacing of the slats on the conveyor; the pinch point at the head of the conveyor; the effect of pump failure; the absence of a single emergency stop button in addition to inadequately trained operators. The coroner noted: 'Each of these obvious hazards posed a risk to the safety of patrons at the ride.' The absence of any kind of engineering risk assessment of the ride was also noted.

The most frequently adopted approach to providing quantitative estimates of risk is to combine the probability or likelihood of an event occurring with the severity of its estimated consequences. This normally involves using the language of probability for the estimated likelihood and some sort of arbitrary numerical scale to represent severity. Commonly this might involve five or ten 'grades' of severity. If the likelihood is also estimated on a simple categorical scale with five or ten levels from say 'Likely' to 'Extremely Improbable', then the results can be represented in the form of a risk matrix. These will be discussed in more detail in the next section.

DREAMWORLD AMUSEMENT PARK TRAGEDY: OCCUPANTS TRAPPED UNDER RAFT

Dreamworld is a popular amusement park on Australia's Gold Coast. The most popular ride was the 'Thunder River Rapids', where patrons sat in heavy rubber rafts and were taken along a water channel, through rapids, and up to a conveyor belt to the final stop. On 25 October 2016, as a water pump failed, a raft containing four adults collided with another empty raft at the end of the ride and was lifted vertically before dropping back onto the still-moving conveyor, which pulled the raft into the mechanism. The four occupants were fatally injured. The water pump had failed repeatedly in recent days including two failures on the day of the tragedy. The operator in charge of the ride was on her first day and had received only 90 minutes of training. There was no single emergency stop button. The official inquest found that the ride posed a 'significant risk'.[27] The company operating Dreamworld was later fined A$ 3.6 m (US$ 2.57 m) for breaches of Australia's Work Health and Safety Act.

In many high-risk areas such as product design (e.g., the development of new medical devices) or large-scale engineered projects, there has been a desire for more quantitative assessments of risk. In the 1960s, the initial risk calculations suggested that the likelihood of the Apollo space missions successfully getting men to the moon and back was quite low. Prior to the Space Shuttle programme, a NASA task group suggested that the shuttle should be designed to have a 95% or greater probability of mission completion and a lower than 1% chance of death or injury per mission.[28] The need to determine such probabilities led to the development of the methodology known as 'Probabilistic Risk Analysis (PRA)' or sometimes as 'Probabilistic Safety Analysis (PSA)'. The greatest impetus to the development of formal risk analysis came from the inquiries into the Three Mile Island nuclear accident in 1979, which recommended that PRAs should be routinely conducted for all nuclear power plants.

The first thorough application of PRA was by the US Nuclear Regulatory Commission in their 'Reactor Safety Study' published four years prior to Three Mile Island.[29] The main modelling tools used in PRA involve event trees and fault trees. Essentially, these are mirror-images of one another: Event trees work forward from potential 'initiating events' which might be a water pump failure or the misreading of a display value by an operator. The event tree builds a branching series of potential outcomes of these events on future events as estimated by the analyst. Fault trees work backwards from a particular failure to estimated possible causes both technical and human. Figure 2.4 illustrates a simplified fault tree representation of the Dreamworld incident.

The fault tree starts with an undesirable event at the top of the tree and works backwards to suggest possible preceding events that could lead to this. In the Dreamworld case, the two possible preceding conditions would be a sudden loss of water, due to leakage or pump failure, for example, or a failure of the conveyor belt due to loss of power or the slats breaking. These two are linked by a logical 'OR' as either one could lead to the rafts colliding and overturning. For either of these events to lead to the undesired outcome requires another event which is the lack of an emergency cut-out. This is a logical AND condition so that one of the failures can only lead to the outcome if at the same time there is a failure of the emergency system. If failure probabilities can be attached to each event then the overall failure probability can be calculated.

Fault tree analysis has become widely used in the development and certification of complex engineered systems. There are sophisticated statistical analytical techniques for calculating probabilistic risk estimates from the fault trees and there is a dedicated website just for fault trees (www.faultree.org). One obvious limitation is the ability of the human analyst to foresee every possible precipitating causal condition. The events on the tree have to be generated by subject matter experts (SMEs) but there is no guarantee that every possible contingency has been considered. In fact, research with undergraduate students who were asked to evaluate a fault tree for a car failing to start were as happy with the tree showing only three branches as the one showing six. What was 'out of sight' was effectively 'out of mind'. Professionals are not immune to this bias.[30]

Another problem for PRA is that whilst there are often reasonably precise estimates of the failure probabilities of mechanical components, such as water pumps, precise failure probabilities for human behaviour are not so easily obtainable.

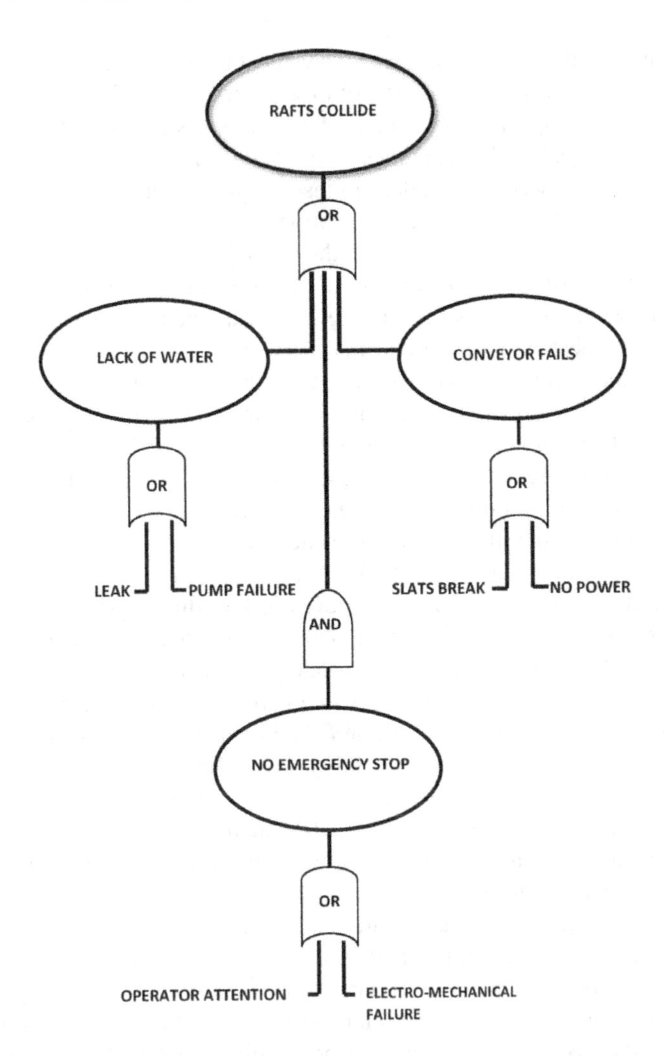

FIGURE 2.4 A partial fault tree for the accident involving the Thunder Rapids Ride at Dreamworld, Australia.

Given the obvious importance of human behaviour in system performance, it has become essential to incorporate estimates of human error likelihood into the PRA. Consequently, a sub-branch of PRA has developed to deal with this and has become known as Human Reliability Assessment (HRA).

The most well-known figure in HRA is **Barry Kirwan** who has published a number of books and articles describing the HRA process.[31] Kirwan was a lecturer in Ergonomics at Birmingham University (UK) having previously been Head of Human Factors for British Nuclear Fuels. He now heads safety programmes at Eurocontrol – Europe's Air Traffic Management authority. Kirwan has described the HRA process in detail. There are essentially three key parts to the process. The first step involves detailing the human involvement in any process or activity. This is usually accomplished with some form of *task analysis* which is simply a method of breaking down an objective into a series of smaller tasks that must be accomplished to achieve that objective. For example, the objective of starting the engines on a jet aeroplane involves a series of smaller sub-tasks requiring the pilot to locate and position certain switches in a particular orientation.

The second step requires identifying the potential human errors – selecting the wrong switch or positioning it incorrectly would be two possible examples. Chapter 3 covers the possible varieties of human errors in greater detail. The theories of human errors outlined there provide the essential framework for enumerating the possible errors that could occur in each task described in the task analysis. Numerous frameworks have been developed in HRA for identifying and labelling human errors, all known by catchy acronyms such as HAZOP, HEART or JEDHI. It is possible to stop the HRA at this point and having catalogued the possible errors that might arise, set about designing ways of mitigating or controlling them.

More commonly, the final step requires quantification of the likelihood of these errors happening. These are generally known as *human error probabilities* (HEPs). A large variety of techniques have been developed for this with the earliest and best known being the *technique for human error rate prediction* (THERP). The technique was developed for use in the US nuclear power generation industry with error rate data obtained from observation of work in operational nuclear power plants and from simulator data.[32] THERP draws on a data bank of these error probabilities to provide estimates of performance reliability. For example, the error probability for an experienced operator entering data on a keyboard is of the order of 1 error per 100 keystrokes. The original handbook of THERP data included an error rate of 0.3% for reading an analogue meter, 1% for reading a graph and 0.3% for choosing the wrong control on a panel of controls.

Although these and similar error rates published by other researchers seem quite low, when considering the number of times these actions are taken across many operators in many locations over extended periods of time, one can expect a fairly high number of errors to occur in absolute terms. In addition, it seems reasonable to expect that other factors such as inexperience or high pressure would increase these error rates quite markedly. THERP contains an additional set of *performance shaping factors (PSFs)* which takes into account factors such as skill level, workload, lighting and fatigue and estimates the impact of each of these factors on the basic error probabilities. Estimate is the operative word here as these are very much imprecise 'guestimates' of the effects of factors which can vary hugely in their impact depending on the presence or absence of other factors. So, for example, the extreme cold would be predicted to increase the error rates of postal sorting office workers but seems only to increase the efforts of highly motivated polar explorers!

Nevertheless, Kirwan reports the results of an extensive validation exercise which showed reasonably high correlations between the predictions derived from several quantification techniques including THERP and observed performance reliability in tasks typically found in nuclear power and process control. Of course, these highly standardised and proceduralised tasks may not be representative of the majority of tasks carried out across a wide range of other occupational and working situations. In addition, the focus in most HRA quantification has been more on omission errors, where something is not done or done incorrectly and somewhat less on commission errors where something that should not be done is done.

A more comprehensive approach to error prediction is known as SHERPA (systematic human error reduction and prediction approach). Following the hierarchical task analysis step, the lowest-level tasks are classified according to a behavioural taxonomy (e.g. 'Retrieval', 'Checking', etc.) and associated error modes leading to a table of predicted errors and error likelihoods associated with performing the task. Empirical evidence has shown that SHERPA can be performed reliably and can successfully predict a high proportion of errors. For example, a team of researchers in the UK used SHERPA to successfully predict a high proportion of the errors observed on automated airline flight decks. The authors conclude that 'SHERPA is the best of the currently available human error prediction techniques'.[33]

The principal problem with HRA techniques, however, is more fundamental. In line with Kurt Lewin's dictum that 'there's nothing so practical as a good theory'; it is the flaws in the underlying theoretical basis that render the practical application of HRA highly debatable. As noted above, the application of HRA requires an activity to be broken down into a sequence of subsidiary tasks arranged in sequence. Whilst some highly routinised and proceduralised tasks such as entering information into the flight management system (FMS) on an aircraft flight deck might be amenable to this form of description, many, if not most, of the activities in the workplace are not. In addition, many of the more devastating failures that crop up in modern high-risk systems arise from the presence of unanticipated conditions requiring complex work-arounds and strategic thinking, rather than from glitches in more routine actions.

In partial recognition of these problems, HRA has developed second- and third-generation approaches such as ATHEANA and CREAM with greater attention paid to commission errors and to the contextual circumstances within which breakdowns in performance can occur. As William Wagenaar noted, these circumstances can be highly bewildering and chaotic, presenting the human operators with huge challenges to overcome. **Nancy Leveson** makes a similar observation noting that major failures often involve two or more very-low-probability events happening together.[34] Sometimes, these have a common cause whereas PRA assumes complete independence, and the resulting probability calculations produce extremely low-predicted risk, thus underestimating the true risk. There have also been some attempts to build databases of human reliability in other domains than nuclear power and process control such as in the maritime industry.[35]

Whilst large-scale high-hazard industries are likely to continue to utilise some forms of probabilistic risk assessment tools incorporating human performance, the approach remains firmly grounded in individualistic approaches to human error and system failure. There is an urgent need to build better practical tools on the theoretical bases provided by systems science which will be covered in Section 3 "Systems" of this book. For most workplaces and organisations, risks still need to be managed, however, and we will look at some of the basic principles of risk management in the next section.

MANAGING RISK

The simplest and most commonly adopted approach to risk management is based solely on the identification of risk sources or hazards and their amelioration using one or more of the approaches shown in Figure 2.5. This 'risk control hierarchy' has been widely applied in a variety of workplaces and organisations. In some cases, it may be relatively easy to eliminate hazards altogether; in other cases, this may not be possible for technical reasons or because the costs of so doing are deemed excessive. In this case, the next approach (substitution) may be tried. For example, one chemical (e.g. a weedkiller), such as glyphosate which is now banned in some countries, might be replaced with household vinegar and salt. Reducing the amount of contact with the hazard would be the next option and the last and, generally, least-effective option would be training and appropriately equipping the worker to provide protection against the hazard. A wide range of PPE can be found in many workplaces from hospitals to construction sites. Logically, the complete elimination of a hazard will be more effective in reducing the impact of the hazard than relying on proper utilisation of PPE by workers.

The next most straightforward risk management technique is to take account of both the nature of the risk source or hazard and the potential severity of the consequences it generates. This is the risk management approach adopted by many large organisations. It is recommended in international standards such as the influential Australian/New Zealand AS/NZ 4360 (2004) or the updated AS/NZS 31000 (2009). Typically, both the likelihood of harm and the severity of the possible harm are graded on a four- or five-point scale. Likelihood might be expressed as 'Unlikely, Seldom, Occasional, Likely and Frequent' and severity as 'Negligible, Moderate, Critical or Catastrophic'. This generates a matrix or grid of 20 cells with each cell corresponding to a particular combination of risk likelihood and risk severity.

The level of priority for management is indicated by the level of shading of the cells in Figure 2.6. The most urgent and important from a risk management perspective are the cells in the upper left of the matrix (indicated by the darkest shading)

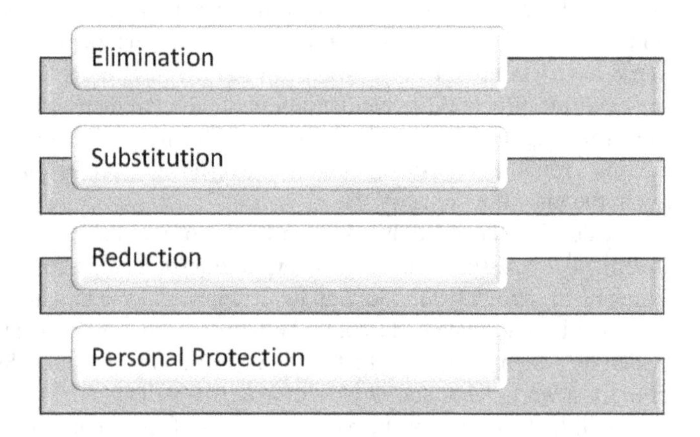

FIGURE 2.5 The basic components of the risk control hierarchy.

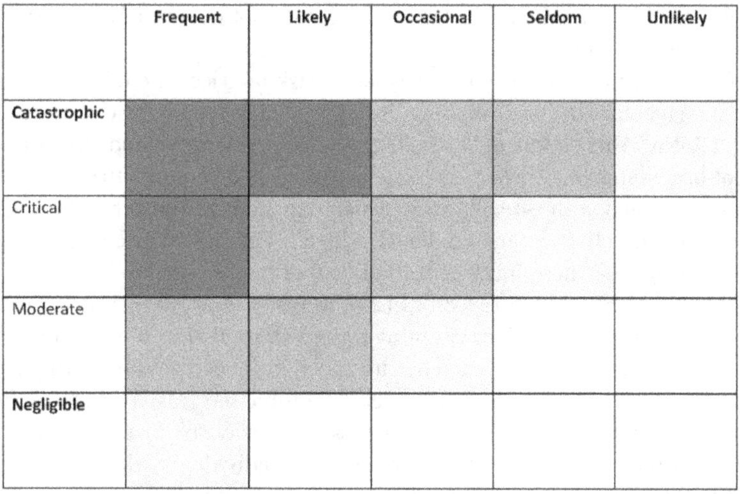

	Frequent	Likely	Occasional	Seldom	Unlikely
Catastrophic					
Critical					
Moderate					
Negligible					

FIGURE 2.6 A risk matrix. The horizontal axis represents the likelihood of occurrence and the vertical axis represents the severity of the outcome. The darker shaded cells represent the areas of greatest risk.

where there is a high probability of major effects. The least important are the cells (unshaded) in the lower right of the matrix where there are low levels of probability of relatively minor impact. The lightly shaded and more heavily shaded cells in the middle of the matrix indicate increasingly important priorities. In this hypothetical example, we have four levels of risk management prioritisation from most urgent (darkest shading) to least urgent. Risk matrices are widely used risk management tools and often form a key component of risk management systems (SMSs) as described below. However, there have been some recent questions concerning their inherent validity and usefulness in risk management decision making.

Tony Cox has looked closely at the mathematical and statistical validity of risk matrices.[36] The question is how accurately a qualitative tool such as a risk matrix can accurately capture the degrees of variation possible in the original quantitative variables of probability and severity of the outcome. The simple answer is 'not very'. Rather dispiritingly, Cox concludes that risk matrices are 'not always better than – or even as good as – purely random decision making'. The only exceptions are where there are plentiful high-quality data on both probability and severity and then several advanced mathematical techniques can be employed to generate useful risk matrices. Unfortunately, even in this ideal circumstance, risk matrices are not valid when probability and severity are negatively correlated. In other words, when high-probability events have low severity and vice versa. This is much more commonly the case in the real world than in the reverse. Finally, if risk matrices are used, Cox has shown that for a 5 by 5 risk matrix to be valid, it must have exactly 3 levels of risk as indicated by the shading or colouring generally used. Typically, four or more levels are used in such matrices to indicate different levels of risk target prioritisation. Cox's final words: 'Risk matrices do not necessarily support good … risk management decisions.'

To further address the question of whether risk matrices are of any actual value in risk management decision making or simply make for interesting décor, **David Bell** and **John Watt** conducted a small experiment where 50 students of risk management or occupational health and safety were asked to judge the likelihood and severity associated with three hazards which depicted the danger of falling from a railway platform, cliff top and medieval bridge.[37] For risk matrices to be useful as a management tool, there must be a high degree of consistency between different judges in how they rate the risks; or else, the whole exercise just becomes highly subjective and inconsistent. The resulting ratings showed almost no degree of agreement between the judges with each hazard receiving every possible rating from low to high and with no differentiation between the three hazards. To account for these findings, the authors refer to many of the issues previously covered in this chapter regarding differences in risk perception between individuals with differing worldviews and the important role of affect or emotion in risk judgment.

It is hard to avoid the conclusion that there are so many different factors affecting risk perception and risk tolerance that attempt to reduce this complexity to a small grid, and using this grid to determine subsequent risk management actions is simplistic at best and downright hazardous at worst. Nevertheless, risks have to be assessed and addressed. Risk assessment lies at the heart of what is known as *SMSs*, which are widely used in areas from leisure activities to space programmes.

SAFETY MANAGEMENT SYSTEMS

In high-risk systems and organisations dealing with hazardous materials or processes, one of the key functions of management is to ensure operations are carried out safely. Traditionally, this has meant the avoidance of unexpected negative events that can lead to losses – whether material, personal or financial. Textbook definitions of SMS refer to management practices and procedures that are concerned with health and safety and the minimisation of losses.[38] The International Civil Aviation Organization (ICAO), as a United Nations affiliated organisation, provides detailed advice to member states about safety management systems and how to design them.[39] The core concept emphasised by ICAO is that of control. SMSs are the means of controlling those variables that can lead to negative events and losses. ICAO does not endorse the idea of 'zero accidents', which is sometimes advanced as the only possible safety goal. Instead, ICAO emphasises relative rather than absolute safety. Management must balance the inherent productive purposes of the organisation (i.e. to deliver the service for which it was designed) with its protective goals (i.e. safety). This dilemma is further addressed in Chapter 3.

As a large international organisation, ICAO sets the framework for individual states to develop their own safety programmes. The advice offered by ICAO follows traditional views of human error (see Chapter 3) and emphasises hazard identification and risk control using tools such as the risk matrices discussed above. ICAO recommends developing a risk matrix for individual hazards and then using the results to generate a 'safety risk index' which prioritises subsequent risk management efforts. We have already noted the limitations of this approach. Each state or nation is required to ensure that every aviation-related organisation from training organisations, airlines and maintenance to manufacturers and air traffic management within its boundaries develops an SMS.

The four key elements of an SMS are shown in Figure 2.7. Identifying hazards is the first process and using the traditional 'likelihood x severity' risk matrix to identify the size of the risks and then implementing appropriate remedial actions constitutes the second process. The effect of these remedial actions needs to be continuously monitored. This step involves the development of 'safety indicators' which are observable events such as number of fatal accidents, number of runway incursions, etc. The fourth and final step involves a philosophy of continuous improvement whereby the SMS is regularly scrutinised for its 'efficacy and efficiency'. A shorthand way of referring to these steps is sometimes referred to as the 'PDCA' (Plan, Do, Check, Act) cycle.

In the United States, the Federal Aviation Administration (FAA) has set out guidelines for an SMS in accordance with the ICAO recommendations and since 2015 has mandated all air carriers to implement an SMS. Similarly, the UK Civil Aviation Authority has issued its own guidance for developing an SMS for the UK aviation industry.[40] Most other nations have similar documents and SMSs have become widely embedded in the world's aviation industry. In Australia, for example, all forms of transportation (air, sea, rail, etc.) are legally required to have a functional SMS. Safety management systems have become established in many other industries. Recently (2015), an occupational health and safety management systems

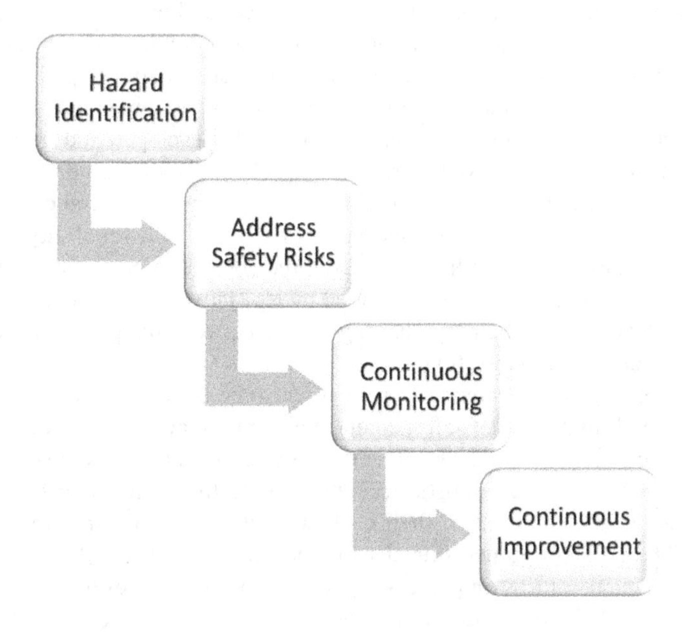

FIGURE 2.7 The four key elements of an SMS.

standard (OHSMS) has appeared as AS/NZS ISO45001 to provide guidance across a broad range of industries and occupations.

The key question, of course, is 'have safety management systems improved safety?' This is an inherently difficult question to answer since there have been many changes in regulatory and operating environments over the past few decades as well as concurrent technological advances in many areas. It is very difficult to design high-quality research studies in complex environments that have multiple influences on the variables (such as injury rates) under investigation. The first systematic evaluation was carried out in 2007 by a team of researchers at the Institute for Work and Health in Toronto.[41] They could only find 13 studies that were not seriously flawed, and few of these looked at the kind of objective safety indicators suggested by ICAO. The best that could be said was that there was no evidence of any negative effects of SMS!

The next thorough review was carried out in 2012 by **Matthew Thomas** on behalf of the Australian Transportation Safety Board.[42] Thomas identified 37 suitable studies mostly conducted with high-hazard industries and manufacturing. Just over half the studies reported data on objective safety indicators. There were some positive findings on injury rates in a variety of industries (construction, manufacturing and chemical processing) but a number of null findings. Again, no negative effects were reported. Because any SMS consists of multiple individual processes that it has proven impossible to disentangle which, if any, are responsible for producing an effect. It may be that other factors beyond the core components of an SMS, such as leadership commitment to safety, have a greater influence on safety indicators than the SMS itself.

The safety management field is a very much larger and more complex area than suggested here and much more information can be found in the sources listed in the Notes section. Almost every regulatory agency in almost every developed country has suggestions and recommendations for best practices in developing appropriate safety management systems. Although SMSs have gradually developed and evolved since the 1970s, their overall effectiveness remains largely undemonstrated.[43] Attempts to pinpoint the processes that might be effective in influencing safety indicators remain at a very early stage. Serious questions can be raised about the reliability and validity of the risk matrix/risk profiling that lies at the core of SMS. On a positive note, the emphasis on SMS, sometimes reinforced by legislative requirements, provides a healthy counterweight in the perpetual tug-of-war between the productive and protective demands of an organisation.

Goals of Safety Management: There is often some confusion and disagreement as to what the ultimate goal of safety management might be. The goal of 'zero accidents' has sometimes been advanced as the only reasonable safety goal. Anything else, it is argued, involves a fatalistic acceptance of death and injury. Some organisations have adopted 'zero harm' as their goal although many have since modified this approach. The counterarguments revolve around excessive costs and over-regulation with an unproductive focus on preventing even the most minor event. If such events do occur, as is likely, then the safety policy can be seen to have failed leading to other undesirable behaviours. In most countries and organisations, a more pragmatic approach to risk has been adopted. This is sometimes referred to as the ALARP or 'As Low as Reasonably Practicable' principle.

In the UK, health and safety regulation is tied to this principle under the Health and Safety at Work Act of 1974. UK law has used the principle of 'reasonable and practicable' in settling employer liability cases for nearly two hundred years. It has been said that with the ALARP principle 'the typically British concept of reasonableness has been made somewhat more precise'.[44] In other words, the benefits of ameliorating a given risk must be weighed against the potential costs of doing so. Safety-at-any-cost is not the goal. What is reasonable and practicable will also vary from one activity or technology to another. Given that the costs include injury and loss of life, this raises the thorny question of how to cost injury and death. A good deal of research and debate has gone into the topic of costing injuries and valuing human life with widely varying results. Typical values range from less than US$ 1 m to more than US$ 25 m with a mean of around US$ 7 m.[45] Few countries maintain comprehensive records of injuries and hospitalisations so accurately costing these is subject to numerous uncertainties.

CONCLUSIONS

Our ability to calculate probabilities and analyse risk can be seen as one of the most significant advances in the evolution of human thinking. Quantitative risk analysis and the development of systematic procedures for the management of risk would seem to provide an exact and objective approach to the goal of avoiding risk and reducing harm in all areas of life from leisure activities to aviation and nuclear power generation. On closer inspection, however, the apparent objectivity and precision dissolve in the face of plentiful scientific evidence as to how people form perceptions of risk and decide what risks they are willing or prepared to tolerate. These decisions are affected by a host of psychological and cultural factors that influence a person's general and social outlook. Risk professionals are not grown in a Petri dish and so are subject to these same influences.

Research has also shown that the assumptions underlying the use of risk matrices as a risk management tool may not be valid. Risk matrices are a core component of safety management systems and as yet the question of the overall validity of SMSs is undetermined and subject to further evidence. As there is such diversity in the composition and makeup of SMSs in any case that it might be more profitable to try and isolate the more effective components of managerial approaches to high-risk technologies and industries. We will examine this question more closely in Chapter 6, where we look at the proposed characteristics of 'high-reliability organisations' or HROs. Of course, if the requirements were perfectly specified and there were no unexpected breakdowns in technological, mechanical or human performance, then there would be no lack of reliability and unwanted outcomes would likely not occur. It is to the thorny question of 'human error' that we now turn to in the next chapter.

NOTES

1. Peter Bernstein was an economist who wrote a number of books on finance and economics including the award-winning *"Against the Gods: The Remarkable Story of Risk"* published in 1996 in New York by John Wiley & Sons.
2. There are 36 possible combinations of numbers that can appear on the faces of two dice (i.e. 6 × 6 = 36). A total of 9 can be thrown with a 3 and a 6; or a 4 and a 5 plus the reverse pairings for a total of 4 chances out of 36 or approx. 11%. A total of 3 can only occur with a 1 and a 2 plus the reverse for 2 chances out of 36 or 5.5%.
3. The International Organization for Standardization (ISO) revised risk management guidelines are contained in *ISO 31000 – Risk Management* published in 2018. Former standards such as AS/NZS 4360 have been superseded by one based on the ISO 31000 referred to as AS/NZS 31000 (2009).
4. The fascinating history of risk is covered in more detail in Peter Bernstein's book (see Note 1).
5. Cohen, B., & Lee, I.-S. (1979). A catalog of risks. *Health Physics, 36*, 707–727.
6. This material is covered in many publications. For example: Slovic, P., Fischoff, B., & Lichtenstein, S. (1979). Rating the risks. *Environment, 21*, 14–20. A good summary can be found in: Slovic, P. (1987). Perception of risk. *Science, 236*, 280–285.
7. At the time of writing (Jan 2020) the new coronavirus originating in Wuhan, China was dominating the world news. See for example: https://www.who.int/emergencies/diseases/novel-coronavirus-2019; Accessed 29th January, 2020.
8. Slovic, P., Lichtenstein, S., & Fischhoff, B. (1984). Modeling the societal impact of fatal accidents. *Management Science, 30*(4), 464–474.
9. The early study showing significantly different risk perceptions by white males was: Flynn, J., Slovic, P., & Mertz, C.K. (1994). Gender, race, and perception of environmental health risks. *Risk Analysis, 14*(6), 1101–1108. An influential later study was published by Kahan, D., Braman, D., Gastil, J., Slovic, P., & Mertz, C.K. (2007). Culture and identity-protective cognition: Explaining the white male effect in risk perception. *Journal of Empirical Legal Studies, 4*(3), 465–505. The link to climate change denial was shown in a study by McCright, A.M., & Dunlap, R.E. (2011). Cool dudes: The denial of climate change among conservative white males in the United States. *Global Environmental Change, 21*, 1163–1172.
10. A good overview of Lopes' account of risk-taking can be found in: Lopes, L.L. (1987). Between hope and fear: The psychology of risk. *Advances in Experimental Social Psychology, 20*, 255–295.
11. Zuckerman, M., Kuhlman, D.M., Joireman, J., Teta, P., & Kraft, M. (1993). A comparison of three structural models for personality: The Big Three, the Big Five, and the Alternative Five. *Journal of Personality and Social Psychology, 63*, 757–768.
12. Roberti, J.W. (2004). A review of behavioural and biological correlates of sensation seeking. *Journal of Research in Personality, 38,* 256–279. A recent meta-analysis of 39 studies showed a strong effect of sensation-seeking on participation in extreme sports along with smaller effects for extraversion and impulsivity. See: McEwan, D., Boudreau, P., Curran, T., & Rhodes, R.E. (2019). Personality traits of high-risk sport participants: A meta-analysis. *Journal of Research in Personality, 79*, 83–93.
13. An entertaining account was published by: Lendrem, B., Lendrem, D.W., & Dudley, J. (2014, 11th December). The Darwin Awards: Sex differences in idiotic behaviour. *British Medical Journal.* The authors explain the difference between regular stupidity ('shooting yourself in the head whilst demonstrating that a gun is unloaded') and truly idiotic stupidity ('shooting yourself in the head whilst demonstrating that a gun is

loaded'). A more academic look at gender differences in risk-taking involving a review of 150 studies can be found in: Byrnes, J.P., Miller, D.C., & Schafer, W.D. (1999). Gender differences in risk taking: A meta-analysis. *Psychological Bulletin, 125*, 367–383.

14. Blais, A.R., & Weber, E.U. (2006). A domain-specific risk-taking (DOSPERT) scale for adult populations. *Judgment and Decision Making, 1,* 33–47.
15. Loewenstein, G., Weber, E.U., Hsee, C.K., & Welch, N. (2001). Risk as feelings. *Psychological Bulletin, 127*(2), 267–286.
16. Slovic, P., Peters, E., Finucane, M.L., & McGregor, D.G. (2005). Affect, risk, and decision making. *Health Psychology, 24,* S35–S40.
17. This incident was discussed by Lopes in the chapter referred to in Note 10 above. Further information can be obtained from the website of the Royal Fleet Auxiliary Historical Society (http://www.historicalrfa.org/rfa-reindeer).
18. A typical study showing no correlation between the questionnaire measures and risk-taking by forestry workers for example was: Salminen, S., Klen, T., & Ojanen, K. (1999). Risk taking and accident frequency among Finnish forestry workers. *Safety Science, 33*, 143–153.
19. Wageaar, W.A., & Groeneweg, J. (1987). Accidents at sea: Multiple causes and impossible consequences. *International Journal of Man-Machine Studies, 27*, 587–598.
20. Wagenaar explicitly discusses the role of risk-taking in accidents in: Wagenaar, W.A. (1992). Risk taking and accident causation. In J.F. Yates (Ed.), *Risk Taking Behavior.* Chichester: John Wiley, pp. 257–281.
21. A brief discussion of the problems with the design of roll-on roll-off ships can be found at: https://www.marineinsight.com/naval-architecture/ro-ro-ship-design-dangers/
22. Department of Transport. (1987). *Mv Herald of Free Enterprise, Report of Court No. 8074 Formal Investigation*. London: Her Majesty's Stationery Office.
23. Wilde, G.J.S. (1982). The theory of risk homeostasis: Implications for safety and health. *Risk Analysis, 2*, 209–225.
24. The study of Munich taxi drivers conducted by Aschenbrenner and Biehl in 1994 is discussed in Simonet, S., & Wilde, G.J.S. (1997). Risk: Perception, acceptance and homeostasis. *Applied Psychology: An International Review, 46,* 233–252. The study of Oslo taxi drivers appears in: Sagberg, F., Fosser, S., & Saetermo, I.-A.F. (1997). An investigation of behavioural adaptation to airbags and antilock brakes among taxi drivers. *Accident Analysis and Prevention, 29*, 293–302.
25. Evans, L. (1996). Risk homeostasis theory and traffic accident data. *Risk Analysis, 6*, 81–94.
26. Numerous studies of vehicle automation and driver performance have been reported by Neville Stanton and colleagues. For example: Stanton, N.A., et al. (2001). Automating the driver's control tasks. *International Journal of Cognitive Ergonomics, 5*, 221–236. A recent overview of highly automated driving can be found in: Navarro, J. (2019). A state of science on highly automated driving. *Theoretical Issues in Ergonomics Science, 20*, 366–396.
27. Coroners' Court of Queensland Inquest: Inquest into the deaths of Kate Goodchild, Luke Dorsett, Cindy Low & Roozbeh Araghi at Dreamworld, October 2016 (courts.qld.gov.au)
28. Bedford, T., & Cooke, R. (2001). *Probabilistic Risk Analysis: Foundations and Methods.* Cambridge: Cambridge University Press.
29. Published by the U.S. Nuclear Regulatory Commission '*Reactor Safety Study: An Assessment of Accident Risks in U.S. Commercial Nuclear Power Plants. WASH-1400 (NUREG-75/014)*' was published in 1975. One of the scenarios covered was the core damage suffered at Three Mile Island in 1979.

30. The study involving undergraduate students was published by Fischoff, B., Slovic, P., & Lichtenstein, S. (1978). Fault trees: Sensitivity of estimated failure probabilities to problem representation. *Journal of Experimental Psychology: Human Perception and Performance, 1977, 3*, 552–564. Professional judges were used in a study by Dube-Rioux, L., & Russo, J.E. (1988). An availability bias in professional judgment. *Journal of Behavioral Decision Making, 1,* 223–237.
31. See for example: Kirwan, B. (1992). Human error identification in human reliability assessment. Part 1: Overview of approaches. *Applied Ergonomics, 23,* 299–318. For a detailed handbook on the field see: Kirwan, B. (1994). *A Guide to Practical Human Reliability Assessment.* London: CRC Press. A more recent summary of the field can be found in: Kirwan, B. (2008). Human reliability assessment. In *Encyclopedia of Quantitative Risk Analysis and Assessment.* London: Wiley.
32. The early development of HRA can be found in the work done for the U.S. nuclear power industry by Swain, A.D., & Guttmann, H.E. (1983). *Handbook of Human Reliability Analysis with Emphasis on Nuclear Power Plant Applications.* NUREG/CR-1278. Washington, DC: U.S. Nuclear Regulatory Commission.
33. Harris, D., Stanton, N.A., Marshall, A., Young, M.S., Demagalski, J., & Salmon, P. (2005). Using SHERPA to predict design-induced error on the flight deck. *Aerospace Science and Technology, 9*, 525–532.
34. See page 34 of: Leveson, N. (2011). *Engineering a Safer World: Systems Thinking Applied to Safety.* Cambridge, MA: MIT Press.
35. A useful critique of HRA approaches can be found in: French, S., Bedford, T., Pollard, S.J.T., & Soane, E. (2011). Human reliability analysis: A critique and review for managers. *Safety Science, 49*, 753–763. An earlier, particularly trenchant critique of HRA was published by Dougherty, E.M. (1990). Human reliability analysis – Where shouldst thou turn*? Reliability Engineering and System Safety, 29*, 283–299.
36. Cox clearly explains the basic mathematical limitations and issues with risk matrices in the aptly titled: Cox, L.A. (2008). What's wrong with risk matrices? *Risk Analysis, 28*, 497–512.
37. Ball, D.J., & Watt, J. (2013). Further thoughts on the utility of risk matrices. *Risk Analysis, 33*, 2068–2078.
38. Examples can be found in Glendon, A.I., Clarke, S.G., & McKenna, E.F. (2006). *Human Safety and Risk Management, 2nd Ed.* Boca Raton, FL: CRC Press. This text covers many of the issues discussed in this chapter in greater detail.
39. International Civil Aviation Organization. (2009). *Safety Management Manual (SMM), 2nd Ed.* Montreal, QC: International Civil Aviation Organization. The manual can be downloaded at: https://www.icao.int/safety/fsix/Library/DOC_9859_FULL_EN.pdf
40. The FAAs Safety Management System guidance document can be downloaded: https://www.faa.gov/documentLibrary/media/Order/8000.369.pdf.

    The UK Civil Aviation Authority (CAA) have issued a manual '*Safety Management Systems (SMS) Guidance for Organisations: CAP 795*' in 2015. It is available at: https://publicapps.caa.co.uk/docs/33/CAP795_SMS_guidance_to_organisations.pdf.
41. Robson, L.S., et al. (2007). The effectiveness of occupational health and safety management system interventions: A systematic review. *Safety Science, 45*, 329–353.
42. Thomas, M. (2012). *A Systematic Review of the Effectiveness of Safety Management Systems.* ATSB Transport Safety Report AR-2011-148. Canberra: Australian Transport Safety Bureau.
43. Li, Y., & Guldenmund, F.W. (2018). Safety management systems: A broad overview of the literature. *Safety Science, 103,* 94–123.

44. Bedford, T., & Cooke, R.M. (2001). *Probabilistic Risk Analysis: Foundations and Methods.* Cambridge: Cambridge University Press.
45. Ale, B.J.M., Hartford, D.N.D., & Slater, D. (2015). ALARP and CBA all in the same game. *Safety Science, 76*, 90–100. Practical guidance on ALARP such as from Worksafe (NZ) can be found here: https://worksafe.govt.nz/managing-health-and-safety/managing-risks/how-to-manage-work-risks/

3 Human Error

The study of error has both a very long and an exceedingly short history. Human failing has been the central driving force in many literary, philosophical and religious accounts of the human condition. The ancient Greeks knew a thing or two proclaiming 'the wisest of the wise may err' (Aeschylus). History provides ample evidence of the prevalence of mistakes in judgement and errors in execution. Consider the decision early in his presidency by the revered US President John F. Kennedy to authorise a covert invasion of Cuba to overthrow the recently installed Fidel Castro. Everything that could go wrong did go wrong.[1] Some errors were based on faulty information and others on strategic and planning mistakes. A much-chastened Kennedy recovered from this fiasco and quickly restored his reputation with his handling of later events including the Cuban missile crisis.

The history of scientific enquiry into human error is rather shorter. The empirical science of psychology developed from its roots in mental philosophy with the establishment of the first experimental psychology laboratory by **Wilhelm Wundt** in Leipzig in 1879. Experimental psychologists quickly became fond of the reaction time (RT) method as a means of measuring the timing of mental processes. Simple RT involves timing the onset of a response to a stimulus such as a light or a sound. Early experimenters did their best to prevent their participants from making errors as these interfered with this primary goal.[2] For example, 'catch tests' were introduced where no stimulus was presented. If the participant erroneously responded, then they were told that data from the whole block of trials would be discarded. As a result '(they) learn to keep their eagerness within bounds' as the experimenters quaintly put it.

It was not until somewhat later (the 1950s) that the strategic nature of reaction time was studied. It was soon found that participants can trade speed for accuracy depending on the importance assigned to one or the other. Importantly, studies showed that optimum performance could be achieved even when a small amount of error (around 10%) was accepted. **Paul Fitts** who discovered this, and later became one of the founding figures in engineering psychology or what has become known as the science of 'human factors' (see Chapter 4), pointed out what may be a key principle of human performance which is that in many everyday tasks such as writing or talking, we have become adapted to performing at a level that includes a small amount of error to achieve optimal efficiency.[3] In other words, we can expect to find errors wherever humans perform everyday tasks. Errors are actually an inherent part of efficient normal task performance. Without some error, we would not be able to be such efficient typists or talkers.

At much the same time as the first experimental psychologists began their laboratory studies in Europe, the American psychologist **William James** published his

DOI: 10.1201/9781003038443-4

influential 'Principles of Psychology'.[4] William James, brother of the celebrated literary figure Henry James, defined psychology as the 'Science of Mental Life' although this included physiological phenomena as both antecedents and consequences of mental operations. James meticulously examined many of the issues still central to modern-day psychology such as memory, attention and the brain. He was interested in habits and the occasional errors associated with habitual actions such as taking out one's door key on arrival at a friend's door: 'Who is there that has never wound up his watch on taking off his waistcoat in the daytime or taken out his key on arriving at the doorstep of a friend?' Who indeed.

Such observations led James to several conclusions that are still valid today. Most importantly he realised that frequent (habitual) actions require little or no conscious attention. For this reason, we are generally unable to report the exact order in which we dress or even how we turn on a tap. We can carry out habitual actions whilst our attention is elsewhere. Were this not the case, James observes, we would have a very limited behavioural repertoire as full attention would be required for every task no matter how familiar. This flexibility in allocating attention gives us enormous power to carry out complex actions more or less simultaneously. Everyday errors thus provide a window into the hidden power of habitual or automated actions acquired through practice.

THE VARIETIES OF HUMAN ERROR

Despite this profound insight, the experimental study of human error remained a strangely neglected field until relatively recently. **Donald Norman**, then a professor of cognitive psychology at the University of California, San Diego, developed some of William James's insights further using then-current concepts in the psychology of memory and learning.[5] Norman, later to become a hugely influential figure in the field of human–machine interaction leaving academia for consulting roles in industry, including Apple and HP computers, and publishing a theory of action slips that built on James's insight that action sequences consisted of an initial 'triggering' condition and a chain of subsequent neuro-muscular 'events'. Norman realised that certain action errors might result from the inadvertent triggering of subsidiary steps in the chain whilst other errors might result from failing to trigger any of the steps at all.

The basic outline of Norman's 'theory of act selection' is shown in Figure 3.1 using one of his examples. The intention to leave work for home activates a higher-level mental structure referred to as a schema. This now-active schema takes control over a subsidiary set of lower-level schemas required to fulfil the intention. Following the required route, maintaining lane position, avoiding obstacles, controlling speed and so forth are examples of lower-level schemas now activated by the higher-level schema. Suppose that on this occasion we have decided to take a small detour midway along our route to call in at the fish shop. This subsidiary schema needs to become active just as we reach the appropriate point along our route. In Norman's terminology, the activation of the schema needs to be triggered upon reaching a particular location.

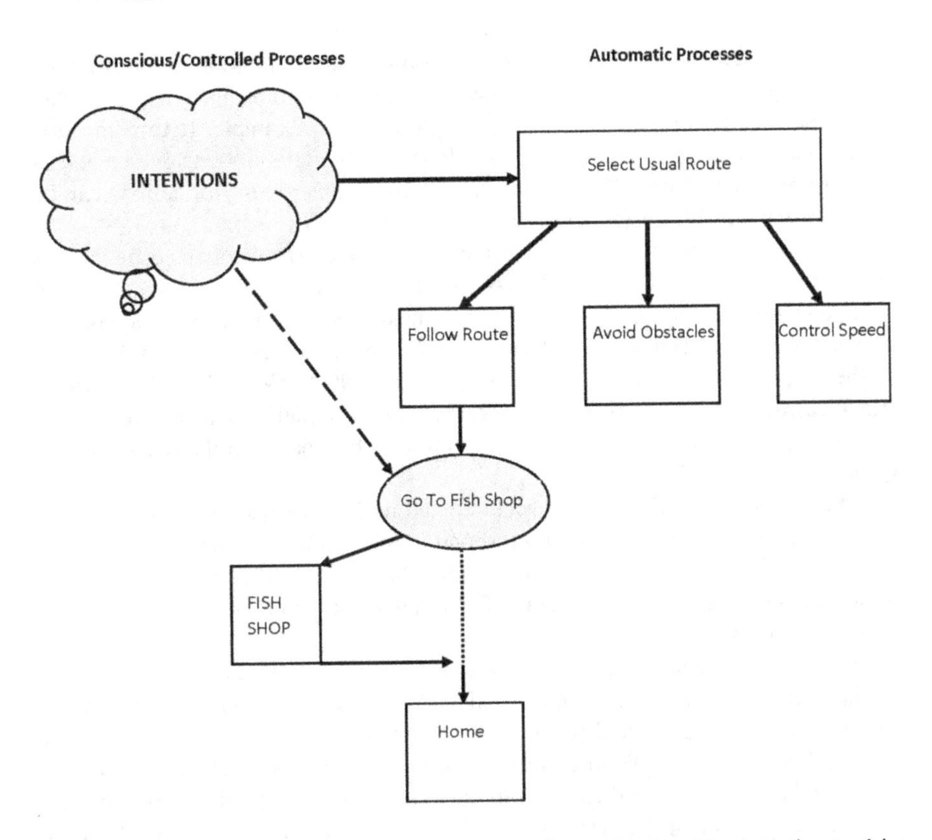

FIGURE 3.1 An illustration of Norman's theory of act selection. The intention to drive home along a regular route automatically activates a set of schemas involved in driving such as avoiding other traffic, controlling speed, etc. If we wish to deviate from the regular route, then the conscious intention to 'Go to the fish shop' must be consciously triggered at exactly the right moment or else we will continue on home as normal.

AIRBUS A320-111 ON DESCENT INTO STRASBOURG-ENTZHEIM AIRPORT, 20 JANUARY 1992

During an instrument approach to runway 05, the crew programmed the Flight Control Unit (FCU) for a standard 3.3-degree angle of descent but did not realise that the FCU was currently in 'HDG/V/S' (i.e. Heading/vertical speed) mode instead of 'TRK/FPA' (i.e. Track/flight path angle) mode. The first officer entered the digits '3.3' into the FCU. In the mode that was set, this actually instructed the aircraft to descend at a vertical speed (V/S) of 3,300 ft per minute instead of the desired 3.3-degree angle that would have been obtained in the flight path angle (FPA) mode. The inadvertently commanded descent rate was more than four times greater than it should have been. This excessive descent rate brought the aircraft in contact with terrain at 2,621 ft above sea level, killing 87 of the 96 on board.[6]

Triggering requires a very precisely timed injection of attention at a time when our attention may well be directed elsewhere – listening to an interesting speaker on the radio or ruminating on some work problem, for example. If this injection of attention is not made on time, the subsidiary schema fails to be triggered and as Norman says: 'I find myself home, fishless!' This simple theory of actions can be used to explain a wide variety of commonly reported action errors.

Norman examined around 1,000 examples of action errors collected by various people through the lens of his ATS (Activation-Trigger-Schema) theory. Many of these action errors will be all-too-familiar to most of us. Examples include opening the refrigerator door and then wondering what we intended to get; accidentally pressing the 'Caps Lock' key on the keyboard AND ENDING UP WITH CAPITALISED TEXT; turning on a kettle or coffee maker without first putting water in it; mistakenly putting a frequently used spice or herb in a dish rather than the one called for by the recipe and so forth.

If Norman's theory provides an accurate and useful explanation of action errors then most observed errors should correspond to one of the three key aspects of the theory: the selection of the intention, the activation or lack of activation of the relevant schema or the faulty triggering of a schema. Examples of each of these are shown in Table 3.1.

Norman's classificatory scheme provides a neat and useful way of bringing order to what may seem to be a haphazard collection of ways in which people's actions can turn out to be faulty and potentially contribute to unintended outcomes with undesired consequences. Whilst most action errors seem to be classifiable in this way there is an absence of direct scientific evidence for the validity of the scheme. Norman suggested that the best way to validate the theory and the associated classificatory scheme would be via laboratory testing of predictions derived from the theory. Surprisingly little empirical research has been conducted in the years since Norman's work.

One person in particular who was greatly influenced by Don Norman's work on action slips was **Jim Reason** a lecturer (and later, Professor) in psychology at the

TABLE 3.1
Sources of Action Errors According to Norman's ATS Theory.

Source	Category	Example
Intention	Wrong mode	Actions inappropriate in the current context such as flight crew acting as though aircraft descending in 'angle of descent' mode when aircraft actually in 'rate of descent mode' as occurred in the Strasbourg crash in 1992
Activation	Over-activation	Capture errors – Actions 'hijacked' by stronger habitual actions
	Under-activation	Omission errors – Step left out of a sequence or procedure
Triggering	Wrong timing	Blends of actions where two schemas are activated simultaneously
	Failure to trigger	Forgetting intentions such as when a momentary intention is later forgotten

University of Manchester in England. Reason published a highly influential book in 1990 prosaically entitled *Human Error*. His four-decade interest in human error began with an observation of his own simple action slip where he recalled finding himself in the middle of making tea but instead spooning cat food into the teapot.[7] This incident occurred when he was looking for a new research area having found himself, unsurprisingly, running out of eager volunteers to participate in his experiments on motion sickness!

Reason decided to look further into the nature of slips and action errors. Good science begins with observation – whether it be animals in their natural habitat or the motion of the stars and planets. Apart from a few anecdotes quoted by Freud and James no scientific body of evidence on action errors existed. With **Klara Mycielska**, Reason conducted two small-scale studies where 98 volunteer subjects kept detailed diary records for one or two weeks.[8] Of particular value was the fact that alongside a description of each unintended action the participants were asked to rate the frequency and recency of occurrence of both the intended action and the unintended action. The two studies yielded a collection of 625 reported errors but the most striking findings were contained in the rating data. More than 80% of participants reported that the intended action was one they performed frequently and had carried out recently. Similarly, for the same proportion of participants, the unintended action was also one that they performed frequently and recently. These data are at the very least highly consistent with a cognitive (rather than say a motivational) explanation of action errors in that the intermingling of highly familiar actions seems to account for the great majority of such errors.

The gathered observational records of action slips are also consistent with William James's account of lapses in attention and habitual actions. Up to this point, we have discussed action errors as those unintended actions that occur during the normal course of events without any conscious awareness of anything abnormal or untoward. Of course, many of these errors are quickly detected and corrected as long as there is sufficient attentional capacity to do so. Obviously, there will also be errors that occur in the course of dealing with a known problem. Reason makes a useful and often-repeated distinction between these classes of error (see Figure 3.2). The first group consists of *slips and lapses* of the kind we have been discussing. The second group can be referred to as *mistakes* where an intended action fails to achieve its intended goals. Reason adds a fourth category of *violations* to refer to a class of

FIGURE 3.2 Reason's four categories of error. Mistakes and violations involve deliberate conscious control whereas slips and lapses generally occur without conscious awareness.

actions ranging from the routine (e.g. not stopping at amber traffic signals) to the exceptional (e.g. acts of sabotage). Whereas slips and lapses can only be defined with reference to internal control mechanisms, mistakes and violations exist by reference to external standards – laws, regulations, normal practices, expected standards, operating procedures, etc.

By 'internal control mechanism' Reason is referring to the level of attentional involvement in guiding action. When we begin learning a skill such as driving a car or playing an instrument, our entire stock of attention needs to be invested in the task at hand. A significant portion of this attention is taken up with monitoring feedback – the sound of the engine, the closeness of obstacles, the relation between our vehicle and others on the road, etc. At this stage, our actions are guided by the gathering of moment-to-moment information or feedback and the task is described as a 'closed loop' (see Figure 3.3). At this stage, we may not know exactly what feedback to pay attention to and what to ignore.

This level of control has been labelled as 'Knowledge-Based' by Danish electrical engineer and safety scientist **Jens Rasmussen**. As well as describing the early stages of learning a complex skill, this also describes the cognitive control required when a familiar task no longer makes sense or there is a conscious awareness of conditions outside normal experience, in other words, 'problem-solving' or 'diagnosis'.

At the opposite end of the spectrum, when we have been practising a task or activity for a great length of time and our actions have become almost entirely habitual or automatic, the level of control can be described as 'Skill-Based'. Moment-by-moment monitoring of feedback has been abandoned and replaced with occasional attentional checks. This level of control can also be described as 'open-loop' (see Figure 3.3). As proposed by Norman and others, most slips and lapses are strongly connected to either a lack of attention at a particular moment when attention is

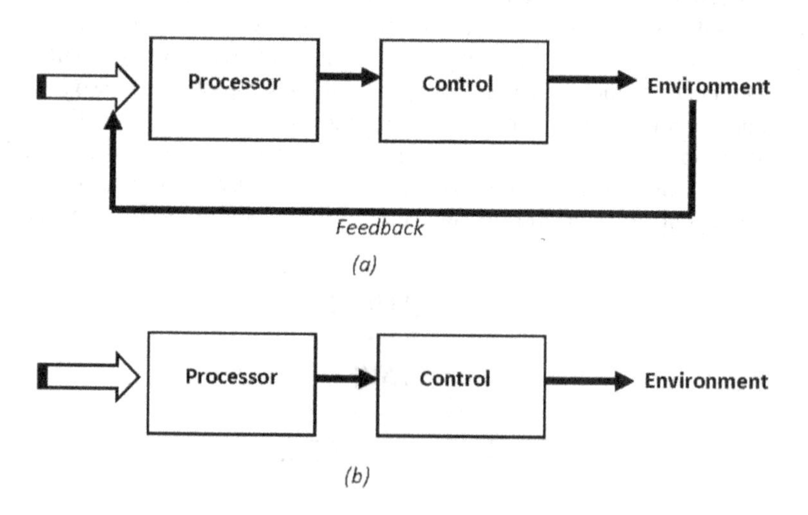

FIGURE 3.3 Two forms of control: (a) Closed-loop control is governed by feedback from monitoring the outcomes of control actions; (b) In open-loop control, action occurs without feedback from prior action.

required (e.g. diverting from a normal route to visit the fish shop) or else over-attention when we focus attention at a time that disrupts the smooth flow of behaviour. Plentiful evidence, chiefly from studies of sporting performance, demonstrate that paying attention to a golf swing or a tennis serve in the moment of execution, for example, can drastically impair performance.[9]

To explain the resulting error forms, Reason described a model which he called GEMS ('Generic Error-Modelling System'), which combines the psychological mechanisms outlined by Don Norman with the theory of the strategic control of action developed by Rasmussen. We will return to Rasmussen's work in greater detail in Chapter 8.

As noted above, Rasmussen proposed that performing any skill can be carried out using one of three levels of cognitive control. Take the automobile driving example for instance. The learner approaching their first driving lesson has some knowledge of cars (having been a passenger), roads and traffic and, possibly, some study of the rules of the road. All of this knowledge is stored in memory. Having buckled in with the ignition turned on, the learner has to draw on this knowledge (as well as the direction of their instructor) to commence controlling the vehicle and eventually negotiating various types of road way and traffic conditions. At first, the process is mentally demanding and requires total concentration. With time, some of the tasks become automatised requiring less and less conscious attention. Essential rules ('give way to the right', etc.) are learned and become established procedures routinely invoked whenever the relevant conditions occur. Eventually, almost all aspects of car control, lane-keeping, collision avoidance, obeying traffic rules, etc. become incorporated into action schema that run almost autonomously, freeing up conscious attention for other activities.

In Rasmussen's theory, similar actions (e.g. driving a car down a road) can involve different levels of cognitive control from the knowledge-based control of the learner to the rule-based control of the intermediate and the stimulus-based control of the experienced practitioner. When confronting unfamiliar situations, the experienced practitioner may revert from skill-based control to rule or knowledge-based control. GEMS provides a descriptive framework for mapping the cognitive control mechanisms (SRK) onto the four error classes (slips, lapses, mistakes and violations) to predict the likely forms of any resulting errors.

At the skill-based level, everything from the examples of William James to the conceptual analysis of Norman to the diary data collected by Reason and Mycielska points to the importance of attentional 'capture' in generating slips, as the focus of attention is somewhere other than on performing the action. At the rule-based level of control, conscious attention is focused on the task at hand with the aim of determining what stored rule is most relevant. Two classes of error could occur. Firstly, the situation might be misunderstood and so an inappropriate rule is retrieved and applied. Secondly, even if the situation is correctly interpreted then the wrong rule might be retrieved or the correct rule might still be misapplied. Finally, at the knowledge-based level of control, many errors are possible. Most commonly, these arise either from lack of knowledge; faulty or incorrect knowledge; or limitations on working with the knowledge at hand duc to constraints on memory and attention. These cognitive limitations will be covered in more detail in Chapter 4.

Reason suggests that underlying this heterogeneous collection of errors are two clear principles. The first principle ('similarity matching') is that the fundamental driver of the whole process is a search for patterns in memory that match the patterns of the problem situation. A good deal of research conducted around this time, as well as more recent findings, all support the notion that skilled performance and expert decision making are largely contingent on similarity matching. People rely on matching the cues provided by the situation to guide a search in memory for stored information that corresponds to the same, or similar, pattern of cues. Ideally, the most relevant situational cues are identified and a correct match is made along with the appropriate action guidance. Unfortunately, there are numerous ways for this process to go wrong, leading to various observed errors in performance.

Reason's second principle is referred to as 'frequency gambling' whereby the search for a stored pattern that matches the current situation is biased in favour of those that are most regularly used. So it is that the behaviours enacted in error are often routines that have been recently carried out in a related context. As Reason notes, these processes are generally efficient and effective in ensuring that things go right but lead to recognisable types of failure when things go wrong.

Through a combination of his writings, extensive consultancy work, as well as his ability as a conference speaker and keynote presenter, Reason became one of the most visible and influential figures in the field of human error for nearly two decades. His work was particularly influential in the aviation, power generation and oil and gas industries and later in the healthcare sector.

COMING TOGETHER: ACADEMIC CONSENSUS ON HUMAN ERROR

The obvious importance of human error both for the psychological theory of skilled performance and for practical purposes began to attract increasing scientific attention. Shortly after the Three Mile Island nuclear disaster in 1979 (see Chapter 4), a small highly selected group of researchers met together in 1980 at what became known as the 'Clambake Conference' and again in 1983 in Bellagio, Italy. The group consisted of all the leading scientists at the time with interests in human error, including both Don Norman and James Reason, as well as Eric Hollnagel, William Wagenaar and Jens Rasmussen.[10]

The group were asked to provide answers to a set of fairly general questions such as 'What is an error?', 'What are the causes?', 'Can errors be predicted?' as well as a long series of rather more specific queries such as 'Is error-free performance possible?' and 'Is error needed for adaptation?' By and large, the consensus views were largely in accord with the distinctions made by Norman and Reason discussed above. An error was considered to occur when there was 'a deviation from intention, expectation or desirability'. This covers both action slips (deviation from intention) and mistakes (deviation from desirability or expectation).

Less clear cut were the discussions on causes and prediction of human error. Causation is a complex topic as we shall see in Section 3 of the book. For the conference participants, the natural framework for considering causation was provided by the generally accepted framework in Western thought since the Renaissance. Causes are conditions that precede the effects they produce. To this way of thinking, accidents can be seen as the result of a chain of causes and effects and to popular notions such as 'breaking the chain' as a way of averting an accident. However, the participants recognised that finding the cause of individual errors can be difficult, if not impossible, although it might be possible to predict errors on a broader basis as in the techniques for human error rate prediction (e.g. THERP) discussed in Chapter 2.

An integration of sorts between Norman and Reason's analyses of action slips was proposed by two German psychologists who focused on the role of intentions in directing and guiding actions.[11] As Norman suggested, actions are initiated by goal-directed intentions but their implementation is directed by lower-level 'initiation, implementation and termination intents'. These three categories were used to classify the same collection of reported slips and action errors analysed by Norman. An example where a bus driver reported finding himself stopping at a bus stop whilst off-duty and driving his car to the supermarket could be classified as a sidetracking of an intention and thus an 'initiation slip'. This corresponds with Norman's classification of the same action error as a 'capture error' whereby something common to both actions (the bus driving and the car driving both passed through the same environmental features) diverted the action from its intended path to a 'stronger' more habitual routine.

One class of implementation errors corresponds with Norman's 'mode error' category where the action is in response to a false specification of the situation of the kind that occurred in the Strasbourg crash (see Table 3.1). Interestingly, as described below, the conscious formation of 'implementation intents' has been shown to be a powerful strategy for avoiding action slips of the kind where an intended future action is forgotten or mistimed thus illustrating the role of implementation intentions in both generating and avoiding action errors.

Whilst there remains some disagreement over the exact specification of the underlying mechanisms, there is broad agreement on the general nature of action errors as the manifestation of our brain's evolved capacity to free our actions from continuous conscious control to run semi-autonomously, allowing our conscious mind freedom to engage with other thoughts and activities. This provides enormous benefits as we can reflect, plan and deliberate whilst carrying out routine skilled behaviour such as driving a car but with an inevitable, if occasional, penalty of the unintended action slip or lapse.

Even these slips and lapses can have an upside. As Paul Fitts noted earlier (see above) some error is necessary for optimum performance. Scientists and artists frequently report occasions when unintended actions (i.e. 'errors') result in serendipitous outcomes such as the discovery of penicillin or an unexpected musical or artistic effect. Rather than trying to eliminate error from human performance, it would be more productive to design and build systems designed to accommodate certain amounts of error whilst minimising the likelihood of errors leading to unwanted and undesirable system states.

FREUDIAN SLIPS

We shouldn't leave the topic of action errors without mentioning another influential view of the same phenomena earlier propounded by **Sigmund Freud** in his *Psychopathology of Everyday Life* published in 1910. Freud presented the classic analysis of slips of the tongue in terms of his own theory of the dynamics and conflicts that govern mental life. In particular, slips of the tongue occur when suppressed wishes manage to break through into conscious behaviour. Freud offers various examples including one where speaking at the opening of the lower house of the Austrian parliament the president of Austria welcomed those present before saying: 'I herewith declare the sitting closed'.[12] In another example, a domineering wife reported to Freud that her husband 'Required no special diet but can eat and drink what I want'. These could be interpreted as instances where the suppressed desire to ignore pesky parliamentarians or control one's partner unexpectedly burst into the plain light of day.

Whilst Freud's theory *could* explain one kind of action error – certain 'sidetracking' or 'capture' errors for example – the theory does not account for the many other kinds of errors which can be explained by the simpler and empirically validated science of attentional capabilities and limitations. From the studies of naturally occurring slips and unintended actions and the theorising of James, Norman, Reason and others we can safely conclude that such behaviours are intrinsically bound up with the way in which skilled actions are organised and developed and, in particular, with their attentional requirements.

REMEMBERING TO REMEMBER

NORTHWEST AIRLINES DC-9-82 FLIGHT 255, CRASHED ON TAKE-OFF, DETROIT 16 AUGUST 1987

The crew were flying their sixth leg of the day. The flight was cleared to proceed to Runway 3C which was the shortest of the available runways. During taxi, the crew were busy reconfiguring the flight computer for the shorter take-off and missed their taxiway exit. After take-off, the aircraft rolled from side to side and failed to gain altitude colliding with a lighting pole 840 m past the end of the runway. Only one small child survived with 154 others fatally injured. It appears that the crew missed a checklist item and failed to ensure the flaps were extended for take-off. Flaps increase the surface area of the wing and generate the increased lift necessary to safely climb after take-off.[13]

Despite the many obvious differences between their tasks, one thing that air traffic controllers, maritime pilots, nurses and power plant controllers have in common is the need to remember to perform some action at some point in the future. Whilst engaged in one stream of action, many skilled tasks depend on the operator 'remembering to remember' to carry out another action either as some future step in the sequence or at some particular point in time.

Academic researchers have only recently (since the 1990s) taken a significant interest in the prospective memory performance. Given its clear and obvious relevance to many practical activities – air traffic control, piloting a ship or aircraft, healthcare delivery, etc. it is even more surprising that it has not received very much attention from researchers in the human factors or safety science communities. An outstanding overview of both the scientific basis and the practical implications of prospective memory research has been provided by **Key Dismukes**, a senior research scientist at the NASA-Ames Research Centre in California.[14]

Dismukes makes the point that like much of our biological and psychological functioning we only really notice prospective memory when it fails. We have already briefly mentioned the forgetting of an intention and the failure to activate an intention at an appropriate time or place. We can also include forgetting whether we have actually performed an action, particularly one of a habitual nature such as washing hands or flushing the WC. Scientists really only started investigating these phenomena in a systematic fashion after **Einstein** (no, not that one) and **McDaniel** developed a simple experimental paradigm that could reliably produce errors in remembering intentions within a laboratory setting in a limited period of time. Prior to that most studies, such as those described earlier, relied heavily on diary records and retrospective memories for slips and lapses.

Research studies have shown that prospective memory can be affected by a variety of factors. People are most likely to remember an intention to perform some action when the action is particularly important to them, when they are able to focus their attention on just one task with minimal interruptions and distractions and when the environment provides highly salient and unambiguous cues related to the prospective task. Conversely, having a variety of tasks of varying priority to accomplish whilst being subject to distractions and interruptions in an environment that fails to provide clear-cut cues, is likely to lead to prospective memory failures. Unfortunately, real-world working environments are almost all of this latter type!

Jim Reason has pointed out that maintenance activities, especially in nuclear power plants and in aviation, possess almost all the characteristics that engender prospective memory failures resulting in omissions of necessary steps or activities.[15] He lists ten characteristics that make a task especially prone to such errors. These include steps that are functionally isolated and are not cued by the immediately preceding step; steps that follow the achievement of the main goal; steps that must be repeated; steps that involve multiple items and so forth. Everyday examples are provided by the early design of ATM machines and the task of photocopying. In the case of the original ATMs, the task was originally designed so that the return of one's card was the final step in the sequence occurring after the cash had been dispensed. Unsurprisingly, a very common error was for people to pocket the cash and walk away leaving their card in the machine. Since the goal of the task was to obtain cash, once that task had been achieved the task appeared completed and the return of the card was overlooked. The problem was easily fixed by redesigning the procedure so that the machine returned the card before dispensing the cash. Similarly, the most common error found with photocopying is to leave the original copy on the glass platen. Again, the goal of the task has been completed when the last copy appears and there is no cue to remove the original before departing the scene. This is especially likely to occur when the user has had to place multiple items on the platen in succession.

For an individual, the best way of avoiding a prospective memory error is not to delay the performance of an intention if at all possible. If this is not possible, then the evidence shows that the best strategy is to think forward and identify some cues or features of the situation requiring the deferred intention and rehearse or visualise these cues together with the intended action. Dismukes notes that these 'implementation intentions' provide 'a powerful and practical way to improve prospective memory performance'. As we shall see in Chapter 4, it is always preferable to design the task or environment to provide the necessary cues, prompts and alerts rather than relying on individuals to always remember to perform an action. In the healthcare system, for example, text or email reminders of forthcoming appointments have been highly effective in reducing wasted appointments involving client 'no-shows'.[16]

Recent research involving an air traffic control task where undergraduate students acting as 'controllers' were tasked with remembering to implement a specific instruction when certain aircraft appeared on their screens showed that an automated cue that flashed when the target aircraft appeared was highly effective in reducing prospective memory errors.[17] To mimic real air traffic control, the participants were engaged in several activities – routine handling, conflict detection, etc. concurrently. The general principle is that a cue, alarm or alert that attracts attention at the moment the intended action is required to be performed can be highly effective in reducing prospective memory failures.

This point has been further reinforced by recent work on the self-detection of errors made in naval aviation maintenance. **Justin Saward**, a Royal Navy air engineer officer, and **Neville Stanton**, a qualified engineer and psychologist and Chair of Human Factors at Southampton University, carried out a series of studies on the commonly experienced, but relatively unexplored, phenomenon whereby sometimes after an event we spontaneously remember that we forgot to do something or did something incorrectly. The authors label this 'Individual Latent Error Detection' or 'I-LED'.[18] The conditions most likely to trigger I-LED were both temporal (less than two hours elapsed since the error) and contextual (encountering cues that relate to the task or being in a similar environment). The authors recommend several simple interventions that are likely to be effective, including placing visual reminders (e.g. a picture of a fuel filler cap) nearby or including verbal reminders alongside data entry sections of the relevant paperwork, and preferably having the paperwork completed in situ rather than another location.

Checklists and Procedures: A widely used tool for reducing the incidence of forgetting errors in complex tasks is the checklist. For many decades, checklists have formed the cornerstone of safety in aviation and are a mandatory activity at several critical stages of flight. More recently positive results have been obtained in reducing healthcare errors through the use of checklists.[19] However, distractions and

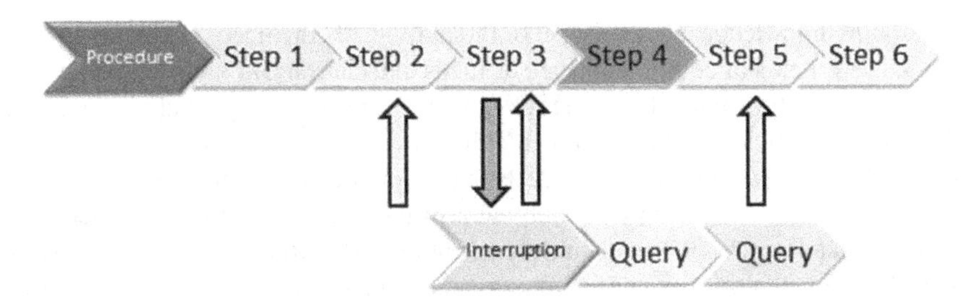

FIGURE 3.4 A schematic of a six-item checklist showing an interruption occurring at Step 3 in the procedure. The arrows illustrate the resumption of the checklist one step early (Step 2), correctly at Step 3 or two steps late (Step 5).

interruptions can easily lead to errors even when working methodically through a checklist as occurred in the 1987 DC-9-82 crash in Detroit.

Figure 3.4 illustrates a schematic six-step checklist. If the crew were interrupted during the execution of the checklist at the third step, for example, to attend to some other task for some period of time and then switched back to the checklist, there are three possible outcomes. One of these is to correctly resume the checklist at exactly the same point where it was discontinued (i.e. Step 3). However, it can be challenging to recall exactly where you were in a highly habitual action amidst other distractions – in the Detroit case, the crew would have completed this same checklist five times already earlier that day.

Of the two other possibilities, one is to resume at an earlier point – say step two. This involves repeating steps two and three. The other possibility is to resume at a later point – say step five which involves omitting step four altogether. When step four is a safety-critical item such as extending the wing flaps then disaster is all but inevitable. Omission errors constitute a large proportion of errors in many fields such as aircraft maintenance for example.[20]

ERROR AND SKILL IN CONTEXT

Psychologists tend to think about memory and attention, skill and error as features of individuals and to seek explanations at the level of individual mental predispositions and neural functioning. **Edwin Hutchins** a colleague of Don Norman's at the University of California, San Diego, has studied memory, skill, attention and error in a broader social and cultural context. Hutchins was trained as a cognitive anthropologist and applied the broader anthropological perspective to the problems he worked on. One of the most notable of these concerned oceanic navigation, which he first studied in the context of traditional Micronesian navigators, and later in the context of US Navy operations.[21] As an experienced ocean navigator himself Hutchins was able to construct extremely detailed accounts of the cognitive, social and computational requirements of navigation.

Central to his account is the notion that the performance of a complex task such as navigation draws on processes and resources distributed amongst a variety of sources – the knowledge and expertise of the individuals involved, characteristics of the task itself, items of equipment (such as maps and charts) and the social organisation of the navigation team. This practice evolves over time to cope with the comings and goings of individual team members. Hutchins' arguments are far too detailed and complex to cover here but he suggests that error needs to be seen in a social and cultural context and not simply as a problem in individual cognition. Furthermore, he argues that since some errors are an essential part of the development of individual skills it is important to design the tasks, equipment and training that make-up work in a way that allows this variability to occur. This is a theme later taken up by Eric Hollnagel and discussed in Chapter 10.

Later, Hutchins built on another of his own areas of skill and expertise, this time in aviation, to undertake studies involving the operation of large commercial jet aircraft. In making the approach and landing the aircraft must achieve certain speeds so that the appropriate configuration of wing flaps can be set. Flaps are physical extensions of the wing surface that can be extended and retracted allowing the aircraft to operate safely at different speeds. Higher speeds are required for efficient cruise and lower speeds are required for landing which must be made at a specific speed.

It is tempting to assume that success (and failure) at setting the right speeds is exclusively the result of the pilots' ability to remember what speeds to set and when to set them. Hutchins painstakingly explains why this is in fact not the case. To ensure there was no misunderstanding on this point, Hutchins entitled the paper 'How a Cockpit Remembers Its Speeds'.[22]

In describing the process by which the pilots interact with various cockpit devices to coordinate flap settings with airspeed, Hutchins draws attention to the multiple ways in which the necessary information is represented in the cockpit – not only in terms of information in the pilots' memory but through various physical means involving information displays and adjustable features of the airspeed indicator. The key point is that much of the process of coordinating speed and aircraft settings takes place externally to the pilot's memory – so much so that we can think of pilots and machines as constituting a 'joint cognitive system'. The important corollary to this is that errors and failures must be accounted for in terms of the functioning of this system as a whole and not just as a function of the crew members' individual mental operations. This important point will be taken up further in Section 3 of the book on the application of systems theory to the problem of safety.

ERROR AS AN INDUSTRIAL SAFETY PROBLEM

Errors are not simply a topic of academic interest but have long been viewed as a critical element in industrial safety. The clearest and most comprehensive approach to error as an industrial safety problem was developed by **Herbert. W. Heinrich** in his classic *Industrial Accident Prevention* first published in 1931.[23] Heinrich worked for Travelers Insurance Company in Hartford, Connecticut, a city known as 'The Insurance Capital of the World' but also for its long association with Mark Twain who wrote two of his best-known novels (*Adventures of Huckleberry Finn* and *Adventures of Tom Sawyer*) whilst living there.

One might wonder why an insurance company would be interested in safety but as professional assessors of risk, insurance companies are also in a position whereby better understanding and management of risks at work represents reduced costs (less claims) and therefore increased profits. Heinrich looked through approximately 75,000 reports and insurance claims for injury at work as the basis for his theory of accident causation. Heinrich proposed that there were five key elements in any accident sequence with an 'unsafe act' or an 'unsafe mechanical or physical condition' at the heart of the sequence. Heinrich illustrated his theory with five dominoes (see Figure 3.5) to drive home the central message that if you successfully address the unsafe act/unsafe condition domino then this nullifies the sequence and prevents the preceding factors from manifesting as an accident. To quote Heinrich: 'In accident prevention the bull's eye of the target is in the middle of the sequence – an unsafe act of a person or a mechanical or physical hazard'.

Heinrich also set out a slightly different model in flow chart form (see Figure 3.6) with four factors or conditions preceding an accident. These are labelled 'Management', 'Man Failure', 'Unsafe Acts of Persons' and 'Unsafe Mechanical or Physical Conditions'. Heinrich states that from his analysis of the 75,000 cases: '98 per cent of industrial accidents are of a preventable kind' and that 88% of accidents are due to unsafe acts of persons which leaves 10% attributable solely to unsafe physical or mechanical conditions. Nevertheless, Heinrich cautions that whilst personal actions cause most accidents, physical and mechanical safeguards are likely to prevent the most accidents.

It's important to note that the vast majority of the reports and insurance claim forms that provided the data Heinrich worked with were completed by plant managers rather than the individual workers. A very well-established phenomenon in psychology is the 'actor-observer effect', which shows that when describing the causes of our own actions we focus outwards on situational and environmental factors whereas when describing the causes of other person's behaviour, we are strongly disposed to seeing them as the result of internal characteristics or dispositions. As a consequence of relying on data submitted by observers (i.e. the managers) rather

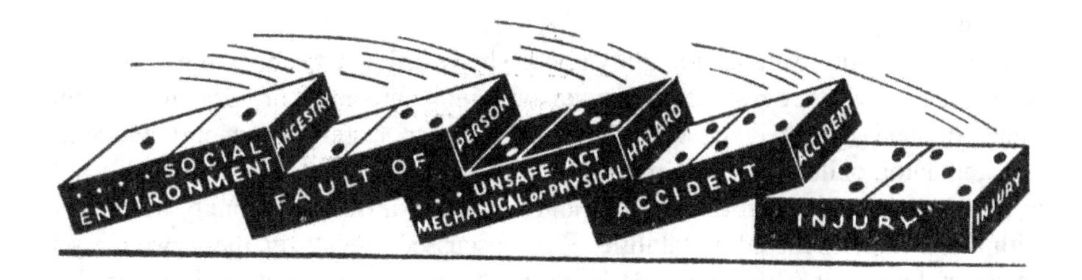

FIGURE 3.5 Heinrich's theory of safety as a sequence of factors depicted as a set of dominoes, illustrating the hypothesised chain of causation from inherited or acquired undesirable traits to unwanted accident and injury. This illustration appeared in the first edition of Heinrich's 1931 book on industrial safety.

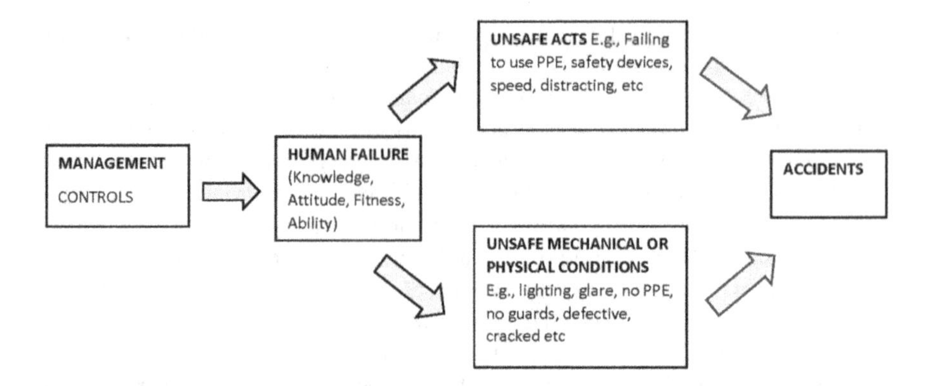

FIGURE 3.6 A flow chart representation of Heinrich's theory of accidents.

than the actors (i.e. the injured workers) themselves, there would have been a strong bias towards dispositional causes which indeed form the essential predisposing factors in Heinrich's domino theory.

Heinrich emphasised the role of personal fault quite strongly. The first two factors in his 'five factor theory' are exclusively devoted to both 'inherited' personal faults such as recklessness, stubbornness and nervousness, and the environments which might encourage these faults. Since Heinrich's reports were almost all filed by external observers representing plant managers and supervisors, it is perhaps unsurprising that their accounts may also have reflected a self-serving function by emphasising the fecklessness of the workers, rather than faults over which the managers may have had control.

As well as setting out in clear and compelling form his theory of accident causation along with a striking visual metaphor of a set of dominoes poised to set off a chain reaction, Heinrich is well known for proposing his 'pyramid' theory of the relationship between minor incidents and major outcomes. He suggested that for every accident involving a major injury there were 29 accidents involving minor injuries and 300 accidents involving no injury whatsoever. Heinrich suggested that accident prevention efforts directed solely towards those events involving injury are misguided and that there should be equal attention to all events irrespective of the outcome.

Heinrich's ratios were later slightly revised by **Frank Bird**,[24] but the premise that all accidents, whether resulting in major or minor outcomes, are drawn from the same population has been heavily criticised. In transportation, in general, and aviation and automobile driving, in particular, slips and lapses are more likely to result in minor outcomes whereas mistakes and violations are more strongly associated with major injuries including fatalities.[25] As discussed previously, these two classes of 'error' have quite different underlying mechanisms and therefore, in contradiction of the Heinrich/Bird theory, represent different populations, requiring different treatments. Heinrich's 'injury ratios' have been labelled the 'least tenable of his premises'.[26]

Heinrich remains an important figure in safety science. He proposed a data-driven approach incorporating psychological principles into safety. Unfortunately, the quality of the data he used to base his principles on was inherently flawed and workplace environments have rapidly evolved since his work in the 1920s–1930s. His overarching focus on flaws in worker behaviour was quickly superseded by the broader focus of work in engineering psychology and human factors discussed in Chapter 4. As we progress further along our 'safety journey' it will become increasingly apparent that event-based causal chain models of accident causation exemplified by Heinrich's dominoes are an inadequate means of understanding all but the very simplest kind of accident. Nevertheless, Heinrich's influence lasted well past his death in 1962 and his domino metaphor enjoyed a major resurgence in a slightly different guise in Jim Reason's work on organisational accidents.

THE 'SWISS CHEESE' METAPHOR

Just seven years after *Human Error* with its focus on individual errors arising from the operation of attentional mechanisms on skilled actions, Reason published *Managing the Risks of Organizational Accidents* with the focus entirely on organisational structures and processes.[27] Reason draws a distinction between 'individual injury accidents' and large-scale organisational accidents of the kind that attract widespread public attention such as the Chernobyl nuclear accident (see Chapter 5). Whilst the distinction between the individual and the organisational accidents might have seemed useful in highlighting the increased complexity of large-scale hazardous activities it is not strictly tenable in the light of evidence on the distributed nature of cognitive and social activities noted by Hutchins, or in terms of our more recent understanding of system dynamics as discussed in Section 3.

Reason's book places a strong emphasis on the concept of *defences*. This is based on the intuitively appealing notion that human behaviour is essentially error-prone and so organisations must create barriers or defences to protect or defend their activity against the inevitability of human failure. An accident can only occur when the defences have been breached. Understanding organisational accidents is, therefore, about understanding how and why the defences failed or were circumvented.

Defences can consist of a range of protections from 'soft' such as legislative requirements, rules and procedures to 'hard' such as physical barriers, alarms, interlocks and escape and rescue devices. The more hazardous the activity the more layers of defence are required. Reason argues that over time the effectiveness of some defences may gradually erode due to overfamiliarity or complacency. Additionally, in the absence of catastrophe, organisational management may become tempted to transfer resources from the protective side of the business to the productive side, further eroding the organisation's defences against hazards.

To illustrate this point Reason developed the 'Swiss Cheese' analogy which has since been very widely repeated. In its original formulation (see Figure 3.7), the organisation's multi-layered defences are idealised as impenetrable barriers preventing hazardous conditions from causing material losses. This is mostly the

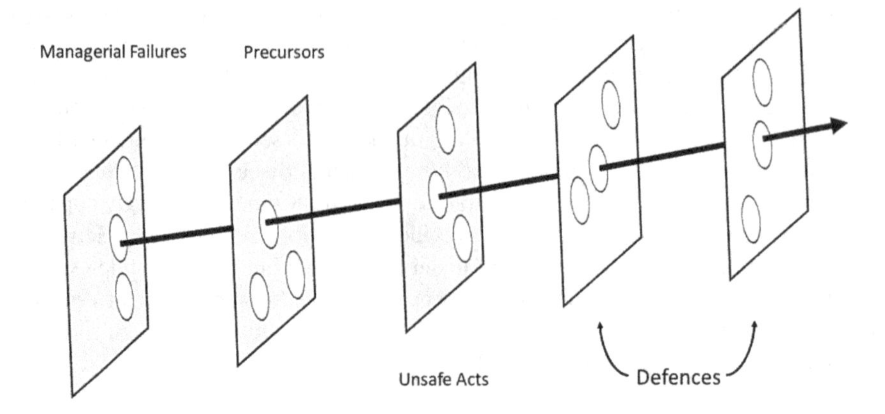

FIGURE 3.7 Reason's initial model of accident causation. Weaknesses in each layer from upper management to individual actions and to the system's defences are represented metaphorically as 'holes'. An accident trajectory is represented as a straight line passing through the 'holes' or weaknesses in each of the layers. This only happens on relatively rare occasions when the 'holes' are in perfect alignment.

view held by management. The reality can be far different. Each defence may have weaknesses – lack of maintenance, poor quality equipment, inadequate training, lack of procedure following and so forth. Often these weaknesses are well known to those at the 'sharp end' of the business but their opinions may be unknown to management or disregarded even if they are known. For example, pleas from crews of roll on-roll off (RORO) ferries such as the *Herald of Free Enterprise*, to install a bow-door open warning device on the ships' bridge were rejected by company management. Over the course of numerous presentations spanning several decades, the analogy with layers of 'Swiss Cheese' gradually grew more and more explicit until the original model became completely synonymous with this mode of representation (see Figure 3.8).

Reason also emphasised a distinction between 'active failures' and 'latent failures'. This distinction refers to the difference between the easily observed actions of those at the 'sharp end' – the misreading of a display, the failure to follow a checklist, etc., and the less observable and often poorly documented decisions of those further away in line management or company boards. Over time Reason changed the terminology to 'latent *conditions*' as a better fit for the idea that high-level organisational decisions create the conditions for failure rather than directly leading to them. In Reason's theory, defences and barriers can be compromised either by unsafe actions of front-line operations personnel or by latent conditions – such as deferred maintenance or inadequate training – resulting from management choices. As noted above, these choices often reflect the trade-off between expenditure on increasing the productive activity of the organisation or expenditure on strengthening unseen defences which after long periods of apparently 'safe' operations can be easily questioned. Combining the active vs latent distinction with the emphasis on defences as

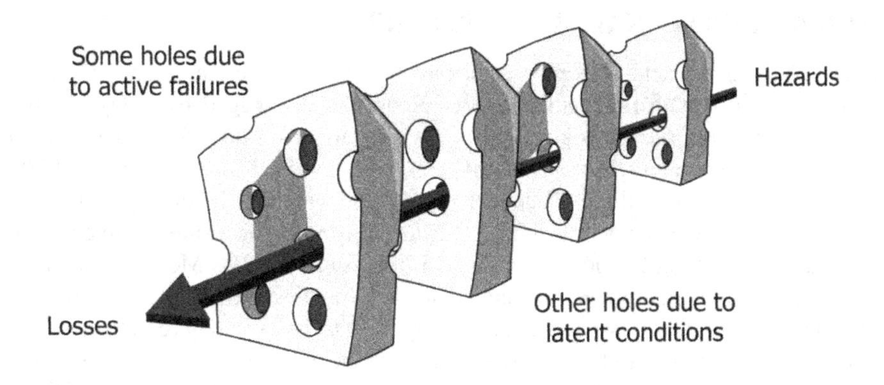

FIGURE 3.8 Two decades later, the Reason model of accident causation became fully synonymous with the Swiss dairy product. ©2008. From: The Human Contribution by Jim Reason. Reproduced by permission of Taylor & Frances Group, LLC, a division of Informa plc.

the primary weapon against accidents yields the model of organisational accidents shown below (see Figure 3.8).

Although the 'Swiss Cheese' analogy was initially used as a means of illustrating breakdowns in defences, it gradually began to take hold of the entire conceptual chain from upper management through front-line workers. Each of these planes of activity was shown with 'holes', often with arrows illustrating the unlikely 'lining up' of faulty processes. Clearly, the analogy has become overextended by this point as it makes little sense to conceive of both a boardroom procurement decision that was taken months or years earlier and a single operator's action as identical 'holes' in hypothetical planes. In this form, the 'Swiss Cheese' analogy looks a lot like Heinrich's dominoes, perhaps illustrating the strong human drive towards simplicity of representation and what the Gestalt psychologists called 'goodness of form'. Interestingly, Heinrich's pioneering work and his domino analogy are never referenced in Reason's work though he surely cannot have been unaware of it.

Reason's work on 'organisational accidents' has had a profound impact on safety thinking in many domains and thousands of references to the 'Swiss Cheese' model can readily be found. The model may have served a useful function at the time as a corrective to an overemphasis on individual fallibility as the 'root cause' of industrial and occupational injury, but as Reason himself later realised,[28] the pendulum may have swung too far in replacing one oversimplified approach based on a human error with another oversimplified approach based on organisational failure. We will look at organisational structure and functioning in more detail in Section 2.

HUMAN ERROR ANALYSIS IN PRACTICE

Researchers in the field of human error have a natural interest in connecting their theoretical work with practical concerns. Reason was engaged by many companies and organisations as a consultant, advising them on how to reduce human error in their operations. As part of his work with the Shell oil company in 1988, Reason, along with William Wagenaar and several others at Leiden University in the Netherlands, developed a technique for use in the field as a safety management tool based on the general ideas espoused in the 'Swiss Cheese Model'. Known as TRIPOD, the tool focused on the 'latent conditions' that were theorised to lie behind later active failures. Assessing the state of an organisation with regard to these latent conditions was mooted to be akin to an individual health assessment based on a few indicators such as blood pressure, pulse rate and temperature.[29] A large pool of these relevant, objectively observable indicators could be assembled and used to generate checklists to periodically audit the current organisational or operational 'state of health'. Responses can be used to build a profile of organisational functioning to best target management efforts towards any areas such as communication, training or maintenance that generate the highest scores. TRIPOD is still available for use but has been largely superseded by other specific techniques as well as the rise of Safety Management Systems (SMS), in general.

By far the best-known application of Reason's error theory, defined as a method to define the 'holes in the Swiss Cheese', was one developed by two psychologists, at that time serving in the US Navy, **Scott Shappell** and **Doug Wiegmann**. Designed as a means of categorising or classifying recorded occurrences, it was initially known as the *Taxonomy of Unsafe Operations*. The taxonomy was designed to allocate factors recorded in the investigation into categories corresponding to four levels of Reason's model, namely organisational influences; unsafe supervision; preconditions for unsafe acts and, finally, unsafe acts of operators. Reason's distinctions between mistakes, slips and lapses were found too difficult to operationalise so were replaced by the categories of decision errors, skill-based errors and perceptual errors. The category of violations was retained. The modified taxonomy became known as the *Human Factors Analysis and Classification System* or HFACS.[30]

An early comparison between US Navy/Marine occurrences with those from the US Army and US Air Force showed a significant disparity in the proportions associated with the category of violations with 33% of Navy occurrences associated with at least one violation compared with 25% (Army) and 10% (Air Force). This stimulated further inquiry into cultural differences between the three organisations and the hypothesis that US Navy culture overemphasised autonomy in decision making leading to a 'can-do' culture involving excess risk-taking. Later interventions designed to increase accountability and adherence to rules were successful in reducing the incidence of certain violations to the same levels as the Air Force. However, more recent comparisons show overall US Navy/Marine violation rates to still be much higher than the US Air Force.[31]

An important advantage of techniques such as HFACS is that by classifying errors in terms of underlying mechanisms rather than by factors specific to a work domain (e.g. 'skill-based error' rather than 'raised flaps too soon') it becomes possible to compare across different domains with one another (e.g. surgery with aviation) and to easily compare trends across time. As noted above, the system also facilitates data-driven interventions rather than relying on current fads and fashions. Tying the categories to clear definitions has enabled analysts to agree on how findings should be classified which is reflected in relatively high levels of inter-rater reliability across numerous studies. HFACS has been used by the US Department of Defence for analysing aviation mishaps across different military and coast-guard services.

However, it is important to note that HFACS essentially classifies findings recorded by the incident investigators so is dependent on the original analyst's opinions about an event and the evidence originally considered relevant. In some cases, where HFACS has been integrated into the entire investigation process it can also serve as a framework for identifying relevant issues. As we shall see later (Chapter 11), accident investigations have tended to focus on front-line behaviours rather than practices further up the organisational hierarchy. If the original investigation goes no further than 'operator error' then methods such as HFACS are less able to reveal anything about underlying factors present elsewhere in the system. This is a significant limitation on the usefulness of error categorisation approaches such as HFACS to provide information on systemic issues in accident investigation.

SUMMARY AND CONCLUSIONS

Error is very much a part of human experience. We have only begun to understand the functions and mechanisms of error relatively recently. An important distinction is between mistakes arising from faulty or inaccessible knowledge and slips of action arising outside conscious awareness. Solutions for the former range from providing information through training to providing information elsewhere in the task or environment. Solutions for the latter depend more on structuring tasks to minimise the effects of these inevitable deviations of action from intention. Internally generated implementation cues along with externally generated alerts and cues can be effective in reducing the impact of prospective memory failures on task performance. The topic of designing tasks and environments to minimise error and maximise desirable performance is covered in more detail in Chapter 4.

Early theories of industrial and occupational safety emphasised the role of individual errors and unsafe acts. The pervasive idea that all accidents irrespective of type or outcome reflect the same underlying process has proven to be misleading. Attempts to balance the earlier emphasis on individual errors with an emphasis on organisational malfeasance has created a false dichotomy pitting one source of error against the other. In later sections, we will look at organisations in greater detail and suggest broader explanatory frameworks that provide a better understanding of safety.

To paraphrase William James, Who is there that has never embarked on a mistaken course of action or found themselves watching the windscreen wipers scraping across a dry windscreen instead of a turn signal flashing? We all have, as human errors are all too real and fairly commonplace. The appearance of error provides information that can be used to devise better equipment, more standardised controls, more salient information and so forth. Improving safety by design will be the topic of Chapter 4. What the human error concept absolutely cannot do, however, is bear the burden of responsibility as a causal explanation of failures and catastrophes. Despite blaring headlines proclaiming 'pilot error' or 'train driver error' the plans or actions of the operating crew (except in rare examples of extreme deliberate violations) cannot carry the full weight of explanation. We will examine this claim in greater detail in the last section of the book.

NOTES

1. Janis, I.L. (1972). *Victims of Groupthink*. Boston, MA: Houghton Mifflin. Both the Bay of Pigs invasion of Cuba and the Cuban Missile Crisis are described in detail along with several other examples of successful and unsuccessful group decision making. The Bay of Pigs was such a fiasco that Janis labels it a 'perfect failure'.
2. Woodworth, R.S., & Schlosberg, H. (1954). *Experimental Psychology*, 3rd Ed. London: Methuen.
3. Fitts, P.M., & Posner, M. (1967). *Human Performance*. Belmont, CA: Brooks/Cole.
4. James, W. (1890). *Principles of Psychology*. London: Macmillan. In an ironic twist I was reading James's chapter on habitual actions in our university library and decided to check the book out and take it back to my office (even more ironically located in the William James building on the Otago University campus). The checkout machines require users to scan their library card which I did but received an error message. On inspection I realised that I had presented the library card I use quite frequently at the public library instead of my less well used university card. Habit rules!
5. Norman, D.A. (1981). Categorization of action slips. *Psychological Review, 88*, 1–15.
6. The information comes from the *Official Report of the Commission of Investigation into the Accident on 20 January 1992 near Mont Saint Odile (Bas-Rhin) of the Airbus A.320 registered F-GGED operated by Air Inter,* produced by the French Ministry of Transport and Tourism dated 26 Nov 1993. The two different descent modes were controlled by a push button on the panel in front of the pilots.
7. Reason has written a short autobiographical account of his career entitled '*A Life in Error: From Little Slips to Big Disasters*' published in 2013 by Ashgate Press.
8. These studies are reported in the book '*Absent Minded? The Psychology of Mental Lapses and Everyday Errors*' by Jim Reason and Klara Mycielska, published by Prentice-Hall in 1982.
9. Many studies have shown that paying attention to skilled movements which have become highly automated can disrupt their smooth execution. See for example; Baumeister, R., & Showers, C.J. (1986). A review of paradoxical performance effects: Choking under pressure in sports and mental tests. *European Journal of Social Psychology, 16*, 361–383.
10. The results of this meeting are described by John Senders and Neville Moray in '*Human Error: Cause, Prediction, and Reduction*' published in 1991 by Lawrence Erlbaum.
11. The paper in Psychological Review carries the names of Heinz Heckhausen and Jurgen Beckmann although the first author had sadly died during the submission process. The reference is: Heckhausen, H., & Beckman, J. (1990). Intentional action and action slips. *Psychological Review, 97*, 36–48.
12. This example was described in Frank Sulloway's '*Freud, Biologist of the Mind* published in 1979 by Burnett Books.

13. National Transportation Safety Board. (1988). *Aircraft Accident Report – Northwest Airlines Inc, McDonnell Douglas DC-9-82, N317RC, Detroit Metropolitan Wayne County Airport, Romulus, Michigan, August 16, 1987. Report No NTSB/AAR-88/05.* Washington, DC: NTSB. The NTSB stated that 'the probable cause of the accident was the flight crew's failure to use the taxi checklist to ensure that the flaps and slats were extended for takeoff'.
14. An excellent overview of the science behind prospective memory is provided by R. Key Dismukes in 'Remembrance of things future: Prospective memory in laboratory, workplace, and everyday settings' published in *Reviews of Human Factors and Ergonomics*, 2010, 6, 79–122.
15. Reason, J. (1998). How necessary steps in a task get omitted: Reviving old ideas to combat a persistent problem. *Cognitive Technology, 3*(2), 24–32.
16. An example can be found in: Macharia, W.M., Leon, G., Rowe, B.H., Stephenson, B.J., & Haynes, R.B. (1992). An overview of interventions to improve compliance with appointment keeping for medical services. *Journal of the American Medical Association, 267,* 1813–1817.
17. This was reported in a paper by Shayne Loft, Rebekah Smith and Adella Bhaskara: 'Prospective memory in an air traffic control simulation; External aids that signal when to act' published in 2011 in the *Journal of Experimental Psychology: Applied, 17,* 60–70.
18. Saward, J.R.E., & Stanton, N.A. (2019). *Individual Latent Error Detection (I-LED), Making Systems Safer.* Boca Raton, FL: CRC Press.
19. Studies have shown reductions in surgical morbidity and mortality. For example a large (> 25,000 patients) retrospective cohort study demonstrated a significant reduction in post-surgical mortality following compliance with the World Health Organization (WHO) checklist. See van Klei et al. (2012). Effects of the introduction of the WHO 'Surgical Safety Checklist' on in-hospital mortality: A cohort study. *Annals of Surgery, 255,* 44–49.
20. Alan Hobbs, at the time an aircraft accident investigator at the Australian Transport Safety Board, wrote a book on aircraft maintenance errors with Jim Reason. See: Hobbs, A., & Reason, J. (2003). *Managing Maintenance Error: A Practical Guide.* Boca Raton, FL: CRC Press. For a recent study involving aircraft maintenance errors see Saward, J.R.E, & Stanton, N.A. (2019). *Individual Latent Error Detection (I-LED), Making Systems Safer.* Boca Raton, FL: CRC Press.
21. His work on US Navy ships is outlined in detail in '*Cognition in the Wild*' published by MIT Press in 1995. Hutchins sets out his theory of 'distributed cognition' in his examination of how navigators carry out their tasks.
22. Hutchins, E. (1995). How a cockpit remembers its speeds. *Cognitive Science, 19,* 265–288. In this article Hutchins uses his expertise as a fully qualified commercial pilot to outline how the task of configuring a jet airliner for landing involves information distributed between pilots and the technological devices in the cockpit.
23. Heinrich published the first edition of '*Industrial Accident Prevention: A Scientific Approach*' in 1931. I have used the second edition which appeared in 1941 published in New York by McGraw-Hill, with the final fourth edition appearing in 1959.
24. The original source of Frank Bird's revised accident ratios is hard to pin down although his work is often quoted.
25. We showed that there were quite different factors behind serious aircraft crashes compared with crashes with only minor outcomes: O'Hare, D., Wiggins, M., Batt, R., & Morrison, D. (1994). Cognitive failure analysis for aircraft accident investigation. *Ergonomics, 37,* 1855–1869.
26. A detailed critique of Heinrich's work can be found in a chapter entitled: 'Heinrich revisited: Truisms or Myths' by Fred Manuele (2003). *On the Practice of Safety,* 3rd Ed. John Wiley.

27. Reason, J. (1997). *Managing the Risks of Organizational Accidents.* Aldershot: Ashgate.
28. Reason, J. (2008). *The Human Contribution: Unsafe Acts, Accidents and Heroic Recoveries.* Aldershot: Ashgate.
29. Hudson, P.T.W., Reason, J.T., Wagenaar, W.A., Bentley, P.D., Primrose, M., & Visser, J.P.E. (1994). TRIPOD Delta: Proactive approach to enhanced safety. *Journal of Petroleum Technology, 46*, 58–62.
30. Shappell, S., & Wiegmann, D. (2003). *A Human Error Approach to Aviation Accident Analysis: The Human Factors Analysis and Classification System.* Aldershot: Ashgate.
31. Shappell, S., & Wiegmann, D. (2004). HFACS analysis of military and civilian aviation accidents: A North American comparison. *Presented at the ISASI 2004 Conference, Gold Coast, Australia.* Downloaded from: https://asasi.org/papers/2004/Shappell%20et%20al_HFACS_ISASI04.pdf

4 Safety by Design

At first glance, the link between pilot errors in the Second World War and the design of modern computer interfaces might not seem all that obvious. However, they are both part of the development of what is variously known as *engineering psychology, human factors or ergonomics*. Nothing seems to galvanise human ingenuity more than a short-term threat or shock such as the outbreak of a novel virus or a war. As we saw in Chapter 2, humans quickly become habituated to risks and threats, and once seen as familiar, they no longer command the same attention or mobilise the same effort. In recent history, war times have most often been the source of the sudden threat that has galvanised national efforts to achieve a highly desired outcome. The additional forces of nationalism and patriotism have also served to harness available resources and stimulate all kinds of innovation.

In the cases of the two world wars of the 20th century, one of the impacts was to bring a variety of scientific efforts to bear on the war effort. In the First World War, with the appearance of aerial warfare alongside traditional sea and ground forces, scientists were brought in particularly to assist with the selection and training of aircrew. On both sides of the Atlantic, psychologists helped design selection tests for aviators as well as other military personnel. Leading academic psychologists of the day such as **Edward L. Thorndike** and **Robert Yerkes** led these efforts to develop tests to improve the 'selection, classification, and assignment of military personnel'. By the end of the war, over 1.5 million US Army recruits had been tested. Similar efforts began slightly earlier in Germany, where the very first modern psychological laboratories had been founded.[1]

Until the outbreak of the Second World War, the efforts of psychologists were largely directed at individual behaviour with topics such as testing and selection, leadership and 'mental disorders' the main concerns. The Second World War brought unprecedented involvement of scientists into the war effort, notably physicists and engineers involved in the design of advanced weaponry as well as new electronic systems designed to detect enemy operations such as Radar and Sonar. At Cambridge University, the first professor of psychology, **Frederick Bartlett**, had become interested in the practical problems raised by the integration of humans into these increasingly complex and sophisticated engineered systems. He wrote an early text on 'Psychology and the Soldier' published in 1927 which covered the three topics listed above. He was later responsible for founding the Applied Psychology Unit which, funded by the UK Medical Research Council, conducted research into numerous questions of practical importance over the 54 years of its existence.[2] However, Bartlett became most famous for his studies, conducted at around the same time, into the remembering of stories which dramatically illustrated the important fact that memory is largely a reconstructive process and not simply a passive recording mechanism. In later life, he became 'Professor Sir' Frederick Bartlett.

DOI: 10.1201/9781003038443-5

One of Bartlett's lesser-known achievements involved the training of a visiting American scholar who spent a year at Cambridge in the 1920s. The experiments they conducted at Cambridge involved testing the effects of anoxia or oxygen deprivation (often referred to as 'hypoxia') on both simple and higher-level cognitive processes. The visiting American returned to the United States and carried out considerable further work on anoxia and other aviation-related matters through extensive field work in the Andes as well as an extended collaboration with Pan American Airways. In 1946 the former student, **Ross McFarland**, also became the author of the first published book specifically on the topic of human factors. His research eventually led to the establishment of the '10,000 foot cut-off', the upper safe limit above which pilots increasingly show significant cognitive declines and must, therefore, use supplemental oxygen.

DESIGNING FOR FLIGHT SAFETY

By far the most influential figure in the development of engineering psychology or human factors at the time was **Paul Fitts**, originally an animal behaviourist who was co-opted into the war effort in 1941 by the US Army Air Forces. Following a trip to Germany to view the war efforts there, Fitts was given leadership of the psychology branch of the Aero Medical Laboratory in Dayton, Ohio, in 1945. Some work on selection and testing had continued between the wars but efforts had accelerated rapidly with the onset of the Second World War. Over 9 million personnel were ultimately given the *Army General Classification Test* (AGCT) before the end of the war. Smaller-scale efforts to test potential air force recruits were also established.[3] Fitts went far beyond the traditional focus on selection and classification of individuals with a couple of ground-breaking studies summarised in reports released in 1947.[4]

For the first time, the focus was on potential deficiencies in the instrument displays and controls themselves rather than deficiencies in the people reading the instruments or operating the controls. Data were obtained from face-to-face interviews with 100 pilots, as well as 524 responses to a written survey. Pilots were simply asked to describe any errors they had made, or personally witnessed another pilot making, in reading or interpreting a display, or in the operation of a cockpit control. This resulted in a collection of 270 display reading errors and 460 control operation errors. The most commonly reported display reading error was to misread the values on a multi-pointer display (see Figure 4.1), most especially the altimeter. The most commonly reported control operating error was confusing one control with another. In one example, the co-pilot of a heavily loaded bomber aircraft inadvertently moved the lever controlling the flaps (devices that increase the lift produced by the wings) instead of the undercarriage. This is an example of the common 'slip' or 'action not as planned' as discussed in Chapter 3.

Instead of concluding that pilots needed to pay more attention, be more careful or were in need of further training, the ground-breaking conclusions were that the equipment needed redesigning to better match human capabilities and limitations. This represented a major shift in thinking away from trying to tailor the individual to better fit their circumstances (whether through selection or training, as had characterised the contribution of psychologists in the First World War) to one of trying to redesign either the equipment, the task or the environment to better fit the individuals who operated them.

Fitts was clear that many of these errors could be avoided by redesigning the displays and controls to better fit with the capabilities and limitations of human perception and motor control. The reports set out a variety of research priorities that would be followed for several decades ahead by psychologists, engineers and designers under the rubric of 'engineering psychology'. This was defined by Fitts himself as follows: 'Engineering psychology is concerned with adapting one important characteristic of the environment, the machines of a technological society, to man's own requirements'.[5] An excellent example is provided by the redesign of the notorious three-pointer altimeter which was the most often reported incident in the wartime study of pilot display reading errors.

The maximum altitude reached by the Wright brothers in their early powered flights at Kittyhawk was only around 10 ft above ground level. However, as aircraft developed in capability an instrument capable of showing the pilot their altitude above the ground became a vital necessity. As early aircraft only flew a few thousand feet high, the design of early altimeters could just follow the familiar design of the analogue watch face with easily readable divisions marked off to show the thousands of feet. Such an instrument could provide the pilot with reasonably accurate readings up to around 10,000 or 12,000 ft. Once aircraft began to attain much greater altitudes the problem of providing a readable display became much more significant. The best solution seemed to once again follow the design of the analogue watch. A single scale showing the 86,400 seconds in a day would obviously not be readable. Instead, one scale marked off in 12 major divisions (hours) with 5 subdivisions within each sector combined with 3 separate hands for hours, minutes and seconds could provide the required information down to the smallest (second) degree of accuracy.

The same solution was adopted for altimeter design allowing a single analogue instrument to indicate altitudes up to 40,000 or 50,000 ft. One hand (the shortest) indicated the tens of thousands of feet, another the thousands of feet and the longest indicated the hundreds of feet. These altimeters appeared in the 1940s and can still be found in some older generation piston aircraft today. Pilots frequently misread the readings on these displays, sometimes with fatal consequences. For example, the altimeter shown in Figure 4.1 indicates 10,180 ft although it takes a bit of practice to figure that out. It would be very easy to misread the value by 10,000 ft (indicated by the small pointer on the '1'). On a dark night with no outside visual reference, the consequences of misreading say 2,500 ft as 12,500 ft and descending several thousand feet could quickly result in tragedy. An example of one airliner crew doing precisely this can be found in **John Rolfe**'s fascinating account of the problem and the search for a solution.[6]

Researchers in the United States, Australia and the UK quickly moved to look at alternative designs for presenting altitude information to pilots. In 1959, the UK formed a committee consisting of civil airline operators, regulators and the RAF Institute of Aviation Medicine (IAM) to look at this problem. Over the next four

FIGURE 4.1 A three-pointer altimeter showing an altitude of 10,180 ft. Based on an image by the US Federal Government in the public domain. Originally from en.wikipedia.

years, a systematic programme of research was undertaken at the IAM that still stands as a model of high-quality research on a safety-critical issue. At the same time, the outcome also provides a cautionary tale for those hoping to see science applied to the critical problems of our time from managing pandemics to dealing with climate change.

The overall strategy was to approach the problem in four steps. Firstly, the existing literature was scrutinised for appropriate criteria to use and expert opinions from subject matter experts (SMEs) elicited. Secondly, the large number of potential display options was reduced to a smaller manageable number for the initial laboratory study using simplified forms of presentation. The results from these trials were then used to design more sophisticated and realistic simulator trials of the selected displays. Finally, the cooperating airlines installed working examples of the most promising design in the cockpits of several of their aircraft for evaluation in actual operating conditions.

The outcome of these studies can be summarised as follows. Firstly, all the studies done at the IAM and elsewhere confirmed the findings reported by Fitts and Jones that misreadings of the multi-pointer altimeter were common and potentially lethal. Secondly, a combination of a digital counter along with a moving-pointer circular display was found to be the most efficient means of accurately conveying altitude information to the pilot. An interesting divergence between expert opinion and experimental findings was also noted. This is sometimes referred to as the *preference-performance dissociation*. Whilst the SMEs preferred the digital display to be located *outside* the moving-pointer display, the measures of accuracy and speed of altitude readings showed the display with the digital counter located *inside* the moving-pointer display to be superior (see Figure 4.2).

The difference between the two was not so marked in the simulator and field trials, although the tendency for the digital outside display to appear as two separate displays was accentuated. Research has shown that we can process multiple features of a single object more efficiently than single features of multiple objects. Having the

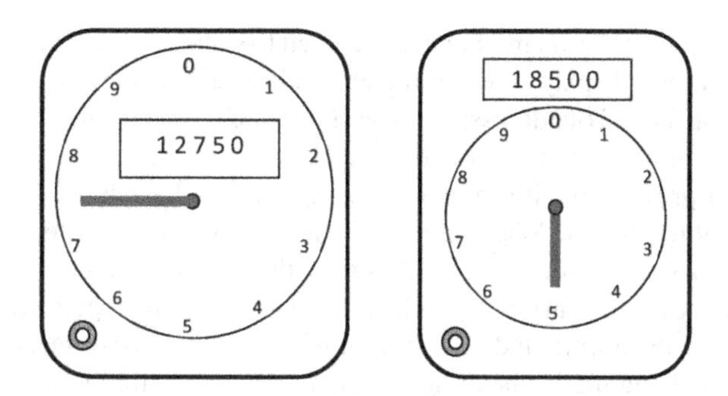

FIGURE 4.2 The two forms of combined analogue and digital representation of altitude information shown to pilots in Rolfe's study of altimeter design.

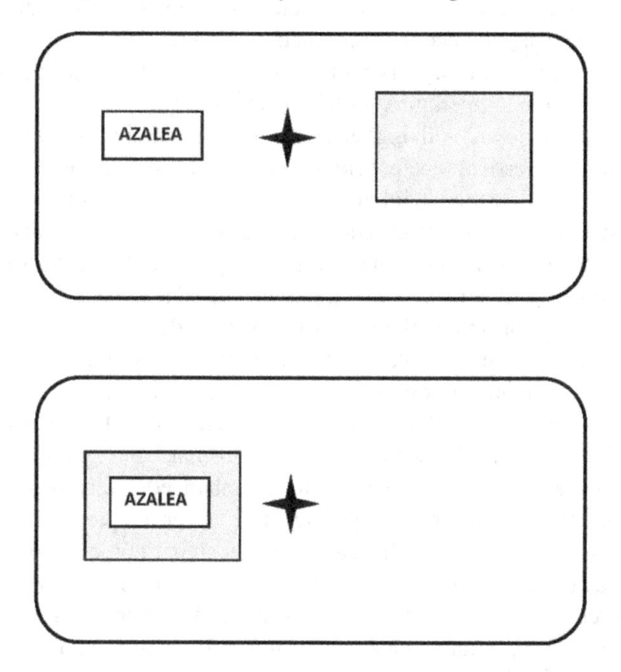

FIGURE 4.3 Processing two features (e.g. a word and a rectangle) as a single object, as shown in the lower part of the figure, is quicker and more accurate than processing them as separate objects, as shown in the upper part. Participants in these studies were asked to fixate on the cross in the centre of the figure and read the word.

altimeter display appear as a single object (with the inside counter) therefore confers distinct advantages in terms of information processing. A simple example (see Figure 4.3) can illustrate this point. It takes longer to read a word when presented opposite a coloured shape, as shown on the upper part of the figure. Presenting them both as a single object improves both speed and accuracy.[7] The fact that separate objects compete for our attention whilst different parts or features of a single object do not, has important implications for the design of displays.

At this point, the question might have occurred to some readers as to why not simply provide a digital display showing the exact altitude? This leads to an important issue not yet considered of addressing the uses to which a display might be put. As an example, you might look at your watch to check the exact time. In this case, the digital format will provide this information most quickly and accurately. Alternatively, you might want to see how long it is until 5 p.m. in which case a quick glance at an analogue watch will be your best bet.[8] Whilst the digital altimeter would be best suited to check readings, particularly when flying level, the analogue altimeter would provide more easily assimilated information when climbing or descending and values are rapidly changing. Embedding a digital counter within an analogue display meets both needs.

The widespread introduction of the improved altimeter design has undoubtedly saved many lives and contributed significantly to improved aviation safety. The design has again changed since the introduction of flight decks with electronic displays which have allowed designers more flexibility in how airspeed, altitude, etc., can be displayed. Like temperature, altitude is inherently a variable that is naturally thought of as 'up and down' with higher values at the top and lower values at the bottom. Thus a moving vertical tape provides a more 'natural' way of representing this variable. In contrast, direction is inherently a circular variable as it exists all around us, so the most appropriate representation remains a circular 'clock' face. On airliner flight decks altitude is now represented by a moving vertical tape whilst direction is represented with map-like displays and a moving circular scale.

John Rolfe, who coordinated the IAM studies, provides a cautionary note regarding the altimeter redesign example. Although civil airlines and many military services adopted the 'counter-inside' altimeter, the old three-pointer design remained in many aircraft and can still be found at the present time. History provides many examples where research evidence on safety benefits has been ignored. As an example, Rolfe cites clear evidence in favour of the enhanced safety provided by rear-facing seats in commercial aircraft. Airlines have never adopted this arrangement. Presumably, this is due to a belief that customers much prefer facing forward towards the direction of travel. Military aircraft are generally configured with rear-facing seating since 'customer' preference is not an issue! In other cases such as climate change, clear and compelling evidence may similarly be ignored in favour of ideological or politically driven decision making.

CONTROLS AND DISPLAYS

Aviation is not the only area where engineering psychology (or human factors/ ergonomics) has resulted in significant safety improvements through redesigning equipment to better match human capacities and limitations. The words of **Charles Hopkins**, a former President of the US Human Factors and Ergonomics Society, are worth quoting directly:

> In some cases, the distribution of displays and controls seemed almost haphazard. It was as if someone had taken a box of dials and switches, turned his back, thrown the whole thing at the board and attached things wherever they landed.[9]

FIGURE 4.4 The Three Mile Island nuclear power plant. Image from: http://phil.cdc.gov/phil/ ID#: 1194.

Astoundingly, Hopkins is talking about the control rooms at the Three Mile Island (TMI) nuclear plant in Pennsylvania following a serious incident that took place in 1979 (see Figure 4.4). Again, John Rolfe's warning seems prescient – you can clearly demonstrate how to improve safety and save lives through human factors engineering of displays and controls but you can't force an industry to take notice. Fortunately, TMI made the US nuclear power industry sit up and pay attention!

THREE MILE ISLAND, PENNSYLVANIA, NUCLEAR MELTDOWN, 28 MARCH 1979

'The accident began about 4 a.m. on Wednesday, March 28, 1979, when the plant experienced a failure in the secondary, non-nuclear section of the plant (one of two reactors on the site). Either a mechanical or electrical failure prevented the main feedwater pumps from sending water to the steam generators that remove heat from the reactor core. This caused the plant's turbine-generator and then the reactor itself to automatically shut down. Immediately, the pressure in the primary system began to increase. In order to control that pressure, the pilot-operated relief valve opened. It was located at the top of the pressurizer. The valve should have closed when the pressure fell to proper levels, but it became stuck open. Instruments in the control room, however, indicated to the plant staff that the valve was closed. As a result, the plant staff was unaware that cooling water in the form of steam was pouring out of the stuck-open valve. As alarms rang and warning lights flashed, the operators did not realize that the plant was experiencing a loss-of-coolant accident. Other instruments available to plant staff provided inadequate or misleading information'. [10]

The control room at TMI illustrates many of the basic problems in designing equipment to fit human capabilities. The same problems appear in other nuclear reactor control rooms as well as almost every industrial and manufacturing operation that one can think of. Disturbingly, many of these problems continued to be found in nuclear plants long after the TMI accident. There are many excellent sources of information on essential design requirements that can be used to create more visible, legible and useful visual displays. Similar guidelines can be found for auditory displays and to a much lesser extent for other kinds of displays – haptic, kinaesthetic, etc.

To display something means to make some information available in a form that can be readily understood by the receiver. Some animals indicate fertility through mating displays or release of pheromones. Humans display emotion through tone of voice, facial expressions or skin tone. For these strategies to work both transmitter and receiver must possess the same sensory abilities – to perceive the same part of the electromagnetic spectrum, or to sense pheromones, etc. In the same way, information is transmitted within and between computers and other machines by electronic digital signals. Since both the transmitter of the information and the receiver of the information can process the same signals then information can be transmitted directly. However, since humans cannot take in electronic or digital information directly, another means must be found to transfer information stored internally by the machine or acquired through its sensors to the human brain. Any of the five human senses may be used, with the visual channel predominant and the auditory channel secondary.

All display design should begin with the fundamental question of what information needs to be made available to the human operator so that they can carry out their assigned role in the operation of the system. No more and no less. This is not as straightforward a question as might be assumed. In some cases, the answer seems fairly obvious – a pilot needs to know their altitude and airspeed; a driver needs to know their road speed; a ship's crew needs to know their precise position on the surface of the sea and so forth. At Three Mile Island, it turned out that the operators needed to know whether the relief valve was actually open or closed and not simply whether it had been commanded to open or close. Could this have been known in advance? In principle, the answer is yes, but only if the displays and controls had been designed following a comprehensive analysis of the operators' tasks, known, oddly enough, as *task analysis*.

Task Analysis: The practice of breaking a job or task down into a series of simpler elements goes back over a century to the origins of 'scientific management'. Physical jobs such as bricklaying were broken down into a series of steps. Eliminating, combining or rearranging these steps could result in faster and more efficient work. More recently this has become familiar as 'time and motion' studies. This becomes a little harder to do with work that consists largely of thinking or other unobservable cognitive activities but in principle this can be done in the same way. As will be discussed later, demands on attention and working memory are particularly important considerations when performing some form of *cognitive task analysis* (CTA). The logic of CTA essentially follows that of basic task analysis for physical tasks but with the emphasis on mental operations such as perceiving, judging, deciding, etc.

The main development in task analysis was that of *hierarchical task analysis* (HTA) formulated in the 1970s.[11] Influenced by the then-current theories on problem-solving, planning and goal-directed behaviour, HTA consists of establishing a task hierarchy from an overall goal down through a series of sub-goals. HTA fits well within an overall systems perspective which was becoming established with the increasingly complex and sophisticated technological developments in aviation, manufacturing, power generation and others. For example, the goal of starting the engines on a jet aircraft can be described as consisting of three sub-goals and their associated actions. The first two sub-goals are to provide power and fuel (as in an automobile) with the addition of a third sub-goal requiring a pressurised air source to start the fan blades turning. Providing power is achieved by turning on the main battery and then starting the auxiliary power unit (APU), a small jet engine situated in the rear of the aircraft. Once running, the APU provides the power to provide the pressurised air needed. To start turning the blades, the air supply to the cabin is temporarily switched off and the air diverted to the engine. Fuel has to be loaded on board and is then provided to the engine by turning on the fuel pumps.

Whereas CTA and HTA techniques developed out of theory in psychology, *functional flow analysis* developed from industrial and systems engineering and serves to identify the sequence of functions or actions that must be performed by the system. The functional flow block diagram represents the sequence of system functions as a series of top-level functions. Each top-level function can be decomposed into lower-order functions in a similar manner to HTA. In our example, 'starting engines' would be one top-level (or 'zero order') function in the system chart for 'operate jet aircraft' along with numerous other functions such as 'program navigation system', 'configure aircraft', 'load fuel' and so on. Each of these zero-order functions would have first-level functions and each of these might have second-level functions and so on.

Whatever the technique, the purpose is to produce a detailed map of all the functions and necessary tasks for the system to achieve its purpose whether that be 'generate power', 'transport passengers' or whatever. Decisions will also be made regarding what is referred to in human factors and ergonomics (HF/E) as 'allocation of function'. This refers to which functions can be achieved by electrical or mechanical means and which functions require, or would benefit from, human involvement. Many of the functions in aviation that used to require human involvement (e.g. astro-navigation) have been replaced by technologies that do not utilise humans (firstly inertial navigation systems and then satellite-based systems). Similar changes are rapidly occurring with automobile technology, initially with systems such as Anti-lock braking system (ABS) that can intervene to control braking, and more recently lane-keeping and collision avoidance systems, all the way to fully autonomous vehicles. In the face of ever-advancing technology, the question of what functions the human pilot, driver or process controller should undertake and what should be undertaken automatically becomes increasingly difficult to answer.

Once the allocation of function decisions has been made and a detailed list of functions and tasks drawn up, we can actually begin to specify what information must be transferred between different parts of the system. Where information must be transferred *to* the human then we must specify the optimal means of doing so in terms of display characteristics. Where information must be transferred *from* the human then we must specify optimal means of doing so in terms of controls and other input devices.

CRASH OF BRITISH MIDLAND BOEING 737-400 NEAR KEGWORTH, UK, 8 JANUARY 1989

The aircraft departed London Heathrow on a scheduled service to Belfast. As the aircraft was climbing through 28,300 ft a fan blade fractured in the No 1 engine resulting in juddering and some smoke and fumes entering the cockpit. The commander took control and disengaged the autopilot. He later stated that 'he looked at the engine instruments but did not gain from them any clear indication of the source of the problem'. After a brief discussion, the first officer closed the throttle for the No 2 engine. The aircraft diverted to East Midlands airport. The No 2 engine was then closed down. On the final approach to the airport, power from No 1 engine abruptly ceased. The commander called for the No 2 engine to be restarted. The aircraft struck an embankment on the M1 motorway just 900 metres short of the threshold of the runway.[12]

Visual displays, from watches to aircraft flight decks, increasingly make use of high-tech systems such as LED screens to provide clear sharp images in all lighting conditions. At the other end of the spectrum, it is not uncommon to find expensive high-tech carbon fibre sailplanes equipped with a display consisting of a single cotton thread attached to the centre of the windshield. Glider pilots need to maintain attention outside the cockpit as much as possible and find the position of the cotton thread an excellent display of the aircraft's balance in a turn. If the rudder and aileron are properly coordinated the aircraft will not 'slip' in the turn and the airflow will keep the thread centred on the windscreen. This example also serves to emphasise the key point that what is critical in display design is the provision of information that will assist the operator (in this case, the pilot) to make the appropriate decisions or execute the appropriate actions (in this case, apply the correct amount of rudder to prevent the nose of the glider pointing away from the direction of the turn). Sometimes 1c of cotton can do just as good a job as US$ 1,000 of LED display. Sadly, the high-tech displays on the Boeing 737 involved in the Kegworth crash seem not to have conveyed the necessary information to the crew of that aircraft to enable them to quickly and accurately identify the damaged engine and shut it down.

A great deal of information about the specific design details of individual displays and controls can be found in textbooks, technical reports, ergonomic guidelines and in the form of widely adopted standards such as those produced by the International Organization for Standardization (ISO), whose proud byline reads 'Developing standards from soap to spacecraft'. ISO provides several standards on ergonomic (human factors) design principles from individual displays, to mental workload and task design. Similar standards are promulgated by the European Committee for Standardization (CEN) many of which also cover HF/E concerns, such as this recent standard (EN 16186 – 2018) for the design of displays and controls in railway train drivers' cabs. This provides a good idea of the typical scope of the HF/E design guidance that is available:

> legibility and intelligibility of displayed information: general rules concerning the layout of information on the displays, including character size and spacing; - definition of harmonized colours, symbols, etc.; - definition of harmonized principles for the command interface (by physical or touchscreen buttons): size, symbols, reaction time, way to give feedback to the driver, etc.; - general arrangements (dialogue structures, sequences, layout philosophy, colour philosophy), symbols, audible information, data entry arrangements.[13]

The latest iteration of the Boeing 737 at the time was the 400 series that came equipped with a 'glass cockpit' where the traditional electro-mechanical instruments had been replaced by a number of LED screens. The designers had chosen to 'update' the round clock style indicators with a large hand used in previous versions of the 737 to portray engine rpm, temperature and vibration levels, with electronically generated circles with a small cursor moving around the outside of the dial. The pilots found these more difficult to read leaving the accident investigators to conclude that 'the detailed implications of small LED pointers rather than the larger mechanical ones ... require further consideration'. Neglect of basic HF/E has unfortunately been widespread across many industries and workplaces, not just aviation and nuclear power generation.

DESIGN FOR ACTION

In fact, numerous basic HF/E design issues can be found right under our noses. No one has written more prolifically or entertainingly on this topic than **Donald Norman** who after training in electrical engineering turned to psychology and established himself at the University of California in San Diego as a leader in the field of cognitive psychology. Increasingly frustrated by examples of poorly designed systems and technologies Norman decided to leave academia to try and make a difference in the design of products for human use by joining Apple computers in 1993. Since then, Norman has combined various roles in industry and academia and is continuing to do so well into his 80s. His most well-known book *The psychology of everyday things* was written whilst Norman was a visitor at the Applied Psychology Unit in Cambridge.[14]

Norman provides many examples of the inherent frustration of not being able to figure out how to use everyday objects such as digital watches, cameras, microwave ovens, etc. Even wash basins, entrance doors and room lighting controls can puzzle and perplex. The example most favoured by writers in HF/E is the 'classic' four-burner cooktop (see the upper part of Figure 4.5). There is simply no way of unambiguously linking each of the controls to the appropriate burner on the conventional gas or electric cooktop. The recent development of ceramic glass cooktops has now made it much easier to arrange the controls to spatially conform with the arrangement of the heating elements (see the lower part of Figure 4.5). Although this particular problem has now been superseded by technological advances, the general

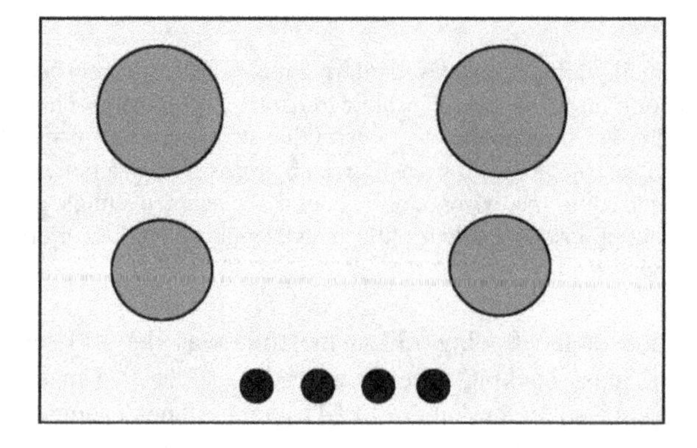

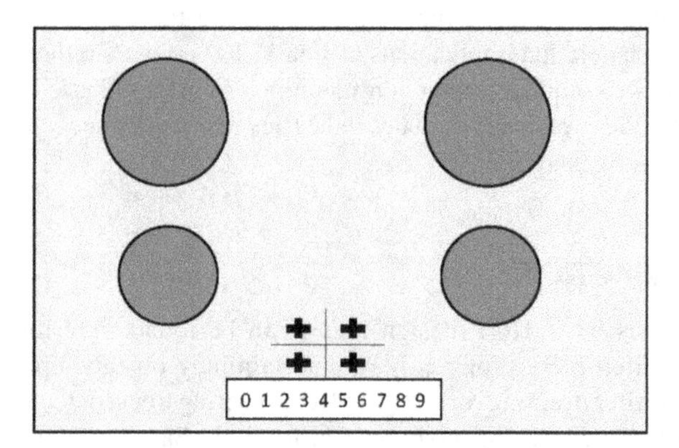

FIGURE 4.5 The design of a multi-burner stovetop is a classic example of ergonomics and human factors. Conventional layouts as shown in the upper example, are potentially confusing (and dangerous) as there is no obvious way of knowing which knob controls which burner. The example in the lower half of the figure shows a modern design of a Bosch ceramic cooktop that overcomes this problem. The selection of a burner is made by pressing one of the four + arranged in the same spatial configuration as the burners. The amount of heat is controlled by selecting a number on the horizontal numerical strip beneath.

problem of *mapping* the relationship between controls and displays (i.e. which control relates to which display) or between the movement of control and the resulting movement in the world (e.g. which way does a boat turn when you move the tiller to the left?) remains a hugely important one in determining the usability and safety of real-world designs.

When we use well-designed equipment our hands just seem to know where to go and the results of our actions are exactly what we expect. This can only happen when our 'natural' expectations have been taken into account by the designers – something that Norman has demonstrated has been somewhat rare. The study of these 'natural' expectations has shown some strong stereotypes – for example, that rotating a knob clockwise will result in a corresponding clockwise movement of the related display. In some countries (e.g. the UK, Australia) there is a similarly strong stereotype that associates downward motion (of a light switch) with 'ON' and upwards with 'OFF'. The existence of the reverse stereotype in other countries (e.g. the United States, Canada) shows that this stereotype is a learned cultural expectation, not a hard-wired neural one.

A version of the four-burner stove problem was present in the Kegworth Boeing 737-400 cockpit. The aircraft has two engines, operated by two power levers, and there are two sets of engine instruments known as the primary displays and the secondary displays. The designers were faced with a mapping problem (see Figure 4.6). Should the primary instruments be displayed together, with the secondary instruments arranged separately on another panel, or should both the primary and secondary instruments for the left engine be portrayed together with a corresponding display for the right engine? The designers opted for the first arrangement which meant that the secondary instruments for the left engine were spatially aligned with the power lever for the right engine. This may have been the key to understanding the confusion the crew had as to which engine was vibrating and needed to be closed down.

This conclusion was confirmed by experiments run at the RAF Institute for Aviation Medicine. Computer-generated banks of dials were arranged to mimic the arrangement used in the Kegworth 737-400. In one condition the dials were grouped by status – primary instruments, in the left-hand panel versus secondary instruments on the right, as shown in Figure 4.6. In the other condition, the instruments were arranged by the engine so that both primary and secondary instruments for the No 1 engine were grouped in the left-hand panel and those for engine No 2 in the right-hand panel. Participants responded to a discrepant dial reading by pressing one of two buttons which they were told represented the left and right engines. The results clearly showed that performance was much better with the instruments grouped by the engine so that there was clear compatibility between the location of the dial showing the discrepant reading and the appropriate response.[15]

As a general principle, the controls for a set of displays should be configured in the same spatial layout as the displays or elements being controlled.[16] Modern touch screens now mean that displays can be touched directly maximising this correspondence. The problem with non-corresponding arrangements such as the classic four-burner cooktop and the arrangement of primary and secondary engine displays

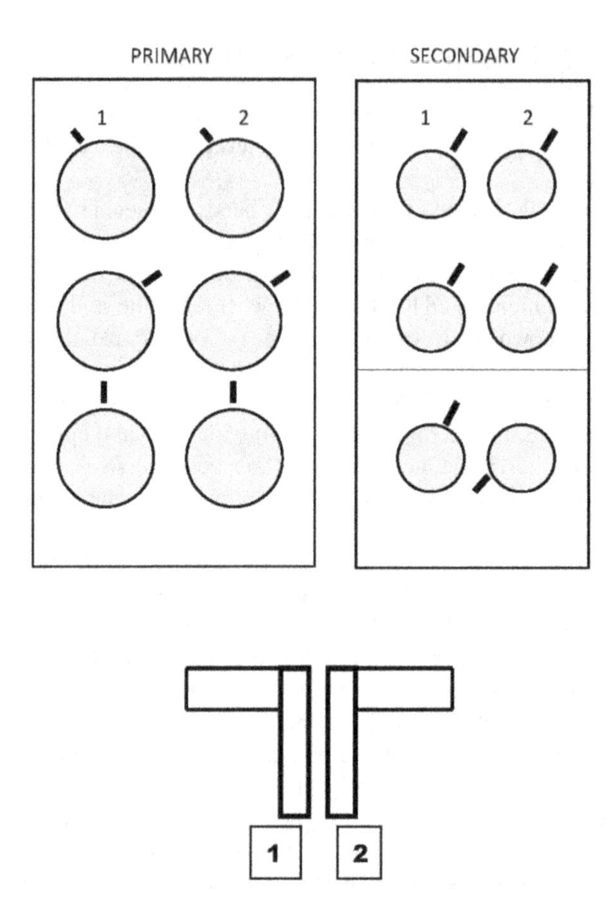

FIGURE 4.6 A simplified layout of the power controls and engine displays on the flight deck of the Boeing 737-400 involved in the Kegworth accident. The Number 1 power lever operates the engine on the left wing and the Number 2 power lever operates the engine on the right wing. The vibration gauges were designated as secondary instruments and are shown as the bottom row of the right-hand panel. The vibration gauge for the No1 engine, on the left, showing a high reading is in line with the power lever for the No2 engine on the right.

in the Kegworth accident lies in the additional number of mental operations that are required when controls and displays do not naturally correspond. This principle, that good design is one that minimises the number and complexity of intermediary processing steps, is one that underlies a good deal of safe and effective design from the aircraft flight deck to the orientation of 'you-are-here' maps and domestic cooktops.

The principle can be extended to designing displays to also be compatible with their associated task demands. Tasks can be characterised in numerous ways but one important aspect is their attentional requirements. These can be thought of as varying along a continuum from highly focused requiring attention to a single element (e.g. in Figure 4.4, locate the narrow tower topped with a spherical object) to broadly divided requiring attention to multiple elements (e.g. Is the power plant situated in the countryside or in the city?). Similarly, when reading nutritional information panels on food products, the task might be focused (e.g. Is this high in sodium?) or divided (e.g. Is this a healthy product?).

Chris Wickens, a leading engineering psychologist and former head of Human Factors and director of the Aviation Research Laboratory at the University of Illinois Urbana-Champaign, formulated the *proximity compatibility principle* (PCP) as a theoretically based guide to the design of complex displays. The principle is that where tasks require high proximity or involve divided attention and/or integration across elements, then the optimal display will be one with high display proximity where the display elements are linked or connected in some way. Conversely, when the task demands are low proximity, as when it is necessary to isolate a specific element, then the optimal display format is one with low display proximity.

The PCP provides a tool for display designers to match their display design to the users' task requirements by manipulating the proximity of display elements. The elements of a display can be said to have high proximity by virtue of their spatial proximity, common colour, connecting lines or by being combined into a single 'object' such as a wheel or rectangle. A good deal of careful research has supported the PCP as a useful design tool.[17] One interesting development has been the validation of *object displays* as a useful means of facilitating decision making in situations where multiple elements must be combined to make an overall judgement. Stress and time pressure can inhibit our ability to search for, and process, multiple elements. An early example was suggested for nuclear power plant operators faced with the task of simultaneously monitoring a large number of indicators showing individual parameters (pressure levels, temperatures, etc.). Arranged in the form of a 'wagon wheel' when all values are normal then the lines linking each point form a regular polygon. Deviations from normal are indicated by deformations of the polygon and are instantly perceptible without any computation or other intervening mental operations. The same 'wagon wheel' format has been used to represent nutritional information for packaged food items and the effects compared with the same information presented in traditional alpha-numeric list form. Results have been in accordance with the proximity compatibility principle. People find it easier to quickly evaluate the overall healthiness of the product with the 'wagon wheel' object display, although the search for specific information (e.g. sodium levels) is easier with the conventional panel information.[18]

Although the 'wagon wheel' object display has not been commercially employed for nutritional labelling, the health star rating system (used in Australia and New Zealand) which 'awards' between one-half and five stars according to the product's overall 'healthiness' has become quite popular. By combining the levels of various key elements (amount of sugar, fat, sodium, etc.) into one object (a star) the health star is more compatible with an integrated task such as judging a product's healthiness but is not compatible with determining the specific levels of individual ingredients. For that, the detailed list of ingredients and quantities listed on the nutrition panel is more compatible. Similar systems (star ratings or 'traffic lights') are widely used in Europe and elsewhere to provide information about energy consumption on cars and household appliances as well as nutritional guidance on foods.

A number of innovative displays have been developed to assist with monitoring patients in surgery. Some were introduced into medical equipment and US patents were taken out on these novel display formats by pharmaceutical companies. Although research has generally supported their use for tasks that involve integrating information from a number of cues, enthusiasm for the displays waned under some criticism. One problem is that to produce a regular shape, variables measuring quite different processes in quite different units of measurement, such as heart rate, blood pressure, oxygenation levels and so forth, must be scaled so that the same line length corresponds to whatever the expected or normal level of that variable might be. This means that some changes necessarily become less perceptible than others when combined in this way. If a change was detected, the anaesthesiologist would then have to review the specific numerical or waveform data to fully diagnose the problem.[19] The main benefit of object displays then seems to be in the initial detection of non-normal conditions. Recent research has confirmed the usefulness of this type of display format in improving performance and reducing workload in vigilance tasks that involve looking for changes over lengthy periods of time. This is a task that human operators are known to struggle with.[20]

Another promising approach to designing displays is known as *ecological interface design* (EID). The approach stems largely from the work of psychologist **J.J. Gibson**, who studied the ability of various animals, including humans, to successfully locomote around their environment and perform complex tasks such as throwing a spear, catching an object, or dodging out of the way of a falling rock. All of these actions are the result of the *direct perception* of information contained in the light reaching the eye. Most of us can move around to accurately catch a ball thrown by another person without knowing anything about the physics involved, or needing to perform any complex mental operations such as calculating a parabolic trajectory! The idea of EID is to make the key environmental information for any task as directly perceivable as possible rather than in a form that requires intermediary processing steps.

The EID approach strongly echoes the point made earlier that display design must first be preceded by an analysis of what information is required to be represented. EID has an established methodology for doing this known as *cognitive work analysis* (CWA). This analyses a set of tasks at several levels from the functional purpose of the work down to its physical components. From this analysis, the information requirements for carrying out the task can be established. The analyst then determines what cues in the environment provide this information and then determines a means of presenting this information directly (i.e. in a way that can be seen without the need for mental computation) to the operator.

Over the past 30 years, EID has been used to develop displays in a wide variety of fields from medical devices to air traffic control, nuclear power plant control, power grid control, driving and rail level crossing design. Further discussion of EID can be found in Chapter 8. Conventional display design has tended to focus on displaying selected basic parameters one at a time – a car would have one display showing speed, one for engine revolutions, one for engine temperature and so forth. Ecological displays provide displays of information that would otherwise need to be derived (by mental computation) from these basic parameters. For example, a power control panel might display energy balance rather than various input and output measures which together could be used to calculate energy balance.

For another example, as part of a UK project designed to promote 'safer and more environmentally friendly driving', researchers at Brunel University in Middlesex developed a prototype vehicle display based on EID principles.[21] The display shows a vehicle moving inside an envelope representing the field of safe travel which is a space defined by appropriate position in time and space. Warnings are generated if the vehicle's headway or lane position deviates from safe limits. Outer rings represent aspects of ecologically responsible driving such as acceleration and braking forces. The display is designed to make these parameters instantly perceptible to the driver. Much more detailed work on specifying the constraints involved in fuel-efficient driving has since been conducted by **Neville Stanton** and colleagues and this may provide a solid basis for future developments of in-vehicle information systems along similar lines.[22]

Whilst displays are typically thought of as engineered instruments that portray information generated by sensors, other forms of display can be provided by environmental features such as road markings. One innovative recent application of EID has been to railway level crossing design to improve safety. A large team of researchers, mainly from the University of the Sunshine Coast in Australia, used EID to create features around a railway level crossing that would more directly convey the information required to ensure safe conduct for both road and rail vehicles. Examples include marker posts, road markings and reflectors that provided both visual and aural warnings of an approaching train to vehicles on the road. An important feature of the design was that these static features require no external source of power and so can be installed in the remotest locations. The dynamic cues are provided by the appearance of the train itself.[23]

Driving simulator studies were conducted to compare the EID design with existing approaches. Results showed that drivers used the EID design more adaptively, slowing more when a train was coming but maintaining closer to normal speed when the crossing was clear. In another study, the simulation was set to include one trial where there was a partial failure of the signalling technology resulting in a late warning of an approaching train. In this situation, 40% of the drivers in the EID condition stopped at the crossing compared with only 4% approaching a traditional (flashing lights) crossing. A large number of comparisons between conventional display representations and EID displays have now been conducted across a wide range of technological systems with results that generally show clear advantages for displays that have been designed to have high ecological compatibility. Two of the leading figures in EID research have recently concluded that 'EID has proven to be remarkably effective in significantly improving performance'.[24]

Conclusions: The application of the scientific method and the development of a better theoretical understanding of human performance has contributed to major improvements in the design of displays with corresponding significant gains in safety. The improvement to aircraft altimeter design alone has undoubtedly saved hundreds, if not thousands, of lives. A display involves communication – animal displays are designed to communicate such things as reproductive status or dominance. These are essential for survival. In the technological world, the same question should drive display design – what information is essential for performance (which in high-risk domains might also be essential for survival)? Methodologies from hierarchical task analysis to cognitive work analysis have been developed to address this question and are essential precursors to all good design.

A good deal of valuable information on the mechanics of display design has been codified in international standards or presented in textbooks on ergonomics and human factors. Specific questions concerning visibility, legibility, depth of menus and so forth can all be clearly answered. Broader questions that address the fundamental relationship between displays and operator actions can be resolved through the application of relevant theoretical principles. Of particular importance is the principle of compatibility. This can refer to the compatibility between displays and controls, the compatibility between displays and task requirements, or the compatibility between displays and the ecology. Our brief discussion of displays has focused on visual displays but similar approaches and principles apply to displays (auditory, haptic, etc.) to any of the senses.

DESIGNING TASKS FOR HUMAN ABILITIES

It would make no sense to display information for human operators using ultra-sound or infra-red wavelengths as these are beyond the human ability to sense. Similarly, it makes little sense to design the work that the human operator has to do in a way that is beyond the limitations and capabilities of the human body and brain. The physiological and physical limits of human performance are fairly easily determined and humans are only tasked with operations within acceptable physical parameters. The mental limits of human performance are less easily determined and we have only recently begun to understand how best to make use of human abilities in systems of work. In many cases, however, there is still a mismatch with humans tasked with responsibilities for which they are not well suited.

A moment's introspection suggests at least three areas where designers of the tasks we are asked to perform might wish to be especially concerned with human cognitive capabilities. Who has not forgotten where they put something (e.g. parked the car, left the keys, phone or wallet); become distracted or preoccupied whilst reading or watching something and not taken in any of the information; or made a rash or unwise decision. These areas of memory, attention and decision making are of particular interest in designing work tasks that best match our abilities.

The Applied Psychology Unit at Cambridge was responsible for much of the UK's wartime and early post-war research into problems originating in the interaction between humans and machines in military services. Following this period, the Unit became more heavily involved in human–machine problems arising in civilian contexts. Much of this work was undertaken by one of the lesser-known of the many talented individuals who worked at the APU at one time or another. Referred to only as **'R' Conrad,** he worked on problems suggested by the General Post Office (GPO) – at the time Britain's largest employer. During the 1960s, the GPO was undertaking many technological changes involving the telephone service (from switchboards to direct dialling and from rotary dial 'phones to push button keypads) and in the letter delivery service with the introduction of machine sorting involving keyboard data entry and the introduction of postal codes.

The design of the UK postal codes – serving a similar function to the US zip code – was based on the research conducted by Conrad into memorability of alphanumeric sequences.[25] Conrad showed that the most memorable six-character code was one composed of four letters and two digits, with the letters and digits arranged in two groups of three, with each group consisting of two letters and a digit. For example, 'RH6 9DD'. The codes were introduced as part of the transition to mechanical letter sorting in the 1960s. The research conducted by Conrad not only led to the design of equipment and tasks that were most compatible with the way in which human memory works but equally importantly contributed to significant advances in our understanding of human memory itself.

In other studies, Conrad showed that when memory for visually presented words decayed, the errors that appeared were largely based on acoustic similarity to the originals. This suggested that there was some kind of temporary memory storage based on the acoustic properties of words rather than on their meaning. Conrad was also responsible for supervising a PhD student, **Alan Baddeley**, who was largely responsible for developing the important concept of *working memory* (WM) which is one of the key concepts in our understanding of how memory works. Working memory is where incoming information is briefly held – for example, looking up a 'phone number or email address – to allow us to perform some task. Following an instruction sheet for assembling a piece of furniture, for example, requires looking at the information on the sheet, holding it temporarily in memory whilst we locate the relevant parts and then taking the appropriate actions before returning to the sheet. As we don't require this information for very long it is normally not transferred to our *long-term memory* (LTM) store of information where other information such as our home address and the names of capital cities around the world is held.

These two very different kinds of memory storage (working memory and long-term memory) are the key concepts in understanding human memory and provide essential information for the design of tasks that best match our capabilities. A schematic outline of human memory is shown in Figure 4.7. Subsequent research by Baddeley and others has established that working memory has two functionally distinct sub-systems – one for verbal memory and one for visual-spatial memory. There is also a kind of overall 'management system' known as the central executive. Similarly, long-term memory has been shown to have two very different components. One where we remember information such as the name of the longest river in Africa, our car number plate and so forth and another where we remember 'how to' information such as how to ride a bicycle or play the piano. The first store is referred to as 'declarative' memory and the second as 'procedural'.

In addition to these two distinct memory systems, there are three essential memory operations: *encoding*, *rehearsal* and *retrieval*. Any information that we want to retain must first be encoded. This involves noticing something about the information – its sounds, surface features or meaningful connections with other material. This latter strategy is one way of ensuring material passes into long-term storage. Simply repeating material, or forming arbitrary associations (as in mnemonics) are other

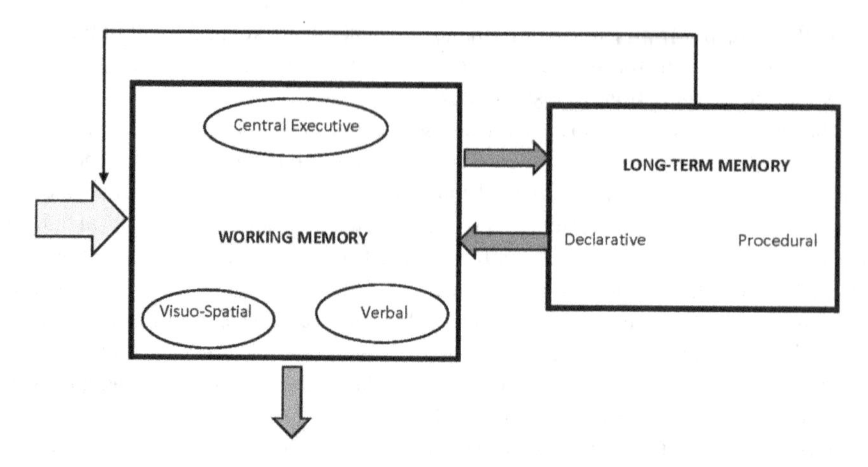

FIGURE 4.7 A schematic outline of the human memory system. The larger arrow on the left of the figure shows that WM receives information from parts of the brain, such as the visual cortex, responsible for processing sensory signals. Information in WM can be both transferred into, and retrieved from, LTM, and also used to select an action response. The smaller arrow in the upper part of the figure illustrates the potential influence of LTM on perceptual processing itself.

forms of rehearsal. Finally, at the point of needing to act – to troubleshoot a faulty electronic panel or an unexpected rise in reactor core temperature – information must be retrieved from memory. The success of this depends not only on a multitude of factors – some circumstantial, such as the conditions (heat, stress, vibration, fatigue, etc.) pertaining at the time, but also on the way in which the material was stored in LTM and specifically the indices or labels that point us in the right direction.

These are all complex topics requiring in-depth explanation. However, at this stage, our interest is purely in the implications of this knowledge for designing the work we ask humans to perform. Memory is obviously critical in tasks such as air traffic control (ATC) and for any industrial or process control task that involves comparing current information (e.g. display readings, written instructions) with one another or with reference values. Since remembering what we have done, or keeping information temporarily in mind, are involved in almost every working task, it is essential to ensure that tasks are designed to ensure that these abilities are not exceeded. For example, it is tempting for designers to build in a wide range of warnings (visual and auditory) for non-normal conditions, but in major events in a control room or flight deck, the number of such warnings can quickly overwhelm memory capacity resulting in stress, confusion and less-than-optimal performance.

RUNWAY COLLISION OF US AIR BOEING 737 AND SKYWEST FAIRCHILD METROLINER, LOS ANGELES INTERNATIONAL (LAX), 1 FEBRUARY 1991

At 06.04 p.m., the LAX Local Controller (LC2) instructed a Skywest Metroliner to 'position and hold' on Runway 24L at LAX awaiting take-off clearance. A US Airways B737 was on approach to the same runway and was given clearance to land by LC2 at 06.04 p.m. At 06.07 p.m., the Boeing 737 collided with the Metroliner destroying both aircraft killing all 12 people on board the Metroliner as well as 22 people on the 737. The official aircraft accident investigation identified a wide range of contributory factors including 'The local controller forgot that she had placed (the Metroliner) into position for takeoff on runway 24 left … because of her preoccupation with another plane'.

The report also noted that the ability of the controller to see 'aircraft on the runways … was complicated by some of the Terminal II apron lights which produced glare'.[26]

So important is working memory in human performance that Chris Wickens suggested that a *working memory analysis* be included in any task analysis prior to system development.[27] Indeed, even an informal analysis could be usefully included in any job evaluation. The analysis might cover questions such as how much information needs to be held temporarily in memory during the performance of this task? One of the most famous studies in modern psychology reported that the answer to this question was '7+/–2 items'.[28] The key word here is 'item' as whilst WM capacity might be around this number for random digits, in general people have little need to remember random numbers but mostly deal with meaningful 'chunks' of information such as postcodes, compass directions, addresses and so forth. Most people will be able to safely hold at least five chunks of information in WM but for how long? Evidence suggests that information decays quite quickly (no longer than 10 or 15 secs) unless it is held for longer by deliberate rehearsal – for example, repeating a 'phone number over and over.

This gives us some idea of the amount of information an air traffic controller, for instance, might be able to retain should the screens in front of them suddenly go completely blank. Measures of the ability to hold information in working memory storage provide one of the better means of predicting performance in a wide range of tasks, including computer programming and air traffic control. This reflects the central importance of working memory in job performance and safety and emphasises the need, especially in safety-critical situations, to ensure there are backups in place to guard against critical failures such as the one experienced by the unfortunate LAX controller. For many years air traffic controllers have used written 'flight progress strips' as a low-tech supplement to electronic screens and human memory.

The fact that working memory includes separate systems for verbal material and visuospatial material has important implications for the design of work. It should be much easier (and indeed it is) to 'multitask' or time-share across these systems than within either one of them. The working tasks expected of pilots, air traffic controllers and others have naturally evolved to take advantage of this ability so that, for instance, pilots are expected to reliably absorb detailed verbal instructions from ATC whilst simultaneously monitoring position with reference to visual cockpit displays. This kind of time-sharing across the two different working memory systems works much better than would trying to absorb two concurrent verbal messages for example.

The role of the central executive in working memory appears to be that of controlling attention. This involves prioritising between competing tasks, directing attention towards the desired source of information whilst minimising distraction and interference from other sources of information, and managing the switching of attention between one thing and another. These are all critical activities in the successful performance of a range of real-world tasks such as performing surgery, piloting a large vessel into a difficult harbour, or dealing with unexpected situations in a control room or aircraft flight deck. Issues of attention management are frequently alluded to in the aftermath of unfortunate events such as those discussed previously in this chapter.

ATTENTION AND PERFORMANCE

With the exception of basic reflexes and orienting responses (i.e. where we turn towards an unexpected stimulus), most of our attention is under conscious direction. This includes determining which of several sources of information related to one task (e.g. driving) to pay attention to (e.g. the speedometer or the road), as well as the management of several demanding tasks being performed simultaneously. The most important distinction is that between *focused* and *divided* attention. In the former, we wish to allocate all our attention to one source of input (e.g. listening intently to a speaker) whereas in the latter we wish to split our attention between more than one source of input (e.g. monitoring speed, altitude and direction).

Attention within tasks: Most people, placed in one of the pilots' seats in a modern airliner cockpit would have little idea of what to look at. Probably most attention would be directed towards the largest and brightest of the displays – one of the two primary LED screens found on both Boeing and Airbus cockpits. Each screen portrays several sources of information and it is only through training and experience that pilots know which part of which screen they should be looking towards at any given moment. Chris Wickens and colleagues have developed a model of the forces that determine where a skilled operator will direct their attention.[29] Known as Salience Effort Expectancy Value (*SEEV*), the model proposes that attention will be determined by four factors. The first is the salience of the stimulus – the larger brighter screens are more salient than a small switch on a panel. Flashing light increases salience and is therefore deliberately used by designers to attract attention when needed (e.g. in a non-normal situation). The second factor is effort. It is easier to stare straight ahead than it is to scan two instruments a metre apart. When we 'zone out' that is exactly what we do as all voluntary effort ceases. Constantly scanning an aircraft instrument panel for information, or the natural environment for possible enemies, is effortful and fatiguing. The third factor is expectancy. Training and experience help us know where the most useful information is likely to be found at any given moment. In recent decades, driver training has placed greater emphasis on training drivers to know where to look in their environment for potentially critical information (e.g. scanning a row of parked vehicles in case a pedestrian suddenly steps out into the road or a door unexpectedly opens). The final factor is value. The value of looking towards nearby traffic is high compared with the value of looking at advertising signs along the roadway.

So, areas that are salient, require little effort, or where there is an expectation that something of value or importance will be found, will attract the most attention. People will spend the least time looking towards anything that has low salience and a low likelihood of showing anything useful, especially if it takes effort to do so. Predictions of the model have been tested in driving, aviation and surgery and have been found to provide a good account of the distribution of attention within a task. For example, a study conducted with surgical scrub nurses in Singapore found the more experienced nurses to have more optimal patterns of visual attention with greater attention paid to the open body cavity allowing greater anticipation of the surgeon's next moves.[30] Gathering information to anticipate imminent demands has also been found to be a skill that marks out more experienced pilots. Colloquially this has been known as 'keeping ahead of the airplane'.

It is important to note that looking towards, or at, something is not completely synonymous with paying attention in the sense of taking in information. Research has found a phenomenon known as *change blindness* (CB) which occurs when we fail to notice a change that has taken place in something we are looking at. Continuity errors in movies provide a good example whereby we sometimes fail to notice that a character is wearing a wristwatch one moment but not the next. The same term is used to describe a failure to detect a change over a slightly longer time period in a relevant feature of the task environment. For example, a traffic sign that changes from indicating one thing (e.g. presence of children) to another (e.g. merging traffic from the right). The majority of participants who repeatedly watched video scenes involving driving along a stretch of road where the traffic sign changed on the last trial failed to notice the change, even though they had been warned to expect such changes.[31] Similarly, in many reported incidents, pilots have failed to notice a change in an instrument display showing, for example, a change in autopilot mode.

A related but slightly different concept is that of *inattentional blindness* (IB) whereby we fail to register the presence of something in plain sight. This can range from failing to see your wallet or keys on the desk in front of you to not seeing an approaching vehicle at an intersection despite looking directly towards it. This is a common feature of vehicle intersection collisions. IB involves the withdrawal of attention from one activity and redirecting it towards another. For example, drivers who are immersed in an involving mobile phone call whilst driving are particularly prone to failing to register significant features of the driving environment such as a vehicle slowing in front of them. Drivers who notice the brake lights of the vehicle ahead, as well as those who don't, are looking in the same places but the latter are not actively 'registering' the presence of this important cue. This finding has been confirmed by later research. There is some suggestion that certain measures of brain activity might be able to indicate whether or not what we are looking at is actually receiving our attention.[32]

Attention between tasks: Most real-world tasks such as driving, piloting or process control require attention to be distributed between a number of different tasks that must take place more or less concurrently. A pilot, for example, needs to monitor instrument displays, plan ahead, listen to air traffic control instructions and interact with other crew members all at much the same time. This is sometimes referred to as *multitasking* or *time-sharing* and draws on the ability to divide attention between different sources of information. How easy is it to do two things at once? Well, it depends on what the 'things' are and how familiar we are with them. In general, we can do two or more things at once if they are both fairly easy tasks. Lyndon B. Johnson's famous put-down directed at Gerald Ford that he 'couldn't walk and chew gum at the same time' is easily interpreted as an insult since these are both simple, routinised actions.[33]

Something can be inherently simple, as above, or become simple through familiarity or practice. After much practice even playing an elaborate piece of music can become an undemanding task through a process known as *automatisation*. In the same way, multiplication tables can become so familiar that answers (9*8=?) can be delivered instantly. If a task has become automatised then it can normally be time-shared easily with another task, even if that task is also quite demanding. However, there are exceptions where performing even two inherently simple tasks together might pose problems and this is when the two tasks share common elements or processes. Once again, Chris Wickens and colleagues have extensively researched such task conflicts and developed *multiple resource theory* (MRT) as a means of predicting when tasks might easily be time-shared and when they may not.[34]

Attention was originally thought of as a single undifferentiated 'pool' which could be allocated to whatever processing was required. Difficult tasks required more attention than easy tasks. Trying hard, or exerting effort, involved more attention than not trying. This 'pool' could be quickly used up when we tried hard to carry out several demanding tasks at once. However, evidence accumulated to show that some tasks that were individually quite undemanding might prove difficult to carry out concurrently. Conversely, some individually difficult tasks could be carried out together. Referred to as *difficulty-structure uncoupling*, this evidence led to the idea that rather than being a single undifferentiated pool, there must be separate pools of attentional resources relating to different processes. For example, Alan Baddeley had already demonstrated that spatial and verbal memory constituted different sub-systems and these can therefore be thought of as providing two different resource pools.

Wickens has added some additional capacities that can also be thought of as relatively separate resource pools. For example, much research has shown that the processes required to initiate and execute responses are quite separate from the processes involved in perception and cognitive processing. Perhaps more obviously, we can easily process information coming in visually and auditorily at the same time. In fact, there is also evidence that we process focal visual information (i.e. whatever we are directly looking at which falls on the fovea) separately from peripheral information (e.g. whilst looking directly at one thing we can become aware of something moving in our periphery). Any task can be characterised in terms of its resource requirement profile on these four dimensions (verbal/spatial; auditory/

visual; perception or cognition/response; and focal/ambient). MRT predicts that the greater the resource overlap the greater will be the difficulty in carrying out the two tasks simultaneously. Considerable research evidence has borne this out.[35]

MRT provides a very useful tool for designing operator tasks and workflow to avoid periods of competition for scarce resources. Of course, people more commonly deal with multiple demands by switching between one task and another. This *task management* involves deciding which task to pay attention to at any moment (i.e. prioritising), interrupting the flow of one task to attend to another and then picking the original task back up again. As we noted in the previous chapter, this can lead to predictable errors – one harmless, whereby some of the original tasks might be repeated, and one potentially consequential, when some part of the initial task is overlooked, as when the flight crew miss a crucial step on a checklist.

Practice and training can improve performance on individual tasks, in some cases through automatisation. Beyond this, however, there seems to be an ability to combine tasks that requires practice in multitasking environments. Individuals who are regarded by their peers as outstanding in their field are often thought to exhibit superior abilities to maintain focus under pressure, correctly prioritising their tasks to maximum effect. The ability to predict which individuals will display such advanced performance would be of enormous interest as it costs many millions of dollars to train surgeons, controllers or pilots. As noted above, individual differences in WM provide relatively good predictors of the ability to perform many real-world complex tasks. However, even this does not fully predict the potential for all-round superior task performance under pressure that is the hallmark of the most superior performers.

Attempts have been made to assess this potential with complex multifaceted tests involving the coordination of multiple demanding tasks under pressure. Some have been developed and used in the selection of military pilots, as training a fighter pilot is a particularly costly exercise with high drop-out rates. One test developed on the foundations of decades of research at the University of Illinois into the ability of individuals to effectively time-share between tasks and manage the demands of coordinating multiple activities has shown some success in predicting high-level performance. Similar tests are now employed in military aviation pilot selection in various countries around the world.[36]

DECISIONS, DECISIONS

Deciding which candidate to select for training as a pilot or doctor is one example of high-stakes decision making. Doctors deciding whom to treat first in an emergency department or a company deciding whether to build a new factory are others. Traditionally, decision making has been seen as a rational process of comparing pros and cons or costs and benefits and choosing an option that maximises the decision maker's overall utility. This 'normative model' of decision making requires considerable cognitive effort to search out all the relevant information, structure it in a way to allow comparisons between alternatives, and calculate overall costs and benefits. Psychologists such as **Daniel Kahneman** and **Amos Tversky** became the most widely cited psychologists of all time through their relatively simple demonstrations of how easily human decision makers could be misled by their intuitions.[37] Kahneman was awarded a Nobel Prize in 2002 for his work, Amos Tversky having unfortunately passed away in 1996. The decision making literature has extensively documented the numerous departures from optimality that can be found in almost every example of human decision making: failing to notice relevant information or seeking out solely confirming evidence; insensitivity to missing information; inability to retain more than a few pieces of information in mind at a time; giving weight to invalid cues whilst overlooking valid ones and many more.

This led to a search for more accurate *descriptive models* of human decision making. **Gary Klein** has become one of the most recognised names in the field of *naturalistic decision making* starting with his studies of decision making by firefighters in the 1980s. Klein proposed an influential model of decision making that is a better fit with the realities of time-pressured, high-stakes decision making in complex dynamic environments. Known as the *recognition-primed decision making* model (RPD), the model describes how decision makers use their experience and expertise to guide their understanding of complex situations.[38] In sharp contrast with the normative models of simultaneous option assessment and comparison, the RPD model shows that expert decision makers typically progress in a serial fashion, considering and then if necessary rejecting hypotheses about the situation until a good fit is obtained. This allows the expert to then progress speedily and directly to an appropriate plan of action. As the name suggests, the RPD model of decision making replaces the calculation and analysis of the normative model, with psychological processes such as similarity-matching and recognition. The RPD model has been hugely influential with applications in numerous fields from medical and aeronautical decision making to the decision making of football (soccer) players and American football quarterbacks!

Although human decision making has often been found wanting, even the most technologically advanced systems such as spaceflight and unmanned aerial vehicles (UAVs) have retained a role for the human being as the ultimate arbiter and decision maker. Half a century ago, Stanley Kubrick's classic *2001 a Space Odyssey* portrayed a rogue sentient computer taking over a spaceship from the human crew. The idea of fully autonomous technology taking over control still remains a disturbing one. Machines could undoubtedly take over the individual processes that constitute decision making from seeking out and gathering information to collating and comparing the information and, ultimately, weighing up the various factors that enter into the most significant decisions. However, as we discovered previously in looking into human judgements about risk, these judgements and decisions frequently reflect emotional and moral concerns as well as purely cognitive ones. We ultimately prefer to have agents that, at least in theory, possess the same emotional and moral sensitivities as ourselves to make decisions with life and death consequences. Whilst pilotless commercial airliners are technically feasible, the lack of an identifiably human presence on the flight deck would create unease, even in those too young to have seen the destructive amoral computer ('HAL') in Kubrick's epic.

LIGHT, HEAT AND REACH

Workplace design is a topic that is comprehensively covered in many ergonomics and human factors textbooks. There are also numerous international standards and guidelines that are available to guide designers and managers towards current best practices in relation to working conditions.[39] The workplace environment refers here to those factors 'external' to the individual that have an impact on their ability to carry out their tasks. Some well-studied examples would include the effects of lighting and temperature, the layout of workstations, the effects of noise and vibration and the effects of extreme environments such as working at high altitudes or at depth (e.g. divers). It would require thousands of pages to cover all these topics in detail so we will simply provide several very brief insights into the scope of this work with three examples: workplace layout; illumination and performance; and the effects of thermal stress on performance.

Workplace Layout: Designing safe and efficient workplaces requires accurate information about the dimensions of the human body. Known as *anthropometrics* (literally: human measurement), the study of human dimensions (height, weight, arm reach, knee height, hip breadth, etc.) provides essential data for the design of airplane cockpits, car interiors, office seating and desks, power station control panels, clothing for firefighters and medics and so forth. Anthropometric data can be used to answer questions such as 'What height should a control panel be located on a piece of equipment so that 95% of the population can see it and use it safely and efficiently?' The answers to this kind of question are especially critical in areas such as military aircraft cockpit design where ejection handles, flap levers, etc. must be within reach of those who fly in them. Research on this topic has been reported from the US Air Force Aeromedical Laboratory at Wright-Patterson AFB, for example. In general, the design goal is to ensure a fit for at least 90% of the user population (i.e. everyone between the 5th percentile and 95th percentile). This can become increasingly difficult and expensive if the user population is broadened, say from male US pilots to include male Japanese pilots who are generally around 40% smaller, US female pilots and so forth.

Unfortunately, many of the values for human dimensions quoted in textbooks and standards were obtained many decades ago from highly selected samples (mostly of European and American military personnel) using simple measuring tapes, calipers and weight scales. To be of current value, anthropometric data need to have been recently obtained (to reflect the continuing changes in body size exhibited by many populations) and specific to the population for whom the design is intended. For example, truckdrivers and firefighters tend to have quite different bodily dimensions from the general population. Often, the necessary data have to be obtained privately by companies engaged in developing their own products, or else by purchasing a commercially available product such as the CAESAR database developed from 3D whole-body scanning of 4,000 participants. However, agencies such as NIOSH in the United States have started making publicly available specialised databases for groups such as emergency medical technicians (EMTs) and firefighters (see https://www.cdc.gov/niosh/topics/anthropometry/default.html). NASA provides extensive anthropometric data on American and Japanese body sizes from the year 2000 for designing workspaces for astronauts (see https://msis.jsc.nasa.gov/sections/section 03.htm).

The International Organization for Standardization (ISO) promulgates a range of anthropometric standards covering machinery design to the specifications of crash test dummies. Many of the problems inherent in such standards – due to their age (the measurements may now be out of date) or due to the population not being rigorously specified or not the same as the target population – are discussed in a report produced by the Australian Safety and Compensation Council (ASCC).[40] An ongoing project known as the World Engineering Anthropometry Resource (WEAR) provides access to over 150 databases (see https://bodysizeshape.com/). This has been designed as an ongoing source of the most recent available data for designers to utilise on a commercial basis. Most data are available on European and North American populations with little anthropometric data available on South American and African populations, in particular. Military populations remain the most studied – for example, the US Army recently completed a 3D scanning study of 10,000 US soldiers.[41]

Lights that Glare: There is a great deal of information available about light and lighting in various kinds of workplaces. As well as general office space, specialised work environments such as mines or air traffic control towers have specific lighting requirements to ensure optimal efficiency and safety of work. Illumination is measured in the SI unit 'lux', which is one lumen per square metre which indicates the intensity of illumination on a given surface. The required illumination depends largely on the nature of the task but also on the age of the workers. The critical task variable is the contrast between task elements. Bold black text on a white background has high contrast, for example, whereas red text on a black background has low contrast. Higher illumination would be required for tasks involving reading red text on a black background or discriminating between wires coloured gold, silver and copper, for instance. Modern standards provide guidance on appropriate illumination for particular tasks and environments. In general, for office work illuminance of 500–1,000 lux is sufficient whereas for visual tasks involving low contrast (such as the wiring example) levels of 1,000–3,000 lux may be required.[42] The age of the workforce is another relevant factor as substantial changes in visual performance occur with ageing. The ability of the eye to accommodate or focus decreases dramatically from youth to maturity. As well as affecting our ability to resolve fine detail, these changes can also entail changes in posture as workers consequently crouch or bend closer to their work. Another significant change with age is sensitivity to glare.

We have all experienced the glare of another car's headlights in a rear-view mirror or the glare of the sun reflected off water or a shiny surface on a summer's day. At best, glare is simply a source of discomfort as we squint or look away, but glare can directly impair visual performance on another task. In the Los Angeles runway collision (see 'Runway Collision of US Air Boeing 737 and Skywest Fairchild Metroliner, Los Angeles International (LAX), 1 February 1991') the controller was experiencing glare from the apron lights that affected her ability to discriminate the shape of the Metroliner lined up on the runway. We have an unfortunate natural tendency to orient ourselves towards a light source ('phototropism') which can induce temporary adaptation to the bright illumination leading to reduced visual sensitivity when looking back towards the work task.

Thermal environment: The effects of heat stress on human performance have also been extensively studied but primarily from a medical and physiological point of view. The legal standards set by most countries are designed to protect workers from heat-related health effects. For example, the NIOSH standards used in the United States draw up relative exposure limits (RELs) and relative alert limits (RALs) based entirely on physiological and medical evidence.[43] From a safety-related point of view, however, it is the effects of environmental stressors, such as temperature, on memory, attention, vigilance, decision making and other cognitive activities that are most critical. In an environment such as air traffic control (ATC), for example, one single error may have significant safety consequences.

British psychologist **Peter Hancock** (a winner of the Sir Frederick Bartlett medal for lifetime achievement in ergonomics), currently at the University of Central Florida, has written extensively on the effects of environmental stressors on cognitive performance. He has summarised the effects of high-temperature stress by distinguishing between three groups of tasks. The simplest, such as simply reacting to a signal, is the least affected, with vigilance tasks (where an operator monitors the environment for a possible signal) the most affected by thermal stress. Dual tasks involving divided attention occupy an intermediate position. To account for these differences in how tasks are affected by environmental stressors Hancock has proposed the *Maximum Adaptability Model* (MAM) depicted in Figure 4.8.

Hancock's theory proposes that dealing with excessive heat, cold, noise or other stressor diverts some attentional resources away from the current task to evaluate the threat and adapt to its additional demands. Diverting attentional resources away from a simple, probably overlearned, reaction (such as stopping for a red traffic light) will have less effect on task performance than diverting attentional resources away from more challenging tasks. The theory suggests a zone of psychological adaptability whereby resources remain sufficient for the task, or else a switch in strategy is sufficient to maintain performance, although this may come at some cost.

A recent study of surgical trainees carrying out a range of tasks used to train surgeons to conduct laparoscopic (i.e. 'keyhole') surgery compared performance at ambient temperatures of 19C and 26C. The participants wore surgical scrubs and performed the tasks under both temperature conditions. No differences in performance were found, but carrying out the tasks at a higher temperature led to elevated perceptions of workload and distraction. In terms of the MAM, it could be argued that the surgical trainees were able to make some psychological adjustments to compensate for the higher temperature but that there was a measurable cost to doing so, reflected in feelings of greater workload. Having to do the task for longer (the study only took 30 min) might have pushed participants past the zone of psychological adjustment into the zone of physiological adaptation leading to observable effects on performance. Physiological changes in response to core temperature increases may result in changes to the body's homeostatic mechanisms draining attentional resources even further. At extremes of temperature, noise, etc., performance will

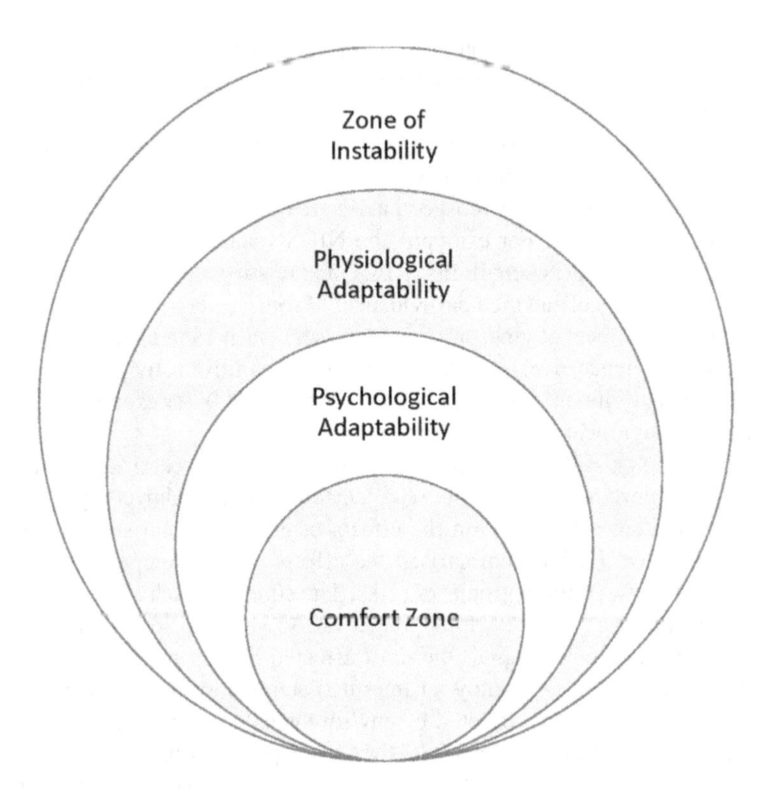

FIGURE 4.8 The Maximum Adaptability Model proposed by Hancock to account for the effects of environmental stressors on human performance. For most tasks, humans operate most comfortably within a range of approximately 15-23C WBGT (Wet Bulb Globe Temperature which takes into account dry air temperature, humidity and radiant energy).

become extremely unstable and may break down altogether as all resources become diverted towards survival.[44]

The MAM is actually a rather more sophisticated theory of the effects of environmental stressors on human performance than described in the simple sketch provided in Figure 4.8. It represents a significant advance on previously prevalent ideas such as the 'inverted u-curve' whereby performance invariably appears to be best in the 'Goldilocks' zone – not too hot, not too cold, etc. Although this is undoubtedly true for most environmental factors it has not proven very helpful in explaining why environmental stressors have the effects they do, or in predicting their likely impacts on human performance and safety. Hancock has recently argued that understanding and predicting the effects of thermal stress, in particular, is becoming ever more important as global temperatures steadily increase.[45]

SUMMARY AND CONCLUSIONS

With the advent of the First World War psychologists were initially brought in from their academic endeavours to provide expertise in measuring aptitudes and abilities so that military personnel could be allocated to areas for which they would be best suited. As military technology rapidly increased in complexity, psychologists and engineers were given a slightly broader remit in the Second World War to find ways to optimise human performance by studying actual behaviour in military systems.

One of the most significant steps in the science of safety began, as does all good science, with observation. In this case, the observations of the Second World War pilots and aircrew of errors and mistakes associated with instrument displays and flight controls. Collating these observations led Paul Fitts and colleagues to show the potential for design-induced error and to uncover the deadly potential of the three-pointer altimeter design. The subsequent appliance of science to improving this display involved careful analysis, experimentation, testing and comparative analysis of alternative designs. This led initially to a great advance on the three-pointer design in the form of the counter-pointer design still used on many flight decks around the world. It is very difficult to precisely estimate the number of lives saved but it would almost certainly be in the thousands.

Within a few months of the serious incident at the Three Mile Island nuclear plant, the Electrical Power Research Institute (EPRI) published a set of reports outlining potential improvements in control room design. The reports catalogue in horrifying terms the extent of the design problems found at the time in the control rooms of nuclear power plants. These include unclear labelling of controls and displays; large arrays of identical undifferentiated controls (including mirror image panels in the same room); controls located so as to be vulnerable to unintended contact and inadvertent operation; glare and reflections bouncing off meter faces; and much, much more. In the present chapter, we have looked at the principles that can be used to ensure that displays provide the appropriate information to guide operators' actions as well as such matters as placement and layout of controls, illumination and so forth.

Subsequent changes and improvements in control room design have been less dramatic than the step forward taken with altimeter design, but, in both cases, technological evolution has rendered some past designs anachronistic. The change from analogue to digital displays has created new opportunities for display design and led to less cluttered instrument panels on flight decks and control rooms alike. The newer digital displays have sometimes created problems of their own as was made evident in the Kegworth accident which involved the pilots misidentifying the malfunctioning engine from their electronic flight displays. Other examples of difficult to read display screens have been reported everywhere from air traffic control centres to more recently designed nuclear power plants.[46]

The most powerful tool in the science armoury, as always, is a good theory. Theory can be misunderstood by some as little more than code for 'guesswork' but this is far from the truth. Theory summarises the essence of what has been discovered through observation, experimentation and testing into succinct form. Theory may be used to generate principles that can be used to guide design, operations and management. Our brief overview of the design of equipment, tasks and of the working environment has highlighted the key role of memory and attention in the safety of almost every kind of work. Theories about memory and attention from modular Working Memory to Multiple Resource Theory (MRT) and the Maximum Adaptability Model (MAM) provide a convenient way of summarising what is known and turning this, whenever possible, into useful practical guidance. Drawing on solid scientific theory provides a much more enduring, stable and empirically justifiable basis for safety management than following currently fashionable management trends.

NOTES

1. Shephard, B. (2015). Psychology and the great war, 1914–1918. *The Psychologist, 28*, 944–947.
2. Reynolds, L. (2004). Wellcome witnesses: The medical research council applied psychology unit. *History of Psychology, 7*, 85–93. Some of the post-war practical problems worked on at the APU included alcohol and driving; keyboard design; a more legible typeface for the Times newspaper; telephone dialling codes and mail sorting postal codes. The UK area postal codes (e.g. 'RH6 9DD') were based on the combination of letters and digits that were found to be optimal in terms of human memory.
3. Alluisi, E. (1994). APA Division 21: Roots and rooters. In H. Taylor (Ed.), *Division 21 Members Who Made Distinguished Contributions to Engineering Psychology* (pp. 4–22). Washington, DC: American Psychological Association.
4. A chapter by Richard Pew focusing exclusively on Paul Fitts appears in the same collection (pp. 23–44) edited by Henry Taylor as cited above.
5. This short definition of engineering psychology comes from a technical report by Paul Fitts in 1947 as cited in the chapter (p. 24) by Richard Pew above.
6. John Rolfe was a Principal Psychologist at the RAF Institute of Aviation Medicine in the UK and was responsible for conducting the studies on improved altimeter design. An excellent overview of this work can be found in: Rolfe, J.M. (1969). Human factors and the display of height information. *Applied Ergonomics, 1*, 16–24.
7. Treisman, A., Kahneman, D., & Burkell, J. (1983). Perceptual objects and the cost of filtering. *Perception and Psychophysics, 33*, 527–532.
8. Miller, R.J., & Penningroth, S. (1997). The effects of response format and other variables on comparisons of digital and dial displays. *Human Factors, 39*, 417–424.
9. Alphonse Chapanis, one of the most important early figures in the development of human factors/ergonomics provides an extended version of this quotation from Hopkins as the opening of his textbook "*Human Factors in Systems Engineering*" published in 1996 (New York: Wiley).
10. This summary is taken directly from "*Background on the Three Mile Island Accident*" produced by the U.S Nuclear Regulatory Commission downloaded from: https://www.nrc.gov/reading-rm/doc-collections/fact-sheets/3mile-isle.html
11. For a thorough review see: Stanton, N. (2006). Hierarchical task analysis: Developments, applications, and extensions. *Applied Ergonomics, 37*, 55–79.
12. Air Accidents Investigation Branch. (1990). *Report on the Accident to Boeing 737-400 G-OBME Near Kegworth, Leicestershire on 8 January 1989*. London: HMSO. The quotation from the aircraft commander can be found on page 3.
13. CEN is the European Committee for Standardization which is one of three European organisations officially recognised by the European Union. Both ISO and CEN standards cover a wide range of industries and activities. See: https://www.cen.eu/about/Pages/default.aspx for information about CEN or https://www.iso.org/home.html for information about ISO.
14. Don Norman's '*The Psychology of Everyday Things*' was published by Basic Books (New York) in 1988. The title was later changed to '*The Design of Everyday Things*'. Norman has since written numerous other books covering design and usability such as

'*Turn Signals Are the Facial Expressions of Automobiles*' (New York: Basic Books, 1993) and '*Living with Complexity*' (Cambridge: MIT Press, 2010). His website is www.jnd.org.
15. Banbury, S.P., Selcon, S.J., & McCrerie, S. (1997). New light through old windows: The role of cognitive compatibility in aircraft dial design. In *Proceedings of the Human Factors and Ergonomics Society 41st Annual Meeting* (pp. 56–60). Santa Monica, CA: Human Factors and Ergonomics Society.
16. This was one of the phenomena investigated by Paul Fitts. See: Fitts, P.M., & Seeger, C.M. (1953). S-R compatibility: Spatial characteristics of stimulus and response codes. *Journal of Experimental Psychology, 46*, 199–210. An excellent coverage of the whole area of compatibility in engineering design is provided by Wickens, C.D., Hollands, J.G., Banbury, S., & Parasuraman, R. (2013). *Engineering Psychology and Human Performance, 4th Ed.* Upper Saddle River, NJ: Pearson.
17. See, for example: Wickens, C.D., & Carswell, M. (1995). The proximity compatibility principle: Its psychological foundation and relevance to display design. *Human Factors, 37*, 473–494. Another good source is: Bennett, K.B., & Flach, J.M. (1992). Graphical displays: Implications for divided attention, focused attention, and problem solving. *Human Factors, 43*, 513–533.
18. US patent 6860266B2 was approved in 2005 for a system for: "*obtaining physiological information from patients and displaying that information in an intuitive and logical format to a physician*". The patent was taken out by Alliance Pharmaceutical Corp of Dartmouth. The development of the 'wagon wheel' for displaying conditions in a nuclear power plant was reported by: Woods, D.D., Wise, J.A., & Hanes, L.F. (1981). An evaluation of nuclear power plant safety parameter display systems. In *Proceedings of the Human Factors Society 25th Annual Meeting* (pp. 110–114). Santa Monica, CA: Human Factors Society.
19. Drews, F.A., & Westenskow, D.R. (2006). The right picture is worth a thousand numbers: Data displays in anaesthesia. *Human Factors, 48*, 59–71.
20. Szalma, J.L. (2011). Workload and stress in vigilance: The impact of display format and task type. *American Journal of Psychology, 124*, 441–454.
21. Young, M.S., & Birrell, S.A. (2012). Ecological IVIS design: Using EID to develop a novel in-vehicle information system. *Theoretical Issues in Ergonomics Science, 13*, 225–239.
22. Stanton, N.A., & Allison, C.K. (2020). Driving towards a greener future: An application of cognitive work analysis to promote fuel-efficient driving. *Cognition, Technology and Work, 22*, 125–142.
23. Read, G.J.M., et al. (2019). From interfaces to infrastructure: Extending ecological interface design to re-design rail level crossings. *Cognition, Technology and Work*, https://doi.org/10.1007/s10111-019-00583-2. There are several other publications on this work including: Beanland, V., Grant, E., Read, G.J.M., Stevens, N., Thomas, M., Lenné, M., Stanton, N., & Salmon, P.M. (2018). Challenging conventional rural rail level crossing design: Evaluating three new systems-thinking-based designs in a driving simulator. *Safety Science, 110 (Part B)*, 100–114. I thank Vanessa Beanland for pointing out to me the important design feature of requiring no external power source.
24. Bennett, K.B., & Flach, J. (2019). Ecological interface design: Thirty-plus years of refinement, progress and potential. *Human Factors, 61*, 513–525.
25. Conrad's work for the post office on push button dialling is reported in: Conrad, R. (1960). Experimental psychology in the field of telecommunications. *Ergonomics, 3*, 289–295. The work that led to the adoption of the alpha-numeric post codes is reported in: Conrad, R. (1962). The location of figures in alpha-numeric codes. *Ergonomics, 5*, 403–406. The work on acoustic confusion in memory for words appears in: Conrad, R. (1964). Acoustic confusions in immediate memory. *British Journal of Psychology, 55*,

75–84. As a testament to the effectiveness of Conrad's recommended format I have no trouble remembering the postal code for my former home address in the UK that I last lived in over 45 years ago.

26. National Transportation Safety Board. (1991). *Aircraft Accident Report. Runway Collision of USAir Flight 1493 Boeing 737 and Skywest Flight 5569 Fairchild Metroliner Los Angeles International Airport Los Angeles California February 1 1991. NTSB/AAR-91/08.* Washington, DC: National Transportation Safety Board.
27. More detailed information on the role of working memory in human performance can be found in: Wickens, C.D., Hollands, J.G., Banbury, S., & Parasuraman, R. (2013). *Engineering Psychology and Human Performance, 4th Ed.* Upper Saddle River, NJ: Pearson.
28. Miller, G.A. (1956). The magical number seven plus or minus two: Some limits on our capacity for processing information. *Psychological Review, 63*, 81–97. One of the ten most often cited papers in psychology.
29. Wickens, C.D., & McCarley, J.S. (2008). *Applied Attention Theory.* Boca Raton, FL: CRC Press. This is an excellent source of in-depth information on SEEV, attention management, multitasking and the design of displays. Coverage of these topics is also provided in the Wickens et al (2013) textbook on engineering psychology (see Note 27).
30. Koh, R.Y.I., Park, T., Wickens, C.D., Ong, L.T., & Chia, S.N. (2011). Differences in attentional strategies by novice and experienced operating theatre scrub nurses. *Journal of Experimental Psychology: Applied, 17*, 233–246.
31. Martens, M.H. (2011). Change detection in traffic: Where do we look and what do we perceive? *Transportation Research Part F, 14*, 240–250.
32. Recording brain activity has become a popular means of investigating the processes of memory and attention. One measure of brain activity arising directly in the wake of the appearance of a specific stimulus is an evoked potential known as P3 or P300 which is found 300msecs after the appearance of a target. The amplitude of this peak correlates with the attention directed at processing the target.
33. President Lyndon B. Johnson has famously been quoted as saying of Gerald Ford that "he couldn't fart and chew gum at the same time". The press at the time 'cleaned' up the quote by changing it to: "he couldn't walk and chew gum at the same time". There is even an academic article with the same title: "Orofacial communication: How do we walk and chew gum at the same time?" that reports studies of coordination between different muscles in the face and jaw: Morquette, P., & Kolta, A. (2014). Published online 3 June 2014. https://doi.org/10.7554/eLife.03235
34. Daniel Kahneman laid the foundations for the study of attention and performance in his 1973 book *'Attention and Effort'* (Englewood Cliffs, NJ: Prentice Hall). The central idea of this, and later work, is that humans have a finite capacity for attention. Attentional capacity is referred to as a resource. The idea of more than one resource pool was first proposed by Norman, D.A., & Bobrow, D.J. (1975). On data-limited and resource-limited processes. *Cognitive Psychology, 7*, 44–64, and further elaborated by Navon, D., & Gopher, D. (1979). On the economy of the human processing system. *Psychological Review, 86*, 214–255.
35. Multiple resource theory was developed in the early 1980s and was explained in several chapters in edited books. For example: Wickens, C.D. (1980). The structure of attentional resources. In R. Nickerson (Ed.), *Attention and Performance VIII* (pp. 239–257). Hillsdale, NJ: Lawrence Erlbaum. A more recent overview of the model and a computational extension to provide specific performance predictions can be found in: Wickens, C.D. (2002). Multiple resources and performance prediction. *Theoretical Issues in Ergonomics Science, 3*(2), 159–177.

36. The background on predicting advanced levels of pilot performance can be found in: Roscoe, S.N., & North, R.A. (1980). Prediction of pilot performance. In S.N. Roscoe (Ed.), *Aviation Psychology* (pp. 127–133). Ames, IA: Iowa State University Press. Later testing confirmed the ability of the subsequent WOMBAT test (see www.aero.ca) to discriminate between experienced and elite glider pilots. See: O'Hare, D. (1997). Cognitive determinants of elite pilot performance. *Human Factors, 39*, 540–552.
37. Their early work was summarised in: Tversky, A., & Kahneman, D. (1974). Judgment under uncertainty: Heuristics and biases. *Science, 185*, 1124–1131. Google Scholar shows well over 35,000 citations for the original article. Danny Kahneman has more recently written a bestseller: *Thinking Fast and Thinking Slow*, published by Penguin.
38. Klein, G.A. (1993). A recognition primed decision (RPD) model of rapid decision making. In G.A. Klein, J. Orasanu, R. Calderwood, & C.E. Zsambok (Eds.), *Decision-Making in Action*. Norwood, NJ: Ablex.
39. Human factors/ergonomics texts such as Sanders, M., & McCormack, E. (1993). *Human Factors in Engineering and Design, 7th Ed.*, published by McGraw-Hill provide useful background information. Waldemar Karwowski has edited several large compendiums in this area and these provide convenient sources of information. For example: Karwowski, W. (Ed.). (2006). *Handbook of Standards and Guidelines in Ergonomics and Human Factors*. Boca Raton, FL: CRC Press. This handbook contains 32 chapters covering 640 pages. Another useful source is provided by Salvendy, G. (Ed.). (2012). *Handbook of Human Factors and Ergonomics, 4th Ed.* Hoboken, NJ: Wiley. This includes a chapter by the ubiquitous Karwowski on 'The discipline of human factors and ergonomics'.
40. Australian Safety and Compensation Council. (2009). *Sizing Up Australia: How Contemporary is the Anthropometric Data Australian Designers Use?* Canberra, ACT: Commonwealth of Australia. The most recent standard produced by ISO is *'Body Measurement: Definitions and Landmarks'* (ISO7250-1:2017).
41. Li, P., Corner, B., Carson, J., & Paqette, S. (2015). A three-dimensional shape database from a large-scale anthropometric survey. In *Proceedings of the 19th Triennial Congress of the IEA*, Melbourne, 9–14th August 2015. Downloaded from http://ergonomics.uq.edu.au/iea/proceedings/Index_files/papers/726.pdf.
42. Helander, M. (2006). *A Guide to Human Factors and Ergonomics*. Boca Raton, FL: CRC Press. Helander gives a couple of interesting examples: one involving safety warning signs for use in mines where the traditional red text/black background provided legibility problems in low levels of illumination, and the second involving problems with using a TV monitor to assist workers doing an intricate manufacturing job with different wires of similar colours.
43. NIOSH. (2016). *NIOSH Criteria for a Recommended Standard: Occupational Exposure to Heat and Hot Environments*. By Jacklitsch, B., Williams, W.J., Musolin, K., Coca, A., Kim, J.-H., Turner, N. Cincinnati, OH: U.S. Department of Health and Human Services, Centers for Disease Control and Prevention, National Institute for Occupational Safety and Health, DHHS (NIOSH) Publication 2016-106.
44. The Maximal Adaptability Model was introduced by Hancock, P.A., & Warm, J. (1989). A dynamic model of stress and sustained attention. *Human Factors, 31*, 519–537. A comprehensive overview is provided in Hancock, P.A., & Vasmatzidis, I. (2003). Effects of heat stress on cognitive performance: The current state of knowledge. *International Journal of Hyperthermia, 19*, 355–372. The study of surgical trainees was carried out at the University of Southern California: Berg, R.J., et al. (2015). The impact of heat stress on operative performance and cognitive function during simulated laparoscopic operative tasks. *Surgery, 157*(1), 87–95. http://dx.doi.org/10.1016/j.surg.2014.06.012

45. Lopez-Sanchez, J.I., & Hancock, P.A. (2019). Diminishing cognitive capacities in an ever hotter world: Evidence from an applicable power-law description. *Human Factors, 61*, 906–919.
46. The reports from the Electric Power Research Institute can be downloaded from: https://www.epri.com/#/pages/product/NP-1118-V1. The reports were produced by contractors from Lockheed Missiles and Space Co. The reports are titled: '*Human Factors Methods for Control Room Design, Report EPRI NP-1118*'. Problems with reading computer displays at a Brazilian nuclear power plant are reported by DeCarvalho, P.V.R. (2006). Ergonomic field studies in a nuclear power plant control room. *Progress in Nuclear Energy, 48*, 51–69. The operators comment on the lack of contrast (pale black letters on a yellow background) as well as light reflections off the screen.

Section 2

Organisations

5 Normal Accidents

A pilot misreads the altitude on their altimeter by 10,000 ft. A nuclear power plant operator inadvertently activates a control on their panel. An air traffic controller forgets about an aircraft they have cleared to line up on a runway. As 'social animals' it is easy for us to see the people in these situations as 'figure' and everything else as 'ground' and to look for information about the individuals (tired? clumsy? forgetful?) as explanations for their behaviours. In every case, however, the individual is also part of an organisation (airline, power generator, air traffic management authority) and the role that organisations play in safety, lightly touched upon in earlier chapters, is now the major focus of this section of the book.

In the second section of the book, our focus will be on the organisation rather than the individual. Whereas our focus on the individual was necessarily informed by psychology and the major advances in understanding human cognition that largely grew out of the 20th-century-war efforts, understanding organisations draws primarily on sociology. Psychologists have drawn heavily on the natural sciences methods of experimentation and control and these have been productive in establishing much of what we currently know about the brain and cognition. These methods no longer represent the 'cutting edge' of psychological science, however. Massive data sets showing the billions of daily social media interactions, detailed genetic analyses and sophisticated statistical modelling are now providing insights into human behaviour in context rather than in the splendid isolation of the psychology laboratory.

PEOPLE IN CONTEXT: SOCIETIES AND ORGANISATIONS

Anthropologists and sociologists have also drawn on a range of methodologies in their efforts to set aside the taken-for-granted appearance of social life to analyse what lies beneath. As members of families, social groups, work and social organisations, regional and national groupings, most of what constitutes these groupings seems too self-evident to require analysis or explanation. In many ways, we have taken more interest in studying exotic tribes than in the social organisation of our own cultures. Sociological analysis has looked at diverse aspects of social organisation. The functionalist perspective has explored the functions served by organisations on the assumption that these must be adaptive and beneficial or else the organisations would not survive. The interactionists focus on the development and maintenance of social relationships that enable an organisation to cohere in a collective sense with a unitary sense of purpose. The more recent concerns with organisational culture, in general, and safety culture, in particular, are offshoots of the interactionist perspective. An analysis of power underlies Marxist, feminist and other perspectives that seek to critique social structures and organisations.[1]

DOI: 10.1201/9781003038443-7

Of most relevance for understanding organisations and the role they play in safety is the perspective developed from the extensive and influential writings of turn-of-the-century German sociologist **Max Weber**. His work greatly influenced many different aspects of sociology despite his life being cut relatively short by the 'Spanish Flu' dying in 1920 at the age of 56. The rapid growth of industrial organisations during the late 19th century created the working world we know today where the large majority of people work for someone else. Many of us now work for large government or shareholder-owned organisations with multiple tiers of executives and managers overseeing the productive activities of those lower down the hierarchy. Weber was interested in the growth of such bureaucracies and their principles of organisation.

It was Weber who spelt out the differences between the traditional forms of authority that governed most societies until the industrial revolution and what Weber called the rational-legal form of authority that characterises the post-industrial revolution form of business capitalism and government in most modern societies. Organisations exist to coordinate the actions of multiple workers whose productive effort must exceed the cost of employing them. In rational-legal bureaucracies, people are recruited on merit and occupy specific roles in the organisation with specific duties and responsibilities and organised schemes of remuneration. Weber saw the development of rational-legal bureaucracy as enhancing efficiency and progress but at the same time containing within it the dangers of dehumanising work in the 'faceless bureaucracy' where individuals become mere 'cogs in the machine'. This can arise when rigid obedience to bureaucratic rules and regulations takes precedence over individual creativity and innovation. This precarious balance between the desirability of strict adherence to rules and procedures, on the one hand, and the desirable exercise of some discretionary variability in work according to circumstances, on the other hand, is particularly relevant to the management and organisation of high-risk activities.

An important concern in the analysis of bureaucratic organisations is with the optimal number of levels in the hierarchy, which can also be expressed in terms of the optimum *span of control*. The span of control refers to the number of subordinate employees for whom each person has responsibility. As a general rule, the smaller the span of control the steeper the organisational hierarchy. Alternatively, if there are few managers each with many subordinates then the organisational structure is necessarily flatter. Management theorists such as **Peter Blau** studied the effects of these different spans of control and levels of organisational hierarchy on the ability of workers to consult and coordinate their actions with others. Blau found that the more specialised and technical the work, the greater the benefits of a smaller span of control. In essence, this allows the front-line workers to deal with most of the problems that may crop up whilst at the same time giving them easy access to a superior with the knowledge to solve more difficult non-routine problems.

The American sociologist **Charles Perrow** worked at a number of universities across the United States culminating in a position at Yale until retirement. His reputation was built on a detailed analysis of power in organisations – both their internal structure and their societal role. As a sociologist with an interest in organisational structure, he noted that the effects of organisational hierarchy and span of control depended significantly on the nature of the work being undertaken. In addition, he noted the presence of informal hierarchies where the 'real power' was wielded by certain individuals lower down the hierarchies. A good example is the power wielded by secretaries and personal assistants serving a 'gatekeeper' function for those high on the formal organisational chart.[2] Perrow might well have remained a distinguished, if relatively unknown, academic researcher and author were it not for several serendipitous events.

The occurrence of the Three Mile Island nuclear reactor accident in Pennsylvania in 1979 has already been mentioned (see Chapter 4). Taking on the role of director of the graduate programme at the University of Wisconsin in 1966 was the other.[3] Perrow set about bringing more female students into the PhD programme, one of whom – Cora Bagley-Marrett – went on to serve on the President's Commission, appointed by President Jimmy Carter, to look into the circumstances of the Three Mile Island accident. Bagley-Marrett recommended that an expert on the sociology of organisations be recruited to advise the Kemeny Commission on the organisational aspects of the accident. Perrow says that as he read over the testimonies of the operators at the plant, he had an epiphany:

> It sounded like the crazy things that went wrong in the universities (and other organizations) I had worked at … but accidents at universities did not cascade into system failures … Second, although my universities were not radioactive, TMI was. Here was catastrophic potential.

But for Three Mile Island (TMI) and Cora Bagley-Marrett the concept of 'normal accidents' might never have been born. The reactor accident seemed to fit with the work Perrow was doing at the time on the relationship between organisational structure and the control of work processes. For several years following TMI, Perrow worked on reviewing the existing literature on accidents and safety, covering much of the ground set out in the first section of the present book. At the same time, his involvement with the TMI inquiry led to high-level (although relatively short-lived) appointments to the Nuclear Regulatory Commission to advise on the optimal organisational structure for nuclear power plants and to the National Research Council to advise on the role of human factors in industrial safety.

Apparently, Perrow's views on the importance of organisational structure for understanding how to better manage complex systems, particularly those with high-risk potential, were not particularly well-received by either the nuclear regulators or the National Research Council. A report submitted to the latter was simply not accepted and subsequently appeared as an article in a high-profile academic journal.[4] The article deals with the perennial complaint of human factors engineers that their input is generally not sought out and if offered is often ignored. This lack of influence has sometimes been attributed by human factors engineers themselves to a reluctance by hard-nosed engineers to accept information that is generally not expressed in hard quantitative terms. As Perrow points out, this has not been universally the case with human factors engineering/ergonomics more readily accepted in some industries such as aviation and less so in others such as the marine transport industry. Some of the differences between these two industries will be discussed further below.

Perrow contrasts 'design' logic, based on embracing the latest technologies and satisfying design criteria in terms of compactness and aesthetic appearance, with 'operational' logic based on ease of use and maintainability. These imperatives often contradict each other. As previously noted, nuclear power plant design clearly emphasised design logic resulting in overcrowded control panels with difficult-to-discriminate controls and unreadable displays. One simple explanation for the preponderance of poor design suggested by Perrow is that the consequences are borne by the operators and may be largely unknown to the design engineers and upper management (unless a catastrophe occurs). When operator concerns have been made known (as in the *Herald of Free Enterprise* case), they can be dismissed as excuses by operators for not doing their jobs properly or as costly additional expenditure of unquantifiable potential benefit.

NORMAL ACCIDENT THEORY

Many interesting and valuable books and papers have been published in recent decades outlining various aspects of the organisational approach to safety. One of the earliest and most influential was the arrestingly titled *Normal Accidents* first published in 1984 by Charles Perrow. After the nuclear accident involving the Three Mile Island reactor in Pennsylvania in 1979, Perrow looked at the organisational structure of the reactor company and proposed that the critical organisational factors that led to the near-meltdown at TMI, as well as numerous incidents elsewhere, were the hitherto unexplored organisational characteristics he labelled as *interactive complexity* and *tight coupling*.[5]

Perrow leads with a homely and easily understood example of interactive complexity similar to the following. Imagine that you and your partner have set the alarm on your 'phone to wake you up in time to get to a job interview at 9 a.m. downtown. For some reason, the alarm does not go off and you oversleep. On waking and realising the time, your partner dresses, grabs coffee and a slice of toast and leaves the apartment. You also make a coffee and put a slice of bread in the toaster. Whilst shaving, you hear the smoke alarm sounding. A crust from your partner's toast has lodged under your bread, burnt and sent smoke into the air. The alarm is high up and you search for a stool to reach it. By the time you have found one and disabled the alarm you are now exceedingly late. You rush out of the apartment forgetting your phone. When you reach your garage you remember that your car is at the dealer's for servicing so you quickly decide to run to the nearest bus stop. No buses appear and there are no other people waiting. It turns out that the bus drivers are on strike over safety issues on the buses. You cannot call an Uber because you have no phone. You miss your interview and fail to get the job. Events that would normally be loosely coupled and linear (get up, breakfast, drive) have become more tightly coupled and complex.

It will come as no surprise to anyone that a nuclear reactor is highly complex. In contrast, a dam is extremely simple – build a wall across a river and a lake will form allowing water to be fed down pipes to turn turbines and generate electricity. Other systems such as an automobile production line can be much more complicated – there are many component processes (and these often involve subcomponent processes) and these must occur in a strict order otherwise the productive goal will not be fulfilled. However, although highly complicated, the system is extremely orderly. The individual parts are easily understood and there are no unexpected interactions between the parts – what happens in the seat assembly room, for example, has little impact on anything happening in the transmission building department.

Complex systems, however, are ones where there is a potential for unexpected interactions between different parts of the system. Perrow illustrates this with a marine example involving a Mississippi oil tanker (the *Dauntless Colocotronis*) that struck a submerged wreck that had been incorrectly charted. The contact resulted in a gash that allowed oil to escape. Some of the oil went into the pump room. The heat in the pump room lessened the oil's viscosity allowing more oil to be drawn in. This oil penetrated a 'packing gland' and began to seep into the adjacent engine room where it evaporated and ignited. Numerous other mistakes followed, including trying to put the fire out with water. Nobody had anticipated that this particular sequence of interactions would ever occur or that failure in one subsystem would affect another unrelated subsystem that simply happened to be located nearby. This is much the same as happened in the simple story about missing the job interview described earlier. On the surface, this seems a simple, linear sequence of breakfast, car and drive to interview. However, the unexpected interactions between these and other events quickly led to mission failure. This is a feature that has characterised many real-world accidents and disasters.

The other critical structural feature is 'tight coupling'. This refers to systems where there is little 'slack' or give and take possible between connected parts of the system. Nuclear and chemical reactions are tightly coupled – once the reaction gets underway it will inexorably continue on its way. The automobile production line can be slowed down or switched off but this is not possible in chemical and power generation. Tight coupling implies invariant sequences (A must follow B and precede C) as well as precision – each component must meet exact specifications in terms of both quality and quantity. Everything must be very precise and timing is all-important. Any system can be characterised in terms of these two concepts and located in a two-dimensional space as shown in Figure 5.1.

Up to this point, Perrow has simply provided an interesting and potentially useful way of looking at the organisational structures that are most associated with errors and disasters. The most controversial part of Perrow's argument, however, was the thesis that for certain systems these characteristics make accidents inevitable or 'normal' for that system. Perrow expresses it perfectly:

> A normal accident is where everyone tries very hard to play safe, but unexpected interaction of two or more failures (because of interactive complexity), causes a cascade of failures (because of tight coupling). The combination of complexity and coupling will bring down the system despite all safety efforts.

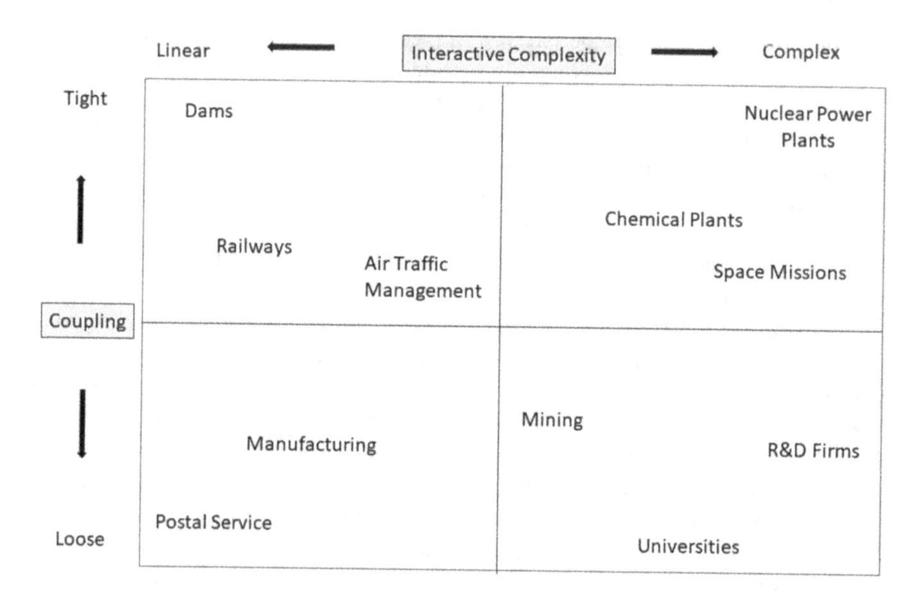

FIGURE 5.1 Technologies and activities as characterised by their degree of interactive complexity along the *x*-axis and coupling shown on the *y*-axis. According to Perrow, the upper-right quadrant involves activities whose high interactive complexity and tight coupling make them vulnerable to 'normal accidents'.

Nuclear power generation is inherently 'accident prone' according to Perrow, not simply because of operator errors or construction weaknesses but because of this inherent potential for multiple interactions of small failures that can rapidly cascade into catastrophic disasters. Perrow then poses the rhetorical question: 'Why then haven't there been more such failures?' and provides the answer himself – because commercial nuclear power plants haven't been operating for long enough! Since TMI, there have, indeed, been many more incidents at nuclear power plants including the devastating accidents at the Ukrainian Chernobyl plant in 1986 and the plant at Fukushima, Japan, in 2011.

EXPLOSION AT NUMBER 4 REACTOR, CHERNOBYL, UKRAINE, 26 APRIL 1986

'On April 26, 1986, the Number Four RBMK reactor at the nuclear power plant at Chernobyl, Ukraine, went out of control during a test at low power, leading to an explosion and fire that demolished the reactor building and released large amounts of radiation into the atmosphere. Safety measures were ignored, the uranium fuel in the reactor overheated and melted through the protective barriers. RBMK reactors do not have what is known as a containment structure, a concrete and steel dome over the reactor itself designed to keep radiation inside the plant in the event of such an accident. Consequently, radioactive elements including plutonium, iodine, strontium and caesium were scattered over a wide area. In addition, the graphite blocks used as a moderating material in the RBMK caught fire at high temperature as air entered the reactor core, which contributed to emission of radioactive materials into the environment'.[6] The International Atomic Energy Association (IAEA) estimates that around 600,000 people were affected by radiation as a result of the explosion.

The events that took place in the control room of the number four reactor of the Chernobyl nuclear power plant were outlined in the official Soviet report of the disaster released in 1986. As part of a scheduled shutdown of the reactor for regular checks and maintenance an experimental test was devised by engineers to determine whether, in the event of an emergency, there would be enough residual energy left in the turbines to keep pumping coolant water into the reactor for the 45-second gap before the diesel generators kicked in. The test was scheduled to take place on 25 March. The shift that took over on that morning inherited a reactor that had already had its power output reduced by half. The manufacturer's recommendation was that when this point was reached the reactor should be shut down completely as the RBMK reactors used in the Soviet Union were known to become highly unstable at low power.

The head shift leader did not shut down the plant as (a) the planned experimental test had not been conducted and (b) it was necessary to get permission from the power grid controller in Kiev to disconnect from the power grid. This failure to shut down could be regarded as a violation, one of several, according to an analysis of the disaster by Jim Reason.[7] Reason lists seven additional violations and errors that were documented in the run-up to the explosion. Reason concludes that the explosion occurred because of 'mistakes on the part of the experimental planners and the management, one serious operator slip … and a series of safety violations'. The initial Soviet report into the disaster also blamed 'an extremely improbable combination of violations of instructions and operating rules committed by the staff of the unit'.

Subsequent research and analysis have given us a much fuller picture of the operation of the Chernobyl plant, the culture surrounding nuclear power generation in the Soviet Union and weaknesses in the design of the RBMK reactor design used solely in the Soviet Union and its satellites. Many Soviet engineers had long-standing concerns about the operation of the RBMK-type reactor. Western designed reactors were mainly pressurised water reactors (PWR) where water is used both as a coolant and as a moderator of the nuclear reaction. These reactors were originally developed by the military to power nuclear submarines and aircraft carriers. The Soviet Union also developed a similar design of reactor known as the VVER. This is now the preferred reactor design in Russia since Chernobyl. In contrast, the RBMK design, used at Chernobyl and elsewhere in the former Soviet Union, was cheaper to build, using prefabricated components, cheaper to operate and produced more power than the PWR/VVER designs. The main design difference was the use of graphite rather than water to moderate the nuclear reaction. This led to a number of technical problems that made the RBMK reactors potentially unstable at low levels of power output and more difficult to control than the water-moderated reactors.

For RBMK-type reactors then, these inherent design features created a far greater potential for accidents than other nuclear power plant designs. If this is the case, then prior to Chernobyl we would expect to find other examples of disaster and near-disaster in these reactors. And indeed we do. According to one analysis, there were at least 11 previous reported accidents between 1966 and 1986 involving Soviet reactors of this type. The most significant occurred at the Leningrad plant where an eerily similar version of events unfolded as would later occur at Chernobyl, where the reactor essentially went out of control when the power was reduced to a low level.[8] For Soviet RBMK reactors, accidents were indeed 'normal' and expected due to the deficiencies of the design itself.

Valerie Legasov was one of the first scientists on the scene after the explosion and was the government's chief scientific advisor on the clean-up operation. He subsequently presented a version of the official government findings at the post-accident conference convened by the International Atomic Energy Authority (IEA). 'He had not spoken about the defects of the reactor itself which helped to turn what would otherwise have been a serious accident into a nuclear catastrophe'. He also felt constrained to refrain from talking publicly about the numerous organisational and political failings that lay behind the event although he had no doubts that Chernobyl was the 'apotheosis and peak of the economic mismanagement in our country over decades'.[9]

The Soviet leaders were much more concerned with managing the political fallout of the accident rather than the actual radioactive fallout. The disaster occurred just as Soviet leader Mikhail Gorbachev was about to embark on a process of opening up the Soviet economy away from the militarised, central command economy of the previous decades. In fact, the director of the Chernobyl plant and long-term party member **Viktor Briukhanov** had only just returned from attending the annual Party Congress in Moscow. Subsequently, Briukhanov was one of six managers charged with criminal negligence, receiving a sentence of ten years imprisonment. Questions about the inherent safety of the RBMK reactor design were not considered at the trial of the plant managers. One of the main issues that subsequently emerged was the failure to learn the lessons of the previous near out-of-control event at Leningrad in 1975. Somewhat ironically, the engineer who drew up the plans for the shutdown and tests at Chernobyl had himself been a trainee at the Leningrad plant at the time of the near-disaster.

The events at Leningrad had revealed a crucial weakness in the RBMK design. For reasons of economy, the boron control rods that were raised or lowered into the reactor to control the rate of the reaction (and thus the power generated) were tipped with graphite. Unfortunately, the initial effect of lowering the rods to reduce the rate reaction was not as intended: 'the tips replaced neutron absorbing water in the top part of the active zone, thus not decreasing but further increasing the rate of the reaction. This was the positive void effect – the deadly design problem of RBMK reactors'.[10] Perrow seems somewhat unaware of the inherent catastrophic potential of the RBMK reactors, attributing the Chernobyl disaster exclusively to 'gross management malfeasance' stretching all the way to the Politburo. It appears that at the time, even the chief designer of the reactor was not fully aware of the significance of the positive void effect.

The development of nuclear power plants for civil use was essentially an offshoot of the military development of nuclear power in the Soviet Union. Consequently, an atmosphere of secrecy and reluctance to disseminate information developed. Information from incidents, such as the near-disaster at Leningrad, was not discussed at scientific meetings and was not used to make improvements in reactor design and operation. Matters of reactor design and operational safety were subservient to the political aims of the programme which were to demonstrate the technological progress and prowess of the Soviet system. These organisational constraints led to the perpetuation of an inherently risky technology, compounded by inadequate documentation and regulation amid a culture of secrecy and limited spread of any 'negative' information. As one nuclear expert put it: 'The Chernobyl accident was practically "planned". Its roots lay in the history of the RBMK development'.[11] Accidents were most certainly normal for Soviet RBMK reactors.

Perrow's coining of an attention-getting oxymoron ('normal accidents') was pivotal in turning a well-regarded organisational theorist into a sought-after authority on a wide range of issues. As he himself notes, 'There's a lot of life after retirement if you have a catchy metaphor that seeps into many parts of society that you never thought of visiting'.[12] In a new 'afterword' added to the republished 'Normal Accidents' in 1999, Perrow tried to steer away from catchy metaphors back to organisational theory, by proposing a tripartite distinction between three different categories of organisation: Those that are inherently *error-inducing*, such as nuclear power and marine transport; those that are *error-neutral*, such as space programmes and chemical manufacturing; and those that are *error-avoiding*, such as air transport and air traffic control. The study of what makes certain error-avoiding organisations capable of engaging in an inherently risky operating environment whilst simultaneously maintaining extremely high levels of safety has become known as the study of *high-reliability organisations* (HROs) and will be the subject of Chapter 6.

ERROR-INDUCING SYSTEMS

The International Maritime Organization (IMO) is the United Nations special agency responsible for shipping and maritime pollution by ships. Their annual report on shipping losses of large vessels (over 100 tons gross) shows an average of around 100 total losses along with around 3,000 incidents each year at an average annual cost of around US$ 2 billion.[13] Safety at sea is subject to myriad influences including the shipbuilders and ship owners; management and port companies; insurers; the state where the vessel is registered; and 'classification societies' which are 'independent bodies which set standards for design, maintenance, and repair of ships'. These issue 'class certificates' which are used by insurance companies. Some classification societies (e.g. Lloyds Register) meet high professional standards whilst others (e.g. Register of Albania) may not.[14] Similarly, widespread differences exist in the regulatory standards of different flag states (i.e. the nation where the vessel is registered) from traditional maritime nations such as Norway and the UK at one end to new entrants such as Equatorial Guinea and Belize at the other. In theory, all maritime activities take place under the umbrella of the *United Nations Convention on the Law of the Sea* (UNCLOS).

In addition to this regulatory patchwork, the maritime industry must deal with economic challenges, congestion in shipping lanes, highly hazardous cargoes of toxic chemicals as well as the challenges of increasingly violent storm systems not to mention seaborne hazards such as ice as well as the roughly 1,500 shipping containers lost overboard each year.[15] For Perrow, this tangled web of influences and interests constitutes an 'error-inducing system: … the configuration of its many components induces errors and defeats attempts at error reduction'. In contrast to nuclear power generation where interactive complexity and tight coupling ensure that the system can never be completely safe, the maritime industry is more heterogeneous. Perrow sees 'some aspects as too loosely coupled (the insurance subsystem and shippers), others as too tightly coupled (shipboard organization), some aspects are too linear (shipboard organization again which are highly centralised and routinised), others as too complexly interactive (supertankers)'. Perrow adds the complex interrelationships between the various legal bodies noted above to the list of complexly interactive influences.[16]

Perrow argues that the error-inducing nature of the maritime system derives more broadly than previously stated. Even though not a tightly coupled, interactively complex system, the maritime industry is nevertheless error-inducing due to three key factors. These are the social organisation of the crew, especially the supreme role of the ship's master; the economic pressures of the industry; and the influences of the tangled set of regulatory and standard-setting agencies.

THE SINKING OF THE *EL FARO*, 1 OCTOBER 2015

The *El Faro* was a US-registered cargo ship of just over 240 m in length and over 30,000 gross tonnage. She was on a voyage from Jacksonville, Florida, to San Juan, Puerto Rico, with a cargo of shipping containers and cars, with a crew of 33. Tropical storm Joaquin with hurricane-force winds of 75–90 knots was tracking across the planned south-easterly track towards Puerto Rico. The Captain's plan to remain south of the hurricane was based on non-current weather data. As the ship sailed closer to the eye of the storm, water entered the car deck which quickly flooded as the inflow exceeded the capacity of the pumps to remove it. Cars became loose. The ship lost all propulsion as it listed heavily. The captain ordered the crew to abandon the ship at 07.29 a.m. The severe weather, the list and the old-fashioned lifeboat design made it unlikely that the lifeboats could be boarded or launched.[17]

Centuries of naval tradition have contributed towards the development of the stereotypical picture of the 'master and commander' as an autocratic and authoritarian figure. Undoubtedly, many ships' captains have indeed been cast in this mould and Perrow emphasises various examples where wilfully inept captains sent their ships to the bottom. Michael Davidson, the Captain of the *El Faro*, had a reputation as a prickly character who liked to remind others who was the boss.[18] Throughout the voyage, the Captain's awareness of the path of Hurricane Joaquin was based on outdated graphical information sent to his personal email from the ironically named 'Bon Voyage System', or BVS. When other crew members expressed their concerns based on more up-to-date information such as from the weather channel, the Captain remained convinced that they were tracking well clear of the storm when in fact, they were heading directly into the eye of the hurricane.

By the 1980s, the aviation industry had begun to realise the significance of poor teamwork on the flight deck as a major contributory factor in many airline crashes.[19] Airlines around the world subsequently introduced teamwork training or *Crew Resource Management* (CRM) – initially as an 'add on' to technical training and later as an integral part of flight crew training. CRM legitimated questioning of authority by subordinates and suggesting alternatives as part of the process of making optimal use of all the resources available on the flight deck. The maritime industry has been much slower to appreciate the value of improving teamwork and communication but since the mid-1990s parts of the industry have introduced what has become known as *Bridge Resource Management* (BRM) training.[20] The US National Transportation Safety Board (NTSB) have been making safety recommendations since the early 1990s that such training is required for all deck officers. All the officers on the *El Faro,* with the exception of Captain Davidson, had attended a BRM course in 2013. The NTSB noted that 'Two members of the bridge team suggested or hinted that they disagreed with the captain's decisions, but the captain disregarded their concerns … the crew deferred to the captain's authority'. The NTSB listed 'ineffective bridge management' as one of the 'probable causes' of the disaster.

Production pressures are a part of any productive activity. The pressure to complete the test plan as prepared by the nuclear engineers played a part in spurring the control room operators at Chernobyl to press ahead with the planned shutdown despite the numerous delays to the schedule introduced by the need to keep supplying power into the grid. Inherent in any passenger or cargo transportation is the need to maintain schedules and avoid delays. The company that owned the *El Faro* expected vessels to arrive within two hours of schedule or else they were considered 'not on time'. The significance of this can be judged against the banner on the company website reading: 'On time, every time'. The Captain's desire to maintain the schedule might also have been increased by the competition for commands on the fleet of brand new ships that the company was introducing to replace the old vessels like the *El Faro.* Davidson had already been overlooked once for a position on the new vessels. In addition, Davidson had just one month previously taken a five-hour detour to avoid a previous storm that turned out to be much less intense than predicted thus compromising the company schedule and increasing running costs. The NTSB's conclusion was that these inherent pressures may well have been a factor in the Captain's decisions to take the most direct route to San Juan despite mounting evidence of the path and intensity of the approaching storm.

As noted above, the complex web of agencies responsible for the regulation of the maritime industry has created many gaps for operators to exploit such as by using Flags of Convenience with lax regulatory standards to register their fleet. In 1998, the International Maritime Organization (IMO) introduced the International Safety Management (ISM) Code, which largely transferred the responsibility for the management of safety to the maritime organisations themselves. This was largely in line with similar trends towards regulated self-regulation in many other industries and activities (see Chapter 1). Whilst some of the major international shipping lines (e.g. Maersk) have taken their responsibilities seriously, far too many operators simply 'regard the ISM Code merely as a paper licence to conduct their business'.[21] As part of the ISM Code, shipping companies were required to develop safety management systems (SMSs), as described in Chapter 2, with the flag states retaining authority to conduct audits to monitor compliance. The owners of the *El Faro* had developed an SMS as required. However, the accident investigators found that the company SMS provided little specific guidance on severe weather or on how to mitigate the risks associated with such events. The NTSB concluded that 'the company's SMS was inadequate and did not provide the officers and crew with the necessary procedures to ensure safe passage, watertight integrity, heavy-weather preparations, and emergency response during heavy-weather conditions'.[22] Despite formal compliance with international regulations as well as with the US Coast Guard – the domestic agency responsible for inspecting and monitoring the US merchant fleet – it was clear that the company viewed the ISM Code more as an exercise in bureaucratic box-ticking than a real exercise in managing safety and risk at sea.

Conclusion: Although systems are designed, built and operated by humans not all the errors that arise can be said to be simply 'human errors'. Interactive complexity and tight coupling are system properties that generate unexpected interactions that could not have been anticipated. Consequently, individuals within the system can find themselves acting reasonably and sensibly within a system that has itself become almost impossible to deal with. Perrow proposes that disasters are therefore inevitable and 'normal' for certain technological systems and activities. However, the catchy metaphor of 'normal accidents' may obscure the deeper analysis of what characteristics contribute towards increasing a system's propensity for catastrophe. Perrow subsequently broadens his argument by characterising certain systems as 'error-inducing'. Chief amongst the contributory factors, according to Perrow, are economic pressures and incentives; regulatory oversight; and the management of the authority. The question of how authority and control should be organised in different kinds of systems is covered in the next section.

THE MANAGEMENT PARADOX

The organisational structure has a direct influence on operator behaviour. The most basic aspects of any organisation are its overall size and shape. As noted above, the span of control or the number of subordinates for whom each person has responsibility is one aspect of shape. An organisation where everybody reports directly to the owner or head has a very flat structure: one where each employee reports directly to an immediate superior, who, in turn, reports to another superior and so forth has a very tall structure. As noted above, this aspect of organisational shape affects the degree of coordination and communication possible. At first glance, the vertical shape of an organisation would seem to dictate the authority structure of the organisation with a broad span of control or flat shape synonymous with decentralised control and highly autonomous employees and a tall structure synonymous with everyone having to obtain permission from someone else above them in the hierarchy. However, equally the broad span of control might equate to centralised authority where everything must be referred to the top boss and the narrow span to a decentralised egalitarian model, with many small, specialised workgroups. The relationship between the span of control and the exercise of authority may also depend to some extent on the size of the organisation.

It is easiest to think of the issue of centralisation and decentralisation of authority in an organisation as separate from the span of control. Authority can be highly centralised or highly decentralised or distributed. In general terms, centralised authority is held to be appropriate where work tasks are highly routine and strict rule-following is essential for performance. Decentralised authority is thought more appropriate where tasks are varied, are unpredictable and require variability and flexibility in responding. It will quickly be realised that few work tasks fall exclusively into one group or the other. Even routine work such as truck driving may occasionally require a creative response such as getting the vehicle out of a ditch. Although academic work is mostly varied and non-routine, strict rule-following may be required for matters involving budgets and ethics.

Nevertheless, Perrow employs a simple centralised versus decentralised dichotomy to determine the appropriate organisational structure for different technologies, characterised by their interactive complexity and coupling. At its simplest, this yields four qualitatively different kinds of systems as shown in Table 5.1. The prototypical loosely coupled system with simple linear interactions would be the industrial assembly line. With this kind of system, either centralised or decentralised authority works well enough, although evidence from organisational psychology suggests greater worker satisfaction and enhanced feelings of control with a more decentralised system. Both Aston Martins and Nissans are produced by assembly line techniques, but the Aston Martin workers have far greater autonomy of operation and greater job satisfaction.

Other processes and activities that are characterised by straightforward linear interactions include railway networks and marine transportation. Simple technologies such as dams are also straightforward and predictable. Unlike the industrial assembly line, however, all these systems are characterised by tight coupling. Once a problem begins to manifest, it is best dealt with by rapid response following preordained procedures to prevent the problem from escalating out of control. Because the system interactions are straightforward and predictable, they can be predicted in advance – protocols prepared, and regular training and drills undertaken to ensure optimal proficiency. In practice, as we have seen, operators can misinterpret the situation leading to erroneous or inapplicable corrective actions. A poorly motivated and ill-equipped workforce coupled with complacent management and a culture of not passing 'bad news' up the authority chain can all combine to create major problems even for systems in this group, however.

TABLE 5.1
Perrow's Classification of the Most Appropriate Form of Authority for Activities and Technologies Varying in Their Interactive Complexity (Linear Vs Non-Linear) and Coupling (Tight Vs Loose).

Interactions → / Coupling ↓	Linear	Non-Linear (Complex)
Tight	Dams, rail and marine transport *Centralisation*	Nuclear, chemical processing, air and space transport *Neither!*
Loose	Assembly line manufacturing *Either!*	Mining, universities *Decentralisation*

Any system with non-linear or complex interactions between its components can produce unexpected and unwanted outcomes. Those systems that are loosely coupled, such as universities, have time to strategise and plan for ways of dealing with these situations. Organisations that allow front-line personnel and others lower in the hierarchy sufficient discretion in their jobs will be better able to identify and respond to adverse conditions generating unexpected interactions. Thus, decentralised control is optimal for these activities and technologies. Perrow's view is that the combination of tight coupling along with the potential for non-linear, complex interactions that characterises a number of transportation, manufacturing generating industries, presents a fundamental paradox for organisational control. The decentralised model, which works for highly professionalised activities such as universities, would be best placed to cope with the potential for complexity, whereas the centralised control model is the one that deals best with activities involving tight coupling. In sum, some activities are best controlled through centralised authority; some with decentralised authority; others can be equally run either way. Activities with interactive complexity and tight coupling present a paradox where neither form of control is satisfactory.

The obvious answer to the paradox would be to combine the best elements of both centralised and decentralised control into a 'hybrid' model. In practice, many large and complex organisations do seem to combine aspects of both centralised and decentralised authority – universities, hospitals and the military to give just three examples. Perrow briefly considers this but remains unconvinced, citing potential ongoing tensions between the two modes along with the alleged expense as reasons for concern. Whilst agencies such as the Nuclear Regulatory Commission sought ways of combining the strengths of following procedures with sufficient operator discretion to deal with the truly unexpected, no clear solution emerged. As Perrow says: 'We could recognise the need for both; we could not find a way to have both'.

Chernobyl also serves to illustrate some of the potential dangers of combining operator discretion within a highly centralised control system to operate technology with inherent critical weaknesses.

THE DYNAMICS OF NORMAL ACCIDENTS

Perrow largely refers to interactive complexity and coupling as static features of organisational structure. However, another well-known organisational theorist, **Karl Weick**, has explored the dynamics of coupling and complexity, specifically in the context of one single event, the world's worst aviation disaster at Tenerife in 1977. Weick, who spent most of his career at the Michigan Ross School of Business where at over 80 years old, he remains an Emeritus Professor, published a detailed analysis of the dynamics of the events that unfolded that day.

TWO BOEING 747S COLLIDE ON RUNWAY AT TENERIFE, 27 MARCH 1977

A Pan American 747 from New York and a KLM 747 from Amsterdam had both been *en-route* to Las Palmas when a bomb exploded leading to both aircraft diverting to Tenerife. By approximately 5 p.m. both aircraft were preparing to leave for Las Palmas. Considerable congestion on the terminal apron led to the KLM aircraft being directed to taxi down runway 12 with the Pan Am aircraft directed along a taxiway with instructions to leave at the third exit. At this point, low clouds reduced runway visibility leaving the control tower with no view of either aircraft. The KLM aircraft turned at the end of the runway and lined up to take off on runway 30. The Pan Am crew were having difficulties locating the required third turn and were still taxiing down the runway as the KLM captain advanced the throttles and initiated the take-off roll. The two aircraft collided on the runway killing everyone on the KLM aircraft as well as most of those on the Pan Am aircraft. The total of 583 fatalities made it the worst disaster in aviation history.[23]

In the years following the Tenerife disaster, there has been much discussion about various facets of the event. Most initial interest was focused on the KLM captain's decision to begin the take-off roll before receiving Air Traffic Control (ATC) clearance. Factors such as the seniority difference between one of the airline's most senior captains and a relatively junior first officer and the fact that the captain spent most of his time in a training role running simulator sessions, issuing his own ATC clearances, were adduced as primary explanations for the decision to take off ahead of ATC clearance whilst the runway was still occupied by the taxiing Pan Am aircraft.

The congested single frequency where an overwhelmed ATC was trying to cope with the unusual influx of aircraft and the terminology in use at the time (subsequently changed in the wake of the disaster) were also highlighted as key factors. The KLM crew had just received their route clearance ('*clearance* to 9,000 ft, right turn after *take-off*') which contained the words 'clearance' and 'take-off' even though it was not a clearance to begin take-off itself. The official Spanish accident investigation placed the blame squarely on the KLM captain and his decision to prematurely take off.

From an individual error perspective (see Chapter 3), the circumstances surrounding the KLM captain's actions in advancing the throttles prematurely were indeed conducive to slips of action which typically occur in highly familiar surroundings, involve established routines and often occur in the presence of much distraction or internal preoccupation. The imminent winding down of the available duty time as the crew rapidly approached the point where they would have to cancel the flight due to the inadequate time remaining to reach their destination was considered by a number of commentators to also have been a significant factor.

As an organisational theorist, Weick's analysis locates these various individual factors 'in the context of a system that is becoming more tightly coupled and less linear'.[24] The airport at Tenerife had become crowded due to diversions caused by the bomb explosion at Las Palmas. The re-opening of Las Palmas initiated a flurry of departing aircraft, leading to the decision to taxi aircraft down the runway as the taxiways were overcrowded. The controllers were forced into improvising means of moving very large aircraft around the single-runway airport. As a small airport, Tenerife lacked ground radar to monitor the positions of these aircraft in the rapidly changing weather conditions which periodically reduced visibility to less than 1,000 m. The Pan Am crew were still on the runway because the 'third left' exit they had been instructed to take involved a switchback of 148 degrees left and right turns that were impossible for a 747 to perform. The disruption caused by diverting multiple flights to Tenerife had significant impacts on crew duty times (flight crews are legally bound to only work for a fixed period of time and cannot exceed this) as well as potential problems with accommodation for passengers and ATC congestion due to the volume of flights inbound to Las Palmas over a short period of time.

Any of these factors could be considered contributory to the genesis of the disaster. Weick considers them over time as part of a gradual change in the dynamics of the situation. Normally, airport operations are relatively linear and loosely coupled. At Tenerife, however, the stress introduced by looming duty time limitations, the physical constraints of a single runway with some exits not designed for the 747, lack of ground radar and poor weather and ATC phraseology limitations combined to increase the coupling and interactive complexity of operations. Other existing vulnerabilities, such as lack of crew communication training that was subsequently introduced to overcome the potential effects of disparities in status, were therefore magnified in their effect. Weick concludes:

> The point of these details is that 'normal accidents' may not be confined to obvious sites of technical complexity such as nuclear power plant. Instead they may occur in any system that is capable of changing from loose to tight and linear to complex.

Weick summarises the processes that might transform a system from linear and loosely coupled to complex and tightly coupled under two headings. The first deals with the breakdown of skilled actions under increasing stress. Human skills are acquired through a series of repetitive practices that can lead to smoother, more accurate and less variable execution. Along the way, moment-to-moment control of the actions becomes increasingly built into the action itself leaving the conscious mind free to wander and engage itself elsewhere. The seeds of error have now been sown. Unwanted interruptions and distractions can derail the action completely causing various types of error. Interruptions or distractions are almost invariably present in any progression from the normal routine to an unwanted catastrophe such as could be observed as the Tenerife disaster unfolded.

One particular pathway leads all the way back to the beginning. It has been shown experimentally and observed anecdotally that one predictable form of skill breakdown involves a return to earlier or more dominant forms of responding. For example, whilst anyone can learn a new skill which involves twisting a control anti-clockwise in order to increase the value of a displayed quantity, when distracted or stressed the likelihood of twisting the control in the opposite, more naturally preferred direction, increases dramatically. When the control in question lowers the nuclear rods into the cooling fluid, for example, then this simple breakdown can be a major factor propelling the system towards catastrophe.

The second set of principles involve breakdowns in communication and teamwork between those actively controlling the system. The three flight-deck crew on board the KLM 747 that day were an exceedingly heterogeneous group. The captain was one of the most senior in the company and head of flight training. The first officer was one of the most junior. The flight engineer was highly experienced but focused on monitoring power plant and systems management rather than crew coordination and communication. Weick suggests that these three acted less like a well-oiled team and more like three separate individuals, thus magnifying the negative effects of increasing stress on performance. Good teamwork and communication can act to reduce task complexity and thus buffer performance against the negative impacts of stress. Weick concluded that Perrow's focus on the structure of technological systems in terms of their complexity and coupling prevented him from looking at the psychological processes that can rapidly alter the system's coupling and complexity.

There are other aviation examples where a design fault turns into a disaster when combined with the kinds of pressures and dynamic system changes Weick noted at Tenerife. One example cited by Perrow was the fatal crash of a Turkish Airlines DC-10 shortly after take-off from Paris in 1974. The DC-10 had a design fault with the latching and locking of the rear cargo door that had resulted in several failures of the door. Rather than (expensively) redesign the whole door, McDonnell Douglas opted for a (cheaper) add on metal plate solution. This modification was not installed on a number of aircraft. Much as at Tenerife, the situation at Paris was one of stress, delay and then a sudden rush to get the aircraft departed. The baggage handler was unfamiliar with the door and failed to latch it correctly. This should have triggered a warning on the flight deck but did not. As the aircraft climbed and pressurised, the door blew out causing the cabin floor to collapse taking vital flight control systems with it. As at Tenerife, the coupling and interactive complexity of the system increased, allowing an existing vulnerability to manifest with disastrous effects.

Although, as Perrow notes, aviation can be considered an error-avoiding system due to its relatively long operating history and widespread public interest and exposure to any adverse consequences, it is still fairly complex and relatively tightly coupled. Unexpected interactions do occur from time to time, as in the case of the loss of a Swissair MD-11 in 1998. The aircraft was operating scheduled flight 111 from New York to Geneva when within an hour after departure smoke began entering the cockpit. The crew decided to divert to Boston but accepted ATC directions to the nearer airport at Halifax, Nova Scotia. Contact was lost shortly after Swissair declared a 'Mayday' and the aircraft shortly crashed into the sea with the loss of all 229 on board.

The aircraft carried a newly installed in-flight entertainment system (IFES) and it appears that arcing in the wiring caused the flammable surrounding insulation material in the cabin ceiling to ignite. As the crew descended towards Halifax, the air conditioning system continued to work (literally, fanning the flames), helping to spread and intensify the fire, subsequently leading to multiple aircraft systems failures causing the loss of the primary flight displays.[25] The nature of the interactions that occurred between the IFES and the flight control systems was unexpected. Regulators had not anticipated the rapid spread of the fire and there were no requirements for such fire-induced failures to be considered as part of the required system safety analysis for aircraft certification.

ASSESSING NORMAL ACCIDENT THEORY

More than 40 years have passed since the events at the Three Mile Island nuclear power plant led Charles Perrow to formulate Normal Accident Theory (NAT). As a sociologist with an interest in organisational structure, Perrow saw the origins of the accident within the internal structure of the organisation. Whereas previous accounts had emphasised the errors of the operators and the poor design of the control room systems, Perrow's analysis pointed to intrinsic problems arising from the design of the system itself. This represented a welcome shift in taking the focus away from the 'errors' of front-line personnel and directing it instead at characteristics of the system within which those individuals operated. Normal Accident Theory thus represents a shift away from thinking about safety in purely individual terms to thinking in terms of organisational and system characteristics. The concepts of interactive complexity and coupling have provided a simple vocabulary for looking into system properties that relate to safety.

Normal Accident Theory has been extended beyond the domains considered in Perrow's book to include areas such as healthcare. Hospital care, for example, tends to be highly complex but loosely coupled. Technological innovations designed to reduce the incidence of errors in medical care may, in fact, result in increased complexity and tighter coupling. One study used NAT as a framework for assessing the effects of computerised physician ordering of medications (known as CPOE).[26] Medication errors have been highlighted as one of the leading factors in adverse events in healthcare. A number of issues from illegible handwriting to confusing drug labelling could be reduced or eliminated by CPOE. Effects on both interactive complexity and coupling were observed. More recently, NAT has also been broadly applied to clinical oncology treatment procedures at a North American hospital with the goal of reducing complexity and loosening coupling.[27]

Despite the continuing enthusiasm for NAT and the ongoing utility of its key concepts of interactive complexity and coupling, there have been some important criticisms. One of the most important is that these key concepts are themselves rather loosely defined.[28] **Nancy Leveson** points out that engineers distinguish between several types of complexity (e.g. structural, dynamic, interactive) and coupling (time-dependent, information-coupling, control coupling, etc.).[29] Different measurement issues arise in each case. It is therefore very difficult to subject NAT to the kind of clear-cut testing required to falsify a scientific theory. Perrow although no doubt pleased with the widespread influence that NAT has enjoyed was more concerned with pushing onwards towards a more thorough theory of organisational structure and safety. To this end, his proposed division between *error-inducing* and *error-avoiding* systems may have greater long-term utility. The question of what characterises *error-avoiding* systems will be taken up in detail in Chapter 6.

Karl Weick has offered some additional insights into why a loosely formulated, essentially untestable theory such as NAT has been so influential in the literature on failures and disasters in modern systems.[30] The two-dimensional grid shown earlier in Figure 5.1 provides a useful starting point from which to frame the discussion of any event or technology. Weick notes that this grid played an important role in the deliberations of the panel investigating the space shuttle Columbia disaster in 2003. Secondly, Weick praises the power of Perrow's conceptualisation in linking together a diverse range of sociological concepts used in organisational analyses. Others have looked at power, communication, diversity, authority and social pressures in organisations but the two dimensions of NAT serve to link these together in a way that has proven heuristically useful. Perrow and others have indicated the need to expand the theory to better address power and status issues as organisations struggle to fulfil their multiple, sometimes conflicting, objectives. Finally, Weick considers the concept of error-avoiding/error-inducing systems to provide a stimulus to further enquiry, particularly into the nature of high-reliability organisations (HROs).

So does NAT provide the key to explaining the failures of so many technological systems resulting in the occasional catastrophic disaster? Perrow himself argues that the answer is negative. Precious few disasters can be attributed to the inherent workings of tightly coupled interactively complex systems. Most of the disasters described in *Normal Accidents* are, according to Perrow, due to mismanagement, regulatory failure or simple component failure. Perrow considers neither of the subsequent major nuclear disasters at Chernobyl in 1986 and Fukushima in 2011 to be examples of 'normal accidents'.[31] In fact, there are very few 'normal accidents' at all in *Normal Accidents*!

All the disasters highlighted in this chapter (Chernobyl, *El Faro* and Tenerife) illustrate the important role that *production pressures* play in catalysing the underlying ingredients that bring about catastrophic failure. These pressures can derive from the commercial imperatives of a commercial business (as in maritime industry) or the political imperatives of a Soviet command economy (as with Chernobyl) but they inevitably emphasise current short-term goal achievement over longer-range considerations such as safety. Whilst Perrow focused on the underlying structural elements (coupling and complexity), Weick showed that systems can exhibit dynamic fluctuations whereby both coupling and complexity may change rapidly over time, so even systems which are not in the upper right-hand quadrant of Perrow's organisational grid (see Figure 5.1) may experience sudden unexpected failure.

In the end, NAT provides a convenient organising framework for looking at various aspects of systems behaviour. We will address this topic in more detail in the third section of this book. In the meantime, we will continue the focus on organisational behaviour by looking at the characteristics of those organisations that manage to operate safely in high-risk environments – the 'error-avoiding' organisation.

NOTES

1. A good overview of the main sociological perspectives can be found in: Goodwyn, M., & Gittell, J.H. (2012). *Sociology of Organizations: Structures and Relationships.* Los Angeles, CA: Sage.
2. Peter Blau's work is cited in one of Charles Perrow's books on organisations: Perrow, C. (1986). *Complex Organisations: A Critical Essay, 3rd Ed.* New York: Random House. In this book Perrow provides an excellent overview of the Weberian model of bureaucratic organisation and its influence on subsequent scholarship.
3. Perrow, C. (2004). A personal note on *Normal Accidents. Organization and Environment, 17*(1), 9–14.
4. Perrow, C. (1983). The organizational context of human factors engineering. *Administrative Science Quarterly, 28*, 521–541.
5. Perrow, C. (1984). *Normal Accidents: Living with High-Risk Technologies.* New York: Basic Books. The book was re-published in 1999 by Princeton University Press, with a new afterword and postscript added to the original text.
6. This is the summary of the Chernobyl accident as provided by the International Atomic Energy Association: (https://www.iaea.org/newscenter/focus/chernobyl/faqs)
7. Reason, J. (1987). The Chernobyl errors. *Bulletin of the British Psychological Society, 40*, 201–206. The article attracted a reply that focused more on the design deficiencies of the RBMK reactor: Baker, S., & Marshall, E. (1988). Chernobyl and the role of psychologists – An appeal to Reason. *The Psychologist: Bulletin of the British Psychological Society, 3*, 107–108.
8. Malko, M.L. (2002). The Chernobyl reactor: Design features and reasons for the accident. In T. Imanaka (Ed.), *Recent Research Activities about the Chernobyl NPP Accident in Belarus, Ukraine, and Russia* (pp. 11–27). Research Reactor Institute, Kyoto: Kyoto University.
9. Rich, V. (1988). Legasov's indictment of Chernobyl management. *Nature, 333*, 285. Further covered by Malko, but also in a recent very comprehensive account of the disaster by a Ukrainian historian and Harvard professor of history. See: Plokhy, S. (2018). *Chernobyl: History of a Tragedy.* New York: Basic Books.
10. https://www.world-nuclear.org/information-library/nuclear-fuel-cycle/nuclear-power-reactors/appendices/rbmk-reactors.aspx
11. Malko (2002), The Chernobyl reactor, 23. See Note 8 for full reference.
12. This quote comes from the article by Perrow (2004); See Note 3 above.
13. Information downloaded from: https://www.agcs.allianz.com/news-and-insights/reports/shipping-safety.html

14. Kristiansen, S. (2005). *Maritime Transportation: Safety Management and Risk Analysis*. Oxford: Elsevier Butterworth-Heinemann.
15. World Shipping Council – *Containers Lost at Sea – 2017 Update*. Downloaded from: http://www.worldshipping.org/industry-issues/safety/Containers_Lost_at_Sea_-_2017_Update_FINAL_July_10.pdf
16. These quotations are from Chapter 6 (Marine Accidents) in Perrow, C. (1999). *Normal Accidents*. New York: Basic Books.
17. National Transportation Safety Board. (2017). *Sinking of US Cargo Vessel SS El Faro, Atlantic Ocean, Northeast of Acklins and Crooked Island, Bahamas, October 1, 2015. Marine Accident Report NTSB/MAR-17/01*. Washington, DC.
18. Slade, R. (2018). *Into the Raging Sea*. London: 4th Estate. Slade's book provides a vivid and compelling dramatized account of the ill-fated *El Faro*.
19. Many of the earliest documents are reports of NASA organised workshops and so forth. An early published collection of papers by many of the leaders in the field at the time is: Wiener, E.L., Kanki, B.G., & Helmreich, R.L. (Eds.). (1993). *Cockpit Resource Management*, San Diego, CA: Academic Press.
20. Following the grounding of a Greek tanker off the coast of Rhode Island in 1989, The NTSB recommends that: 'The U.S. Coast Guard: Require Bridge Resource Management training for all deck watch officers of U.S.-flag vessels of more than 1,600 gross tons'.
21. Bhattacharya, S. (2012). The effectiveness of the ISM code: A qualitative enquiry. *Marine Policy, 36*, 528–535.
22. See Note 17 above.
23. The Wikipedia entry for Tenerife provides a detailed coverage of the event (see: https://en.wikipedia.org/wiki/Tenerife_airport_disaster). The FAA also has useful information on their 'Lessons Learned' section: (https://lessonslearned.faa.gov/ll_main.cfm?TabID=1&LLID=52&LLTypeID=2).
24. Weick, K. (1990). The vulnerable system: An analysis of the Tenerife air disaster. *Journal of Management, 16*(3), 571–593.
25. Transportation Safety Board of Canada. (Undated). *Aviation Investigation Report: In Flight Fire Leading to Collision with Water, Swissair Transport Limited McDonnell Douglas MD-11 HB-IWF Peggy's Cove, Nova Scotia 5nm SW 2 September 1998. Report Number A98H0003*. Gatineau, QC: Author.
26. Tamuz, M., & Harrison, M.I. (2006). Improving patient safety in hospitals: Contributions of high-reliability theory and normal accident theory. *Health Services Research, 41*, 1654–1676.
27. Chera, B.S., et al. (2015). Improving safety in clinical oncology: Applying lessons from normal accident theory. *JAMA Oncology, 1*(7), 958–964.
28. Hopkins, A. (1999). The limits of normal accident theory. *Safety Science, 32*, 93–102.
29. Leveson, N., Dulac, N., Marais, K., & Carroll, J. (2009). Moving beyond normal accidents and high reliability organizations: A systems approach to safety in complex systems. *Organization Studies, 30*, 227–249.
30. Weick, K. (2004). Normal accident theory as frame, link, and provocation. *Organization and Environment, 17*, 27–31.
31. Perrow, C. (2011). Fukushima and the inevitability of accidents. *Bulletin of the Atomic Scientists, 67*, 44–52.

6 High-Reliability Organisations

Charles Perrow proposed that two key characteristics of organisations – their interactive complexity and their degree of coupling – were highly predictive of their propensity to suffer catastrophic failures. So likely, in fact, Perrow coined the memorable term 'normal accidents' to capture the inevitability of failure in systems with these characteristics. At first glance, it seems not too difficult to think of systems that do have these characteristics, such as nuclear power generation, which is obviously highly complex and tightly coupled. What about other systems such as airline transport and air traffic management that also seem to be both highly complex and tightly coupled with little room for error?

Figure 6.1 shows a snapshot of airline traffic over the United States just before the Thanksgiving holiday (26th November) in 2019. The density of air traffic and its concentration on a relatively small number of airports is remarkable to view in this way. Add in the fact that each symbol represents an airliner carrying up to several hundred people moving at speeds up to 850 kmph flying along narrow pathways in the sky connecting these few airports and the potential for catastrophe seems all too readily apparent. And yet, none of the aircraft shown in the graphic collided with any other aircraft. Day after day, the same pattern is repeated with the same result achieved. A system that, on the face of it, should meet Perrow's criteria of high interactive complexity and tight coupling, and which ought, therefore, to contain significant catastrophic potential seems to be remarkably safe. Enduringly so.

AVIATION AND THE MID-AIR COLLISION

Commercial aviation has indeed become remarkably safe. In 2018, for example, there were over 38 million flights and a total of 98 recorded accidents of which 11 involved a fatality. Over a five-year period (2014–2018) the average number of fatal accidents was just over five per annum, equating to a rate of just over 2.5 for every million flights.[1] The majority of these accidents (and almost all the fatal accidents) fell into just three categories: runway safety, loss-of-control in-flight and controlled flight into terrain. Runway safety groups together a heterogeneous set of events involving aircraft running off the runway, tail strikes and ground collisions. Loss-of-control in flight refers simply to events where, for some reason, the flight crew lost control of the aircraft during the flight and 'controlled flight into terrain' or 'CFIT' involves aircraft striking the ground, most commonly during the approach to landing. Colliding with other aircraft during flight does not feature anywhere despite the enormous number of scheduled flights and the high concentrations of aircraft around certain airports at certain times.

DOI: 10.1201/9781003038443-8

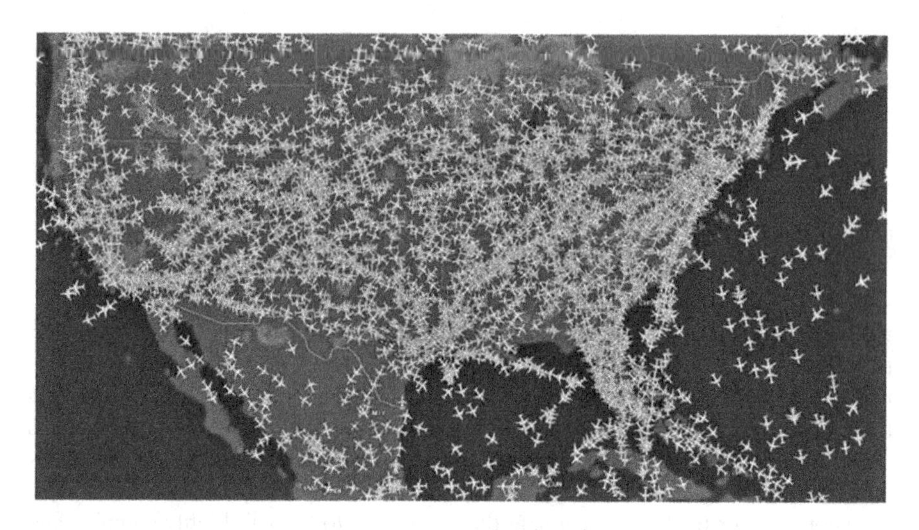

FIGURE 6.1 A snapshot of commercial flight activity over the continental United States. Courtesy of FlightRadar24.com.

After the fatal mid-air collision between two passenger airliners over the Grand Canyon in 1956, efforts were made to find a technological solution to assist pilots in detecting and avoiding other aircraft in the skies around them. The worst ever mid-air collision occurred in 1996 when a Saudi Arabian airline B747 collided with a Kazakhstan Airlines Ilyushin IL-76 just to the west of Delhi, India, killing 349 people on board the two aircraft. The search for technological aids to help avert such events began in the 1970s. First known as the Traffic Alert and Collision Avoidance System (TCAS), these are now referred to as Airborne Collision Avoidance Systems (ACAS) with the International Civil Aviation Organization (ICAO) mandating their installation on all passenger aircraft with more than 19 seats since 2003. An updated ACAS system has been mandated for aircraft flying in European airspace since the start of 2005. Eurocontrol, who manage air traffic services in Europe estimate that ACAS 'reduces the risk of mid-air collision by a factor of about 5'.[2]

ACAS has been a major contributor to the decline in mid-air collisions to the point where it has been suggested that 'A person who flew continuously on a jet transport aircraft in today's environment could expect to survive more than 11,000 years of travel before becoming the victim of a mid-air collision', This equates to one mid-air collision for every 100 million flight hours.[3] The last recorded such collision occurred almost 20 years ago in the skies over southern Germany.

MID-AIR COLLISION ABOVE UBERLINGEN, GERMANY, 1 JULY 2002

On the 1st of July 2002 at 09:35 p.m. a collision between a Tupolev TU154M, which was on a flight from Moscow, Russia, to Barcelona, Spain, and a Boeing B757-200 on a flight from Bergamo, Italy, to Brussels, Belgium, occurred north of the city of Uberlingen (Lake Constance) in German airspace. Both aircraft were operating under Instrument Flight Rules (IFR) and were under the control of the air traffic control centre (ACC) of Zurich. After the collision both aircraft crashed into an area north of Uberlingen. There were a total of 71 people on board the two airplanes, none of whom survived the crash. The air traffic controller in Zurich was working two workstations at the time and did not realise that the two aircraft flying at 'flight level' FL360 were on a collision course until shortly before the accident Both aircraft were equipped with a Traffic Collision Avoidance System (TCAS) which automatically warns pilots if they are on a collision course with another aircraft. The TCAS generated simultaneous warnings in both cockpits a minute before the collision, instructing the Tupolev to climb and the Boeing to descend. Just seconds after the TCAS first alerted the crew to the approaching traffic, the air traffic controller instructed the Tupolev crew to descend to FL350. The crew reacted immediately, disconnecting the autopilot, reducing thrust and pushing the control column forward. At this same moment, the TCAS generated a 'climb climb' message. Both aircraft continued to descend and collided.[4]

The Uberlingen mid-air collision revealed an unexpected interaction between the ACAS technology and the conventional air traffic management system. In this event, one of the aircraft was directed by the controller to descend whilst simultaneously receiving an instruction from the ACAS to climb. At this point, the dynamics of the air traffic management situation changed from being moderately complex and tightly coupled to highly interactively complex and tightly coupled. By and large, air traffic management has reduced complexity and coupling by creating defined aerial corridors along which all traffic is channelled by following highly routinised and standardised procedures and by dividing the overall management task into a number of separate modules – ground control, airfield control, area control, arrival control and so forth, thus limiting the spread of any errors or failures that may occur.

As noted in the previous chapter's discussion of the Tenerife air disaster, even systems that have been carefully set up to reduce overall complexity and coupling may experience temporary dynamic changes that increase vulnerability to catastrophe. This is essentially what happened over Uberlingen as the controller issued an instruction to the Tupolev crew to descend which directly contradicted the instruction provided by the aircraft's automated collision avoidance system to climb. The subsequent uncertainty as to which instruction to follow may have fatally tightened an already tightly coupled situation.

THE BERKELEY PROJECT

A great many organisations are involved in any transportation activity. Equipment manufacturers, airlines and shipping companies, cruise lines, logistics and traffic management authorities, international and national regulatory bodies, insurers and certifiers and so forth. Just as humans come in assorted sizes and characters so do organisations. Some organisations have good reputations, attract and retain workers and maintain good safety records, whilst others are viewed far more negatively, exhibit rapid staff turnover and experience health and safety records far below the norm for their industry. It is probably fair to say that the perfect organisation does not exist and that even in high-risk activities and technologies, exceedingly well-run organisations are few and far between.

Berkeley is part of the vast public University of California system. Located north of Oakland on San Francisco Bay, Berkeley, is one of the world's top-rated universities with numerous Nobel laureates and alumni including Steve Wozniak and Gregory Peck! In the mid-1980s a group of researchers from several different disciplines banded together to conduct field research into several different organisations that shared an apparent ability to operate highly risky technologies and activities with remarkable levels of safety. The leading member of the Berkeley group was political scientist **Todd La Porte**, who had previously written about organisational complexity as a reflection of the number of individual units or sub-groups involved, the degree of differentiation of their tasks and responsibilities and the interdependencies between them.[5] The other key members of the research group were **Karlene Roberts** from the School of Business Administration and **Gene Rochlin** from the Energy and Resources group. Some of the various seemingly risky activities and technologies that exist in the upper right-hand quadrant of Perrow's 'Interactive complexity/coupling' classification chart (refer to Figure 5.1) where nuclear power generation is located, seemed not to be as prone to 'normal accidents' as others. As La Porte puts it: 'Our tacit assumption was that such systems' performance is quite unusual, very difficult to sustain and theoretically inexplicable'.

Systems which *ought* to be unsafe and yet remain stubbornly safe represent an interesting paradox and one which invites further investigation. Do systems which seem structured to fail but which remain alarmingly safe contain some mysterious 'X factor'? And if so, what exactly is it? In an effort to find the answers, the Berkeley team carried out detailed investigations of air traffic control operations at the local San Francisco control centre; air operations on naval aircraft carriers at sea; and the local Pacific Gas and Electrical Company's electrical supply grid. The method used by the investigators was essentially *ethnographic* whereby the researchers participate in the activity as observers, document the actions and social interactions taking place and attempt to put these in some sort of context. The investigations undertaken by cognitive anthropologist Ed Hutchins, on oceanic navigational practices described earlier in Chapter 3, were also ethnographic studies. The result can be very detailed and rich descriptions of the activities linked to interpretations reached by the researchers with reference to the literature on organisational behaviour theory.

The studies carried out on board two US nuclear-powered aircraft carriers provide a rare insight into air operations ('Flight Ops') on board an aircraft carrier.[6] The US Navy has ten ships of the *Nimitz* class, each one a floating city of approximately 3,000 sailors under the command of the ship's captain, as well as a Carrier Air Wing consisting of a similar number of aircrew and support personnel and another small group known as 'Flag' under the command of an admiral. One of the researchers described the carrier as 'A city of 6,000 men with an airport on its roof'.[7] Each group has its own command structure, specialised roles and tasks, and lines of communication but all three groups must cooperate and coordinate to manage extremely complex aircraft and electronic warfare systems under extreme conditions of weather, time pressure and operational stress. The technologies involved are undoubtedly complex in themselves, and there is considerable organisational complexity in the interrelationships amongst and between members of the three groups just mentioned. A seemingly simple instruction from an admiral, such as to have F-14 aircraft standing by ready for a 5-minute launch the next day, involves an intricate set of logistical and administrative operations.

The carriers have a variety of aircraft on board which must be moved up from the hangar below to the flight deck for launch. Space is at an absolute premium in both locations. The launch itself is an extremely hazardous process involving attaching the aircraft to a catapult hook and hurling it forwards so that with full engine power it can achieve sufficient velocity to take off. The aircraft, with engines running, and the numerous deck personnel must coordinate their movements with utmost precision to avoid injury and damage (referred to in the US Navy as the 'crunch rate'). In battle conditions upwards of 50 aircraft may be unpacked from the hangar deck and launched in a short period of time. Despite the dangers, high tempo and limited space on a carrier, the 'crunch rate' can be as low as one for every 10,000 movements.

To achieve prescribed operational goals whilst maintaining such high levels of safety requires individual personnel to make rapid, accurate decisions in concert. A rigidly hierarchical command structure, such as that exists in the military, would simply not have the flexibility to do this. And yet, the formal command structure on a naval aircraft carrier is a typically rigid authoritarian system. Based on observations of operations on the two ships, the Berkeley researchers noted that there existed another kind of parallel structure which they labelled 'organisational self-design'. This consists of more informal networks where the specialised knowledge of those lower down the hierarchy is acknowledged and used to facilitate responding to shifting operational requirements during periods of high-tempo activity. This temporarily bypasses the normal assumptions in a hierarchical system of authority, that those in superior positions have superior knowledge to those in inferior positions. High-reliability organisations (HROs) recognise that often those lower down the hierarchy who are closest to the operational front line have the most accurate and relevant knowledge.

Thus, the inherent conflict between centralised and decentralised control, raised in the previous chapter, is resolved in the form of a highly decentralised latent structure of informal task-related networks that comes into being as and when required. This decentralised structure is widely recognised and passed down from generation to generation so it does not rely on being spontaneously re-invented whenever the need arises. These networks form a fully recognised essential part of operational proficiency and are rehearsed during training in preparation for full-scale operations. With this observation, the Berkeley researchers have noted one significant feature of *high-reliability organisations* (HROs) that enables them to surmount the paradox of control authority noted by Normal Accident Theory. Rather than being mutually exclusive, organisational self-design can produce a decentralised authority structure that comfortably co-exists within a centralised authority structure. HROs have a hierarchical structure like other organisations, but the hierarchy can be made flexible to meet urgent operational requirements.

A helpful but old-fashioned metaphor is provided by the transparency overlay. A transparent acetate sheet with the informal networks drawn upon it can be temporarily overlaid on another sheet showing the formal hierarchical structure, and then withdrawn, leaving the original structure unchanged. The US Navy, and potentially other agencies, have effectively recognised the need for such organisational flexibility by creating an additional small unit ('Flag') alongside the much larger Ship and Air wings to essentially manage the interdependencies that arise when trying to coordinate the complex and tightly coupled interactions that arise in air carrier operations.

In parallel with the naval aircraft carrier studies, the researchers spent time observing activities and social interactions in approach control and en route air traffic control centres in the San Francisco Bay area. Since the point of these ethnographic studies is to provide rich and detailed accounts of human behaviour in its natural context, such accounts naturally resist concise summarisation. One significant feature of air traffic control is the unevenness of workload and pressure. Operational intensity varies widely between off-peak hours and the demands of handling peak-time traffic. At these peak times individual controllers may be responsible for handling up to 25 aircraft in their sector. Bad weather, or simply a change in wind direction as frequently occurs at San Francisco International Airport (SFO) where the parallel runways run pretty much north–south, can entail significant additional pressure as all airborne traffic must be redirected from a pattern leading to one runway direction and reoriented 180 degrees to line the aircraft up with the reciprocal runway. As well as SFO, there are two other commercial airports, five general aviation airports and two large military airports within the same area. The tight coupling and capacity for unexpected surprises rank highly in air traffic control in this area.

The researchers observed similar patterns of adaptive flexibility amongst the civilian air traffic controllers to those shown on naval air carrier operations, as a means of coping with these surges in workload and the additional pressures created by rapidly changing weather. As on the aircraft carriers, the strictly bureaucratic hierarchy of controllers is supplemented by more informal patterns as when, in high-workload conditions, senior supervisors gathered alongside the desks of less senior controllers to provide additional support. Should a complete re-working of air traffic be required to facilitate a change in runway direction, then informal teams rally 'round the controllers in the 'hot seats'. It will be the experienced controller virtuosos who dominate the decision train'.[8] Once the high-tempo periods return to normal, these informal groupings disperse and the traditional bureaucratic hierarchy of roles is resumed. Both the air traffic control centres and the intricate combined operations of air and sea forces on the naval air carriers exhibit what the researchers refer to as 'multi-layered, nested authority systems'. In both cases, operational experience is what counts most in these fast-paced, hazardous activities and during certain periods of activity, expertise-based networks take the place of the more formally instituted hierarchy of roles and ranks.

Whilst observing the air traffic controllers, the researchers noted another important feature of HROs, which is their constant concern with potential failure. These organisations retain a constant awareness of the possibility of disaster and so devote considerable time to preparing for these eventualities by drawing up detailed contingency plans and, most importantly, training and practising their responses. Air traffic controllers and airline pilots have detailed manuals with pre-prepared responses to any situation that has been envisaged by the designers or trainers as a possible eventuality. Pilots and controllers regularly practice these responses in the safety of simulators. However, as we noted in Chapter 2, humans frequently underestimate what they don't know, so that plausible events can easily be omitted from otherwise apparently exhaustive lists. In addition, there is always the possibility of a genuinely new and previously unseen event arising for which no pre-planned response protocol exists.

Karlene Roberts, one of the original key team leaders, has recently summarised the key findings of the Berkeley project.[9] The question that motivated the research was: 'What accounts for the exceptional ability of some organizations to continuously maintain high levels of operational reliability under demanding conditions?' As noted above, the most important finding was that these organisations manage to combine their normal hierarchical decision-making structures with devolved decision making under off-normal or high-tempo conditions. In these circumstances, those closest to the decision who have the greatest expertise are the ones who make the decisions. One result of this is that personnel have high expectations of themselves and take considerable pride in a job well done. Personnel in such organisations are more than usually engaged in their work and enjoy mutual peer group respect.

Closely allied with this is that these organisations engage in continuous training to maintain operational readiness. These organisations are acutely aware of the possibility of failure and maintain a constant state of preparedness. To optimise decision making and prepare for failure, these organisations maintain an appetite to learn from errors and mishaps. Reporting of errors is encouraged, both collectively and individually. The organisations all operate within an environment that is easily recognised as high-risk to both the participants and others so that both the benefits of the activity and the costs of failure are highly visible and salient. Consequently, HROs are as concerned with safety (as this is intrinsic to their operational goals) as they are with the productive goals of the organisation. Whilst most organisations will espouse a concern with safety, in reality, most CEOs are more driven by profitability and returns to shareholders, as well as enjoying considerable personal distance from the effects of failure. Whereas shipping companies can lose a few ships with minimal publicity and carry on operating 'as normal', the high-risk HRO organisations are more publicly exposed to the consequences of any mishaps or failures.

The Berkeley researchers have expressed concern at the tendency to adopt a list of these characteristics as exclusive hallmarks of HROs and turn the achievement of HRO status as 'a marketing label'. Although the acronym 'HRO' became synonymous with the Berkeley group work on effective organisational characteristics, a better term might be 'reliability seeking organisations' which has similar connotations to Perrow's 'error avoiding organizations'.

DEFINING A HIGH-RELIABILITY ORGANISATION

As the influence of the Berkeley group's research began to permeate through the literature on organisational management and safety, questions about the nature of what qualifies as an HRO became more salient. The three organisations originally selected for the intensive study were not the result of a careful and rigorously defined search for organisations meeting some pre-defined standard of safety but were quite simply three highly conveniently located (all on the west coast of the United States within travelling distance of the Berkeley campus) complex organisations, operating sophisticated technology posing an obvious degree of risk to the operators and to others, that appeared to do so relatively safely.

Safety is a difficult concept to define and measure. Most often recognised by its absence, safety is generally thought of in terms of not producing injuries and damage to the workforce or to others who might be involved (e.g. passengers, customers, bystanders or local inhabitants). Comparisons between different organisations and activities require the use of some common unit of measurement (e.g. per capita, hours of exposure, etc.) but this can be problematic as different activities involve different outputs measurable in different ways. The Berkeley group's criteria for selection as an HRO were exceedingly vague and imprecise. High reliability was demonstrated for Air Traffic Control by lack of mid-air collisions; for power generation by the low rate of system 'outages'; and for the US Naval operations by the 'crunch rate'. Critics have pointed out that using such criteria mean that 'it is difficult to think of any low reliability organizations'.[10]

There is also some debate as to whether or not the organisations studied as HROs actually involve high degrees of interactive complexity and tight coupling. Since Perrow himself concluded that actually very few organisations do possess these characteristics, then strictly speaking there can be very few HROs as well. Some critics have claimed that the three original subjects of research (electrical power, air traffic control and naval flight operations) are all linear, well-understood operations with varying degrees of coupling. Whilst this might be largely true, the *Uberlingen* mid-air collision, described above, illustrates the potential for interactive complexity and unexpectedly tight coupling in a system that might otherwise be considered relatively straightforward.

It seems clear that it is essentially impossible to set up appropriate criteria that can be used across a wide range of industries and technologies that would allow us to assess the comparative 'safeness' of one with another. The problems with judging safety performance on the basis of 'output' measures such as days lost to injury, number of collisions and so forth are well known. As well as being totally reactive in nature, they can encourage a focus on small, individual details at the expense of the wider picture of system functioning. The alternative approach is to study a wide range of organisations in an effort to determine what characteristics define the capacity to effectively manage high-risk technologies. In one sense, this simply replaces the problem of defining 'high reliability' with a similar definitional problem in relation to 'effective management'. In some ways, this is an easier problem to deal with along the lines of 'I may not be able to define what a work of art is, but I know one when I see it'. In the same way, some researchers have felt able to recognise a highly effective organisation when they saw one. Debates around the definition of HROs continue but can be put to one side as we continue the search for organisational and psychological characteristics that appear to facilitate effective performance by organisations operating risky or complex technologies.

PERFORMANCE UNDER TRYING CONDITIONS: THE IMPORTANCE OF 'COLLECTIVE MINDFULNESS'

Organisational researchers have continued the investigation of the characteristics of high-functioning organisations in comparison to the relatively dysfunctional in order to determine the ingredients of good organisational performance. In many respects, the same characteristics that are found in high-functioning individuals are those found in safe and effective organisations. These include resilience, flexibility and openness to learning from experience.

Karl Weick, a former Rensis Likert Collegiate Professor of Organisational Behaviour and Psychology at the University of Michigan Business School, has been another key figure in elaborating on some of the key characteristics of highly effective and safe organisations. Although not a member of the Berkeley group, Weick has been a significant figure in HRO theory. Weick's approach has been to re-interpret the existing studies in the light of organisational and psychological theory rather than collecting novel data himself. For the Berkeley researchers, highly reliable organisations were identified through their performance in recording very few failures in their operations despite there being innumerable opportunities to do so. In contrast, Weick's interpretation of the literature is based on identifying 'a distinctive set of processes that foster effectiveness under trying conditions'.[11]

Weick, along with another professor of organisational behaviour at the University of Michigan Business School, **Kathleen Sutcliffe**, and several others, has outlined five important processes that he believes enable the organisations labelled as HROs to function effectively. The first of these is 'preoccupation with failure'. Everyone in the air traffic control centres and on the naval aircraft carriers is permanently aware of how close to the edge (quite literally in the case of the naval flight deck operations) they are. The consequences of failure are very salient. A variety of processes are utilised to foster successful performance and minimise the likelihood of unwanted events. One of these is openness about errors and a willingness to learn from them. Weick cites several examples where openly confessing to an error was rewarded rather than punished. Provided an error was not a malevolent act, then organisations stand to gain more from fostering a culture of open reporting than from one of rooting out and punishing 'bad apples'. Sometimes referred to as a 'just culture', organisations that treat imperfections as a valuable source of learning and improvement ultimately do better than organisations that take a more punitive approach.[12] This is one factor that has differentiated the traditional HROs from those involved in healthcare, for example, which in the past has tended to operate under a culture of punishment involving punitive sanctions and resultant individual secretiveness about failures. It is perhaps not surprising to find that healthcare has been one of the areas that has recently become most enthusiastic about adopting HRO practices. This is discussed further in the following section.

The second process they identified has been labelled 'reluctance to simplify'. Powerful politicians have been known to promote grossly over-simplified pictures of geopolitical reality through 'good guy – bad guy'/'axis of evil' rhetoric. Whatever the merits or otherwise of different political systems, the reality of world affairs is much more complex than can be captured by a simple binary distinction. Similarly, HROs recognise the complexity of their operating environment and strive to match this external complexity with similarly complex and nuanced internal understandings. This requires a degree of openness and heterogeneity to provide different viewpoints, as well as a measure of healthy scepticism regarding sources of information. One key is to effectively manage these differing viewpoints as interpersonal friction could easily develop. The introduction of crew resource management (CRM) training into aviation (see Chapter 4) provides a successful example of encouraging the expression of sceptical and critical viewpoints whilst at the same time managing the potential threats to hierarchy and individual ego that could easily result.

'Sensitivity to operations' is the third proposed characteristic of these effective organisations. This is somewhat similar to the concept of 'situational awareness' which has gained some currency in human factors/ergonomics as a label for having a comprehensive and accurate understanding of what is happening around you in the current moment, as well as having a good grasp on what is likely to occur in the near future. Having good situational awareness depends on being vigilant and attentive and knowing what to look for. Having tired, stressed, distracted or overloaded front-line personnel will inevitably lead to less operational sensitivity and poorer situational awareness. The general theme here is one of active sense-making where organisations display an appetite for enquiring further into their operations, both past and present, to develop a deeper understanding.

The fourth characteristic is 'commitment to resilience'. Most organisations devote at least some of their resources towards trying to anticipate foreseeable risks and where possible developing safeguards and defences. HROs realise that in their complex operating environments not all risks are foreseeable, so devote a sizeable portion of resources towards developing the ability to cope with unexpected risks. This has been referred to as 'resilience' which has become an increasingly widely used term, referring to individuals, communities and organisations. The common thread is a capacity to deal effectively with unanticipated shocks through flexibility and adaptability in behaviour. Preparedness, training and the embracing of flexibility in problem-solving are all characteristics of resilience.

The final characteristic was initially referred to as 'underspecification of structures' but later changed to 'deference to expertise'. This refers to the phenomenon documented by the Berkeley researchers on the aircraft carrier flight decks where at certain times the rigid military hierarchical structure is overlaid with informal authority networks. Those most 'in the know', even if lower down the organisational hierarchy, are temporarily given the right to make important decisions. Investigations of disasters often reveal instances where the safety concerns raised by front-line personnel were downplayed or ignored by middle and upper management or dismissed as impractical or too expensive. The warnings provided to management by crews on the roll-on roll-off ferries that preceded the *Herald of Free Enterprise* tragedy (see Chapter 4) provide one such example.

Eurocontrol, the agency responsible for air traffic management in European skies has recently proposed a set of ten principles for advancing safety. The first principle 'acknowledges that those who do the work are specialists in their work and a vital partner in improving the system'.[13] This embodies the 'deference to expertise' characteristic of HROs and acknowledges that workers frequently need to show flexibility and adaptability however well prescribed the work may be.

Weick proposes that these various characteristics link together 'by their joint capability to induce a rich awareness of discriminatory detail and a capacity for action. We label this capability mindfulness'.[14] This capacity for *'collective mindfulness'* enables HROs to survive 'the dreaded combination of interactive complexity and tight coupling associated with normal accidents'. The concept of mindfulness has become increasingly popular in the context of personal health and wellbeing. From its use in individual meditative practices to refer to a full awareness of the present moment, Weick extends the meaning to cover a range of activities involving an inquiring and critical approach to anything and everything. In essence, it involves constantly being 'on the ball' and being alert to potential differences and changes in the task environment that might indicate a problem. An outline of Weick's theory of collective mindfulness is shown in Figure 6.2.

The ultimate impact of HRO theory might have been more muted were it not for the publication by Weick and Sutcliffe, in 2001 of a less academic, more application-oriented book *Managing the Unexpected.* The book explained the five key HRO principles and the concept of mindfulness in more of a 'how to' manner aimed at appealing to managers and executives looking for ways of transforming their

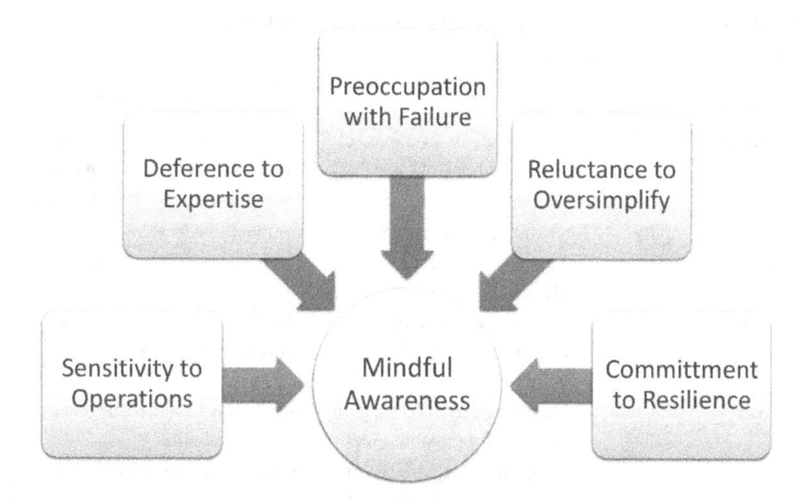

FIGURE 6.2 Weick's five organisational characteristics leading to high reliability. See the text for an explanation of each characteristic.

organisation into an HRO. The book is now in its third edition and has been influential in spreading the idea that 'high reliability' is a state that can be achieved through the application of processes that any organisation could follow.[15]

Summary. The carefully conducted programme of ethnographic research developed by the Berkeley researchers provided a unique set of insights into hitherto little-known worlds – particularly that of naval flight deck operations. Initially chosen because of apparently high levels of operational safety in quite risky environments, the label 'high reliability organisation' quickly became the established term of reference, although 'reliability-seeking organisation' was later thought to be a more appropriate term. Elaborating on the early work and integrating much of it with traditional organisational theory, Karl Weick has been highly influential in developing what is widely referred to as *high-reliability theory.* At least five characteristics have been outlined that might account, in part at least, for the sustained effectiveness of some organisations (e.g. air traffic control) compared with the relatively poorer records of other industries and activities (e.g. maritime industry).

A brief comparison between the generally agreed characteristics of HROs with an organisation that possibly was the very antithesis of an HRO provides a convenient comparison. Table 6.1 summarises the typical HRO characteristics described previously with those of Pike River Coal Ltd who were the owners of the Pike River Mine in New Zealand which suffered a catastrophic explosion on 19 November 2010 as described in Chapter 1.[16] This example serves to illustrate the point that the processes ascribed to HROs are by no means inevitable and that other organisations may well differ markedly on one or more of these characteristics. Other mining operations have also shown similar characteristics to those at Pike River.

TABLE 6.1
Comparing the Five Principal HRO Characteristics with Those of the Pike River Coal Company.

Characteristic	HROs	Pike River Coal
Preoccupied with failure	Constant state of awareness of dangers and risk of complacency	Management preoccupied with investors and production targets. Little attention to mining risks
Reluctance to simplify	Recognition of the complexity of many problems. 'Just' culture	Simple (unworkable) solution to mine escape problem. Culture of blame
Sensitivity to operations	Reporting welcomed to build an accurate picture of operations	Hundreds of incident reports awaiting investigation
Resilience	Stable organisation focused on preparing for possible shocks	Six different mine managers in the past 26 months
Deference to expertise	Incorporate the experience of workers. Overcome hierarchy	Miners' assessments of faulty, badly designed equipment ignored. High turnover of experienced staff

CAN WE RELY ON HEALTHCARE?

A large cohort (over 30,000) of patients who were admitted to New York hospitals in 1984 became the unwitting vanguard of a modern revolution in healthcare. Over the next few years, their records were screened by specially trained nurses looking for any possible evidence that the patient might have been harmed or injured by the treatment they were given. Eventually, 1,133 of the patients were deemed to have been harmed by what the researchers labelled an *adverse event*. When **Lucian Leape**, a Professor of Medicine at Harvard Medical School and his colleagues published their findings in 1991, there was a considerable public outcry at the hitherto unrevealed extent of unintended medical harm.[17]

Extrapolating from the 1984 sample, the researchers estimated that around 1.3 million patients admitted to US hospitals each year were being unnecessarily harmed by the very treatments intended to help them. The number actually dying from these adverse events was estimated to be approximately 98,000 annually – approximately twice as large as the highway road toll. As the researchers pointed out, given the staggeringly large numbers of interventions and drug administrations carried out in hospitals every year anything short of perfection would inevitably lead to sizeable numbers of examples of adverse events, such as administering the wrong dose of a drug. Nevertheless, the picture revealed by the Harvard researchers was highly unsettling. In contrast to aviation safety which periodically bursts fully into the public limelight, particularly following a well-publicised air crash, medical safety had been completely hidden away and not publicly discussed.

BRISTOL ROYAL INFIRMARY, PAEDIATRIC CARDIOLOGY DEATHS

'Between 30 and 35 children undergoing heart surgery at Bristol Royal Infirmary died between 1931 and 1995 who would probably have survived if treated elsewhere … . These "excess" deaths took place in a unit where mortality at the time for children aged under 1 was probably double that for England as a whole, and even higher for neonates. Around a third of children who underwent open heart surgery received less than adequate care.

The inquiry report … painted a picture of a flawed system of care with poor teamwork between professionals, "too much power in too few hands", and surgeons who lacked the insight to see that they were failing and to stop operating. The physical setup was "dangerous", with surgeons on one site – at the Royal Infirmary – and paediatric cardiologists several hundred metres away at the children's hospital. The operating theatre and intensive care unit were on different floors, and children had to be transported by a lift that could be called at any time by others.

The senior management was close to the "old guard" of clinicians and supported them. There was a "club culture" with insiders and outsiders. The style of management had a punitive element and the environment did not make speaking out or openness safe or acceptable'.[18]

In the late 1990s, the US Institute of Medicine (IOM) organised a committee to look into the quality of healthcare in the United States. The IOM is a branch of the federal National Academy of Sciences with responsibility for medicine and healthcare. The IOM's report was published in late 1999 and confirmed the dramatic findings of the Harvard researchers.[19] The report, provocatively titled 'To Err Is Human', in fact, pointed beyond recklessness and negligence and in the direction of broader organisational and systemic issues. As **Mathilde Bourrier**, a French HRO researcher noted, the IOM report argued that a proportion of the deaths that occur in hospitals are totally preventable and that 'Bad communication, bad preparation, lack of feedback, great discrepancies among services, wards and experts are error-inducing'.[20] All these factors, and more, were clearly evident in the multiple cases of preventable death at the Bristol Royal Infirmary in the 1990s. The IOM report recommended four key goals to improve safety in healthcare which as well as certain regulatory changes included the introduction of robust voluntary reporting schemes, prioritising safety as a goal and establishing a national centre for patient safety within the government's health department.

Within a few months, the influential *British Medical Journal* published a special issue devoted to patient safety with invited articles from many American researchers and one (James Reason) from the UK. The title of the editorial ('Facing up to Medical Error') suggested a degree of relief that something long known to medical insiders was now out in public and that efforts could now be made openly to address the various problems. The editors noted that only a decade previously they had been criticised by a prominent physician for trying to draw attention to medical error.[21] It would seem likely that an industry searching for guidance on how to perform more safely in a high-risk environment would sooner or later connect with the growing literature on HROs.

The first concrete sign of healthcare and HROs coming together was the first of a series of conferences 'High Reliability Organization Theory and Practice in Medicine and Public Safety' held in 2003.[22] HRO concepts increasingly gained traction in healthcare organisations, most especially in the United States. To some degree, this represented the kind of aspirational/motivational agenda of management and business consultants looking to promote the latest trend. However, a good deal of published research has appeared since the early 2000s on HROs and healthcare. A 2015 review found 333 relevant published articles, indicating a strong and continuing interest in applying and evaluating HRO theory to aspects of healthcare.[23] The authors note a strong reliance on the early HRO literature from the Berkeley group and the absence of significant subsequent theoretical development. The authors conclude that 'the transition from theory to practice seems to be a struggle'. Seemingly, 'HRO' has become a bit of a catch-all around which various discussions of organisational culture and management now revolve, with less focus on specific applications of lessons learned from the original studies of reliability-seeking organisations.

It's difficult to say whether any measurable progress has been made in the last two decades in reducing harm from the unintended consequences of healthcare. Almost all the research on this topic is US-based and other countries have vastly differing healthcare systems. The original estimates by the Harvard researchers, later incorporated into the Institute of Medicine report, were just under 100,000 preventable deaths from adverse events in American hospitals each year. A more recent study which collated data from four separate studies done in US hospitals produced an estimate of 210,000 annual preventable deaths in the United States from adverse events.[24] All such studies are based on reviews of patient medical records and it is known that these are not complete and accurate in every respect, especially where errors of commission are involved. In all likelihood, the actual numbers are most probably higher than these estimates. Healthcare presents unique challenges as treatments become more complex, populations become older and the extent to which unknown genetic and other individual differences moderate the effects of treatment becomes more apparent. These are all factors that are not present in the studied population of HROs and may limit the possibilities of rapid progress.

THE CONTRIBUTION OF HRO THEORY

Charles Perrow's theory that systems that had certain structural characteristics were inherently prone to disastrous failure, captured a great deal of attention in the late 1980s. The theory was generally taken to mean that certain activities and technologies could never be safely operated. However, several researchers at Berkeley in California, home to thriving power companies, busy airports and massive naval bases, were struck by how safely organisations in these sectors were seemingly able to operate despite having some of the characteristics identified by Perrow. For all of these organisations, the risks they ran were public and highly visible, with consequences for multiple groups in the event of failure. The Berkeley research focused on identifying the characteristics of such organisations that might contribute towards their apparent success, through a broad range of ethnographic and sociological research. Karl Weick added significantly to the HRO literature as well as making HRO theory more widely accessible.

From the start, there has been much criticism of the way in which HROs were identified. Clearly, the original selection was subjective rather than based on any statistical basis for comparing one operation with another. Following Perrow's criteria, the researchers wished to look at complex, tightly coupled systems operating highly hazardous activities or technologies. None of the systems selected were truly interactively complex and tightly coupled, but as Perrow himself noted, very few systems fully qualify. To be considered an HRO, an organisation had to have not failed despite tens of thousands of opportunities to do so. As some critics have pointed out, by that criterion 'it is difficult to think of any low reliability organisations. Any organization that did not have at least this type of safety record would be shut down immediately'.[25]

Weick's work led to HROs becoming defined more by their way of going about their business than by any statistical criteria. This has inevitably led to the appearance of checklists that can purportedly be used to assess whether an organisation is conforming to the HRO operating characteristics.[26] The investigation board into the Columbia Space Shuttle accident in 2003 used the checklist approach to weigh up NASA's organisational functioning and found it to fall well short of the ideal HRO model.[27] This example well illustrates the development of HRO theory from its original beginnings as an empirical research project into the functioning of supposedly super-safe organisations, to its later manifestation as a desirable 'ideal' against which current organisational functioning can be compared via a checklist approach.

Weick's proposal that HROs exhibit 'collective mindfulness' in their careful attention to the possibilities of failure has received a number of criticisms. Most importantly, the question has been raised of how HROs distinguish between minor and inconsequential signs of failure and more important and meaningful ones? In hindsight, working backwards from the known outcome, it is invariably possible to locate warning signs of the disaster to come. How are organisations to distinguish the importance of these signs at the time? This is an important issue that has not been fully resolved. Certainly, there are cases, as with the *Herald of Free Enterprise* tragedy described in Chapter 2, where ferry crews had repeatedly and clearly expressed their concerns to management but had been deliberately ignored. Voluntary safety reporting schemes have a key part to play in promoting the ability to separate out 'signal' (i.e. the meaningful information) from 'noise' (i.e. the background random information).

HRO theory turns up in a wide variety of areas. Karlene Roberts lists over a dozen, including offshore platforms, wildland firefighting and submarine operations.[28] In the latter case, three academics – one a former officer on US Navy nuclear submarines and another previously involved in nuclear submarine power-plant design, compared the United States and Soviet nuclear submarine programmes in relation to HRO theory.[29] The question of interest was the way in which HROs seem to manage to innovate and learn whilst simultaneously retaining their ability to continue to perform their core functions safely and reliably. The study compared the development of fast-attack nuclear submarines in the United States and Soviet fleets between 1970 and 2000. The key difference was the use by the United States of a single 'platform' for all its designs compared with multiple platforms used by the Soviet Union. A platform is a common design framework that can be used to generate multiple products. The aviation industry provides familiar examples. For instance, the Boeing 777 airliner was initially a new platform, different from other Boeing models such as the 747 or 767. Customer demands for a longer range and greater payload resulted in four additional models from the same platform. Whilst the basic design was retained (identical cockpits, fuselage cross section, etc.) both the length and wingspan were extended to increase payload and operating range. Keeping the same underlying platform throughout these developments meant that operating experience with the early models could be generalised to the later models.

Whilst the US Navy based all its submarine designs on one platform, the Soviet navy developed five quite different designs. Where a design offered improvements in one area there were generally deficiencies in others. Some of the designs proved difficult to operate safely, with at least 15 reactor incidents or hull losses for the Soviet designs compared to zero for the US design. The authors suggest that HROs utilise the 'platform' strategy to allow innovation whilst retaining the core of their success by controlling key underlying processes. One implication would be that whilst whole-scale reorganisation or revolutionary new design can sometimes lead to success and sometimes to failure, for high-risk technologies, an incremental, more modular approach provides essential innovative possibilities alongside ongoing stability.

The use of the term 'reliability' to refer to organisations that operate safely in high-risk environments has been criticised. **Nancy Leveson**, Professor of Aeronautics and Astronautics, and Professor of Engineering Systems at MIT, has argued that operating reliably and operating safely are not the same thing.[30] Reliability is the property of consistently performing in a specified way. This need not necessarily be a safe way. More importantly, unwanted outcomes can occur when perfectly reliable components interact in unspecified and unpredicted ways. Leveson cites the December 1999 example of the Mars Polar lander crashing onto the surface of the planet when a software module interpreted the signals from the landing gear as indicating that landing had occurred. The software shut down the engines whilst the vehicle was still descending about 40m above the surface, causing it to crash. According to Leveson, everything functioned perfectly reliably, but safety was compromised by unexpected interactions between components.

One of the major contributions of HRO theory has been to re-direct attention away from an exclusive focus on structural deficiencies and individual errors, towards what an organisation can do well. This change undoubtedly laid the groundwork for the recent developments *in resilience engineering* and 'Safety II' that focus on the ways in which workers almost invariably manage to work safely even in the face of unsafe conditions. These developments will be fully described in the third section of this book. Even the most well-intentioned organisation faces great difficulties in maintaining its 'mindful awareness' over extended periods of time. Vigilance is an extremely challenging state to maintain for anything but short periods of time, and people and organisations that rarely encounter failure will be subject to insidious pressures towards complacency and satisfaction with a job being done well. In the next chapter, we turn our attention towards these social pressures and their role in major technological failures.

NOTES

1. International Civil Aviation Organization. (2019). *State of Global Aviation Safety, 2018*. Montreal, QC: ICAO.
2. See: https://www.eurocontrol.int/system/acas
3. Kuchar, J.K., & Drumm, A.C. (2007). The traffic alert and collision Avoidance system. *Lincoln Laboratory Journal, 16*, 277–295.
4. Bundestelle fur Flugunfalluntersuchung. (2004). *Investigation Report AX001-1-2/02*. Braunschweig: Bundestelle fur Flugunfalluntersuchung. This is an English translation of the official report of the German Federal Bureau of Aircraft Accidents Investigation.
5. La Porte, T. (Ed.). (1975). *Organized Social Complexity*. Princeton, NJ: Princeton University Press.
6. Rochlin, G.I. (1989). Informal organizational networking as a crisis-avoidance strategy: US naval flight operations as a case study. *Industrial Crisis Quarterly, 3*, 159–176.
7. Roberts, K.H., Stout, S.K., & Halpern, J.J. (1994). Decision dynamics in two high reliability military organizations. *Management Science, 40*(5), 614–624. https://doi.org/10.1287/mnsc.40.5.614

8. From Halpern, J. (1989). Cognitive factors influencing decision making in a highly reliable organization. *Industrial Crisis Quarterly, 3*, 143–158.
9. Roberts, K.H. (2018). Advancing organizational reliability. In R. Ramanujam & K.H. Roberts (Eds.), *Organizing for Reliability: A Guide for Research and Practice* (pp. 3–16). Stanford, CA: Stanford University Press.
10. Marais, K., Dulac, N., & Leveson, N. (2004). Beyond normal accidents and high reliability organizations: The need for an alternative approach to safety in complex systems. Engineering Systems Division, MIT. Downloaded from: http://klabs.org/DEI/lessons_learned/papers/marais-b.pdf
11. Weick, K.E., Sutcliffe, K.M., & Obstfeld, D. (1999). Organizing for high reliability: Processes of collective mindfulness. *Research in Organizational Behavior, 21*, 81–123.
12. Dekker, S. (2007). *Just Culture: Balancing Safety and Accountability*. Aldershot, UK: Ashgate.
13. European Organisation for the safety of Air Navigation (EUROCONTROL). (2014). Systems thinking for safety: Ten principles. A white paper. *Eurocontrol*. Downloadable from: 2882.pdf (skybrary.aero)
14. See Weick et al. in note 11.
15. Weick, K.E., & Sutcliffe, K.M. (2001). *Managing the Unexpected: Assuring High Performance in an Age of Complexity*. San Francisco, CA: Josey Bass. A second edition was published in 2007 and the latest (third) edition appeared in 2015.
16. MacFie, R. (2013). *Tragedy at Pike River*. Wellington, NZ: Awa Press. Other examples include the explosion at the Moura No 2 mine in 1994 which is described by Weick and Sutcliffe (2001). See Note 15.
17. The original publication to reveal the extent of harm from adverse events was by Brennan, T.A., et al. (1991). Incidence of adverse events and negligence in hospitalized patients: Results from the Harvard Medical Practice Study I. *New England Journal of Medicine, 304*, 634–637.
18. Dyer, C. (2000, 28 July). Bristol inquiry condemns hospital's 'club culture'. *British Medical Journal, 323*(7306), 181. The whole report can be downloaded from: http://www.uhbristol.nhs.uk/media/2930210/the_report_of_the_independent_review_of_childrens_cardiac_services_in_bristol.pdf
19. Institute of Medicine. (1999). *To Err is Human: Building a Safer Health System*. Washington, DC: National Academy Press. Downloaded from: https://www.nap.edu/catalog/9728/to-err-is-human-building-a-safer-health-system. This has been downloaded more than 84,000 times.
20. Bourrier, M. (2011). The legacy of the high reliability organization project. *Journal of Contingencies and Crisis Management, 19*, 9–13.
21. British Medical Journal. (2000, 18 August). Editor's choice: Facing up to medical error. *British Medical Journal, 320*.
22. Martelli, P.E. (2018). Organizing for reliability in health care. In R. Ramanujam & K. H Roberts (Eds.), *Organizing for Reliability: A Guide for Research and Practice* (pp. 217–243). Stanford, CA: Stanford University Press.
23. Tolk, J.N., Cantu, J., & Beruvides, M. (2015). High reliability organization research: A literature review for health care. *Engineering Management Journal, 27*(4), 218–237.
24. James, J.T. (2013). A new evidence-based estimate of patient harms associated with hospital care. *Journal of Patient Safety, 9*, 122–128.
25. Marais, K., Dulac, N., & Leveson, N. (2004). See Note 10 above for full reference.
26. Checklists for 'assessing your firm's reluctance to simplify' or 'deference to expertise' are included in Weick and Sutcliffe's '*Managing the Unexpected*' (see Note 16). Each scale contains 8 to 12 questions. Examples include: '*People feel free to bring up problems and tough issues*' and '*We trust near misses and errors as information about the health of our system and try to learn from them*'.

27. Hopkins, A. (2007). *The Problem of Defining High Reliability Organizations: Working Paper 51.* Canberra, ACT: National Research centre for Occupational Health and Safety Regulation.
28. Roberts, K. (2009). Managing the unexpected: Six years of HRO-literature reviewed. *Journal of Contingencies and Crisis Management, 17,* 50–54.
29. Bierly, P., Gallagher, S., & Spender, J.C. (2008). Innovation and learning in high reliability organizations: A case study of the United States and Russian nuclear attack submarines. *IEEE Transactions in Engineering Management, 55,* 393–408.
30. This is discussed in: Leveson, N. (2011). *Engineering a Safer World: Systems Thinking Applied to Safety.* Cambridge, MA: MIT Press.

7 Normalisation of Deviance

Around 2,500 years ago, the Greek philosopher **Aristotle** observed that 'Man is by nature a social animal'. Many, if not most, animals on the planet, from ants to apes, live socially. Humans live in complex societies, occupying multiple social niches involving family, tribal, occupational, ethnic, religious and other affiliations. All of these groupings and social identities have an influence on our attitudes, feelings and behaviours. In a classic early textbook of social psychology, named *The Social Animal* after Aristotle's observation, **Elliot Aronson**, one of the most well-known psychologists of the 20th century, defined social psychology as the study of social influence.[1]

SOCIAL INFLUENCE

Modern social psychology has explored the myriad ways in which we are influenced by one another. In the digital age, social influence has even become an occupation in which large sums of money can be earned by posting promotional material on social media. Social influencers work by affecting the purchasing patterns of other people who pay attention to them and their opinions about virtually any consumable from travel and hotels, to fashion and music. This starkly emphasises the role that information provided by others can have on people's judgements and preferences. All social norms are based on observation of what other people do in similar circumstances. Norms that govern greetings (for example, handshake, bow, kiss, etc.) are developed within a culture or sub-culture and exert a guiding influence on everyone within those groupings. Norms are social guidelines that provide information as to what is acceptable and permissible or not in a particular setting. As 'social animals' our wellbeing and survival are closely linked to our ability to harmonise and blend in with others.

A particularly vivid demonstration of this was provided by early social psychologist **Solomon Asch** in 1951 during his time at Swarthmore College in Pennsylvania. Asch had participants enrol in a study of 'perceptual judgment' where the participant entered a room with four others and was shown a series of simple drawings of straight lines of varying lengths. The task was to say out loud which of the three comparison lines was the same length as a reference line. In each case, there was an obvious correct answer. However, each of the other participants in turn announced that one of the obviously incorrect lines was the same length as the reference line. The question was what would happen when we come to the genuine participant who is the fifth and last one to give their response? Asch found that three-quarters of these participants denied the evidence of their own senses and gave an incorrect

DOI: 10.1201/9781003038443-9

answer at least once, with around a third of all their judgements being incorrect and made in conformity with the expressed opinion of the group.[2]

Even though participants were not actually convinced that the normative judgement was correct, most people were prepared to go along with it for the sake of social harmony. In most situations where there are no objectively right or wrong answers, the opinions of others provide the only information that we may be able to obtain. Even in the absence of social pressure to conform to a group judgement, the 'information' that one's judgements are incorrect exerts an influence. In the internet and social media era, there is almost always some 'information' source that can influence the opinions of others. Traditional news outlets now compete with outlandish conspiracy websites as sources of social influence. The pioneering work of these early 20th-century psychologists has highlighted the important role that both informational and normative influences have on our judgements and behaviour.

Normative influences can just as easily be seen in workplace settings. On joining a new workplace, we quickly acquire a sense of what is normal in terms of style of dress, patterns of social interaction and quality of work. A dramatic early illustration of the power of norms to shape working behaviour was provided by researchers at a pajama factory in Marion, Virginia, in the United States. The female workers were paid a piece rate depending on their productivity. The rate was based on an expected output of 60 units per hour with very few employees producing more than this. The researchers concluded that groups of workers acted to suppress production above the expected levels through social influence on workers who started to approach or exceed the expected 60 units per hour. In one case, where the group was broken up, the remaining single worker began producing more than 90 units per hour.[3]

The roles of normative and informational social influence on safety can be seen in one of the most studied disasters of the 20th century – the destruction of the US Space Shuttle *Challenger* a little more than a minute after lift-off. The event had a particular impact as it was broadcast live on television to a worldwide audience as it happened. Within days of the disaster, US President Ronald Reagan initiated a Presidential Commission to investigate the event headed by the then US Attorney General William Rogers. The 12 other members of the Commission included astronaut Neil Armstrong, pilot Chuck Yeager and eminent physicist Richard Feynman.

DESTRUCTION OF THE SPACE SHUTTLE *CHALLENGER*, CAPE CANAVERAL, 28 JANUARY 1986

'Flight of the Space Shuttle *Challenger* on Mission 51-L began at 11:38 a.m. Eastern Standard Time on January 28, 1986. It ended 73 seconds later in an explosive burn of hydrogen and oxygen propellants that destroyed the External Tank and exposed the Orbiter to severe aerodynamic loads that caused the complete structural breakup. All seven crew members perished. The two Solid Rocket Boosters flew out of the fireball and were destroyed by the Air Force range safety officer 110 seconds after launch.

The ambient air temperature at launch was 36 degrees Fahrenheit measured at a ground level approximately 1,000 feet from the 51-L mission launch pad 39B. This temperature was 15 degrees colder than that of any previous launch.

The consensus of the Commission and participating investigative agencies is that the loss of the Space Shuttle *Challenger* was caused by a failure in the joint between the two lower segments of the right Solid Rocket Motor. The specific failure was the destruction of the seals that are intended to prevent hot gasses from leaking through the joint during the propellant burn of the rocket motor. The evidence assembled by the Commission indicates that no other element of the Space Shuttle system contributed to this failure'.[4]

THE LAUNCH DECISION

The failure of the rubber seals (known as 'O-Rings') was quickly established as the precipitating event that led to the breakup of the *Challenger*. The attention of the Commission then became focused on the decision making that led to the decision to launch the shuttle on an unusually cold Florida morning. *Challenger* was originally slated to launch on 22 January but delays to the launch of another shuttle (*Columbia*) led to the *Challenger* launch being pushed back to the 23rd, then the 25th and again to the 26th and 27th, where launches were abandoned due to weather conditions on the 26th and technical faults and weather on the 27th, and finally to the 28th. The forecast for the 28th was for very cold temperatures, down to low 20s (°F) or around –5 (°C). At launch, the temperature was only just above freezing (+2.2°C) and lift off was delayed for several hours to allow ice to melt on the launch pad.

The launch didn't just happen but was preceded by teleconference meetings between NASA managers and engineers at the Marshall Space Flight Center in Huntsville, Alabama, and the launch site at Cape Kennedy, and managers and engineers from the company that built the solid rocket boosters (SRBs), Morton Thiokol in Utah. The weather forecasts for cold temperatures had raised some concerns as no previous shuttle had ever been launched at any temperature below 51 (°F) or 10.5 (°C). A meeting was held at 5.45 p.m. on the 27th where engineers from Morton Thiokol expressed reservations about the performance of the 'O' rings at the low forecast temperatures and suggested that the launch be postponed. Another meeting was to be held later that night (at 8.15 p.m.) and Thiokol were asked to fax relevant charts and written data on 'O' ring performance and temperature through to Kennedy Space Center. One of the Thiokol engineers explained the problem with the 'O' rings to the Commission through an analogy: 'it would be likened to trying to shove a brick into a crack versus a sponge'. At the meeting, Thiokol were asked to quantify their concerns, although clearly no data were available as no shuttle had been launched at these lower temperatures. The Thiokol engineers presented a handwritten note recommending a cut-off for launches below 53°F (12°C). After nearly two hours of discussion, the meeting was recessed and a private discussion was undertaken between the Thiokol personnel.

The senior Thiokol manager present said 'we have to make a management decision. He turned to (engineer) Bob Lund and asked him to take off his engineering hat and put on his management hat'. From this point, the atmosphere at Thiokol seemed to change to one where the engineers were asked to prove beyond a shadow of a doubt that it would be unsafe to launch the next day. When the teleconference re-convened at 11.00 p.m., Thiokol managers stated that they had reassessed the problem and, whilst concerned at the lack of temperature data, were now recommending that the launch proceed. The Rogers Commission concluded that 'the Thiokol Management reversed its position and recommended the launch of 51-L, at the urging of Marshall and contrary to the views of its engineers, in order to accommodate a major customer'.

The idea that a relatively small sub-contractor might bow to the pressure from a large and powerful customer, keen to see its operational schedule maintained, seems plausible enough. *Production pressures*, where operational goals dominate safety goals are commonplace in most industries and activities. Production pressures certainly drove the rapid turnaround of cross-channel ferries, such as the *Herald of Free Enterprise*, discussed in Chapter 2, and were evident in the need to keep the Chernobyl reactors contributing to the Ukraine power grid (Chapter 5). Most recently, a report produced on behalf of the US Congress on the Boeing 737 Max crashes concluded that in their anxiety to produce a viable competitor to the European Airbus A320Neo, Boeing subjugated internally expressed concerns about the aircraft's safety and performance in order to achieve a planned production schedule.[5]

The Rogers Commission and others reached much the same conclusion – that the pressure to maintain the launch schedule of 24 flights per annum was a major driver of the eve-of-launch decision to go ahead with the *Challenger* mission despite the unprecedented low temperatures and concerns expressed by the rocket engineers. Balancing the political and organisational imperatives of keeping the missions going on schedule with the uncertain possibilities of a launch failure, the NASA managers gambled on a successful outcome.

Edward Tufte is a distinguished professor (now Emeritus) from Yale University with a particular interest in the representation of information in visual displays. One of his books, *Visual Explanations*, deals with representations of mechanism and cause and effect.[6] The most famous example was **John Snow**'s work in 1854 on a cholera epidemic in London. By plotting each outbreak on a detailed map of the affected area, it became immediately obvious that the cases all centred around a specific water pump in Broad Street. Once the pump was disabled, the epidemic ended. Tufte analysed the 13 charts faxed by the Thiokol engineers to support their case that 'O' ring failure was linked to temperature and concluded that 'the chartmakers had reached the right conclusion. They had the correct theory and they were thinking causally, but they were not displaying causally'.

Most of the information submitted by the Thiokol engineers was in the form of handwritten notes containing lists of launch numbers, joint temperatures and observed 'O' ring damage following post-launch recovery of the rockets. Figure 7.1 shows one of these charts. The instances listed all had significant 'O' ring damage. The first four were test rockets on a test bed in Utah and the last three were from

HISTORY OF O-RING TEMPERATURES
(DEGREES - F)

MOTOR	MBT	AMB	O-RING	WIND
DM-4	68	36	47	10 MPH
DM-2	76	45	52	10 MPH
QM-3	72.5	40	48	10 MPH
QM-4	76	48	51	10 MPH
SRM-15	52	64	53	10 MPH
SRM-22	77	78	75	10 MPH
SRM-25	55	26	29 27	10 MPH 25 MPH

1-D THERMAL ANALYSIS

FIGURE 7.1 One of the handwritten documents submitted by Morton Thiokol engineers listing the seven data points with 'O' ring issues. From the Report to the President of the Presidential Commission on the Space Shuttle Accident (Washington, DC: Government Printing Office, 1986).

actual Shuttle launches. Both the ambient temperature and the temperature of the 'O' rings are shown. NASA engineers immediately noticed that 'O' ring damage occurred at both the higher end of the temperature range (73°F) and the lower ends. These tables failed to demonstrate a convincing correlation between temperature and 'O' ring damage. Even when set out as a scatterplot, no correlation between temperature and 'O' ring damage can be seen (Figure 7.2).

As Tufte reveals, there *was* in fact a significant correlation between temperature and erosion to the 'O' rings. It was just that the Thiokol charts failed to represent it. Tufte's scatterplot shows data from all 24 previous shuttle launches with joint temperature and 'O' ring damage. The display makes it immediately obvious that every single launch below 66°F had problems and that the forecast launch temperatures, which were another 37°F below this, posed a great leap into unknown territory. It is of course impossible to know if the launch would have been postponed had the full set of data shown in Tufte's figure been presented at the pre-launch teleconference. What seemed at first sight to be a case of a powerful and influential customer steamrollering over the ineffectually expressed concerns of a sub-contractor, might have had more to do with the failure to consider all the available data and to represent that essential data in a visually compelling manner Figure 7.3).

The discussions that took place during the eve-of-launch teleconference discussions have provided fertile ground for social psychological analysis. One commonly explored theme has been that the discussions were subject to a phenomenon known as 'groupthink'. This term was coined by another Yale academic **Irving Janis** to

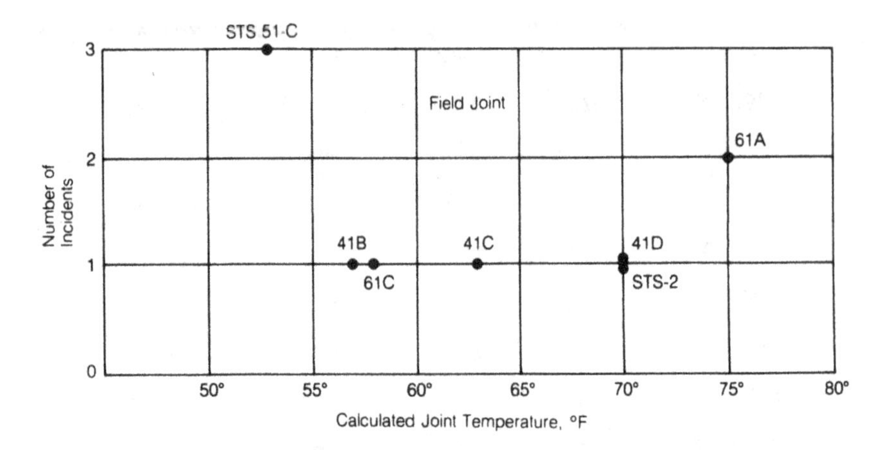

FIGURE 7.2 Scatterplot of the seven flights with known 'O' ring issues as a function of temperature. From p. 147 of the Report to the President of the Presidential Commission on the Space Shuttle Accident (Washington, DC: Government Printing Office, 1986).

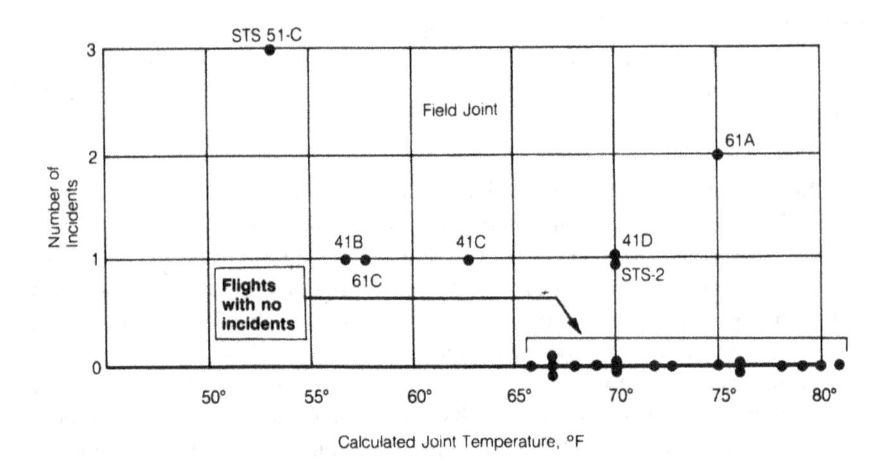

FIGURE 7.3 Scatterplot of all 24 flights with and without known 'O' ring issues as a function of temperature. From p. 147 of the Report to the President of the Presidential Commission on the Space Shuttle Accident (Washington, DC: Government Printing Office, 1986).

refer to a deterioration in decision making often shown in high-level political groups. Janis looked at a number of historical examples including the disastrous attempted US invasion of Cuba in 1961 at the 'Bay of Pigs'. During the cabinet meetings presided over by President John F. Kennedy, the strong desire for group cohesion and the overly directive leadership style shown by Kennedy combined to create the conditions for 'groupthink' where the desire for cohesiveness and solidarity overwhelms the desire for critical thinking. As a result, the shaky basis of the whole invasion plan was never examined.[7]

The more detailed theory set out by Janis included structural faults in an organisation leading to insulated groups of relatively homogeneous members, lacking appropriate norms of impartial leadership and methodical decision making. Add in challenging, stressful circumstances and these cohesive groups were predicted to exhibit the symptoms of 'groupthink'. According to one analysis, examples of all eight groupthink symptoms can be found in the transcripts of the eve-of-launch teleconference meetings. For example, 'self-censorship'. As NASA's opposition to another launch postponement became evident, the Thiokol engineers began to become less vigorous and clear in their opposition to launching below the suggested cut-off of 53°F. 'Pressure on dissenters' was evident in the comment from a NASA manager that he was 'appalled' that Thiokol would recommend postponing the launch. The request by a Thiokol manager to an engineer to 'put on his management hat' similarly suggests pressure on dissenters.

Whilst it is interesting to speculate on the extent to which group dynamics might have affected the eve-of-launch decision making, the previous observations are all subject to alternative interpretations. However, some research evidence on the question was provided by two university researchers in Texas.[8] They analysed the teleconference conversations as reported in the Presidential Commission Report, often referred to as the 'Rogers Report'. Of the 88 statements analysed, 58 were consistent with the suggested 'antecedents and consequences of groupthink' and 30 were not. More interestingly, dividing the statements chronologically over the 24 hours prior to launch showed a statistically significant increasing tendency towards statements consistent with groupthink as time went on. In other words, the demise of the *Challenger* may not have simply been a matter of combustion but of communication.

One particularly troubling point, however, was the lack of any direct evidence that the eve-of-launch decision-making group constituted the kind of highly cohesive group with strong 'we-feelings' that Janis discussed. This is the necessary precursor for 'groupthink' to emerge. To begin with, the discussions were held between three geographically distributed groups that could not even see each other (this was decades before Skype and Zoom). It may be assumed that each of the three groups may have held their own in-group identities and *esprit de corps* although the divergent membership of the groups (both engineers and managers) somewhat mitigates against that. It is somewhat difficult to maintain that the 34 members at three locations could have collectively held the kind of strong collective cohesive spirit of the members of President J.F. Kennedy's cabinet for example.

Diane Vaughan, a Professor of Sociology at Columbia University in New York, has taken issue with the 'groupthink' explanation of the decision-making process just prior to the launch of the *Challenger*. Vaughan's previous interests had been in sociological analysis of organisational 'misbehavior'. In an extraordinarily detailed (575 pages) book, Vaughan looked in detail at the eve-of-launch decision-making and the organisational context that surrounded it.[9] Considering the popular 'groupthink' explanation she pointed out that the conditions required by Janis's theory were not present – there is no evidence that the collective assembly was highly cohesive and bound together in some way. If anything, the opposite was true. Neither were they entirely homogeneous and isolated from outside influences. In contrast to the high-level political groups featured in Janis's 'Groupthink', there were clear norms of evidence and decision making that existed amongst the participating engineers as well as in the demands typically exerted by NASA for recommendations to have strong supporting evidence: 'The discussion did not lack norms requiring methodical procedures for decision making. It was guided by norms and rules of the organisation about how technical discussions must be conducted'.

If groupthink does not provide an adequate explanation of the decision to launch *Challenger*, and the lack of compelling data representation can only take us so far, then what does account for one of the world's most prestigious organisations (NASA) exposing the Space Shuttle programme to such unprecedented risks? To find out, Diane Vaughan undertook an exhaustive examination of the Rogers Report and the documents and evidence contained within it. As in the previous chapters in this section, her approach could also be described as *ethnographical*, although the opportunity to interview participants was restricted by the ongoing legal issues surrounding the disaster. Vaughan set about finding out all that she could about NASA's organisational culture in the months and years leading up to the disaster.

NORMALISATION OF DEVIANCE

Vaughan's 'ethnographic history' of the evolving interpretation of the pieces of evidence of erosion and 'blow back' from the booster rocket joints shows how what was initially viewed with concern came to be accepted as 'acceptable risk'. The first evidence of 'O Ring' problems was discovered in a recovered booster rocket in 1981. This was the second of four 'developmental' flights before the Space Shuttle programme became officially 'operational' from the fifth flight onwards. Although concerned about the problem, the Thiokol engineers considered the redundancy provided by the second 'O Ring' sufficient to ensure that flight safety was not threatened. This was interpreted as the system working in the way it was supposed to.

Vaughan describes the 'normalization of deviance' as a five-step process beginning with the initial appearance of a problem that signals increased risk – in this case, the first observation of an 'O Ring' problem. Step two involves official acknowledgement of the problem – in this case, one of a huge number of anomalies observed on the early 'developmental' stage of the Space Shuttle programme. Step three involved further investigation of the problem with some remedial measures undertaken to improve the situation (in this case, altering the putty-like compound that surrounded the 'O Ring'). Calculations of the extent of erosion that could occur were undertaken and showed that this would still be within 'acceptable limits'. Step four involved 'official acceptance of the situation'. At the pre-launch meetings before the next launch, the Thiokol engineers' recommendations were discussed and accepted. These did not include any specific reference to the 'O Ring' work, however. In the context at the time, the 'O Ring' issue was just one of many anomalies that cropped up during the early stages of the Space Shuttle programme. It did not have the significance at the time that hindsight now imbues it with. The final stage was the successful launch of the next Shuttle which produced a confirmatory cycle that an identified problem had been appropriately analysed and corrected (Figure 7.4).

Expectations surrounding an 'operational' programme are clearly different from those around a developmental/test programme. In response to this public status change as declared by US President Ronald Reagan, internal changes also occurred

FIGURE 7.4 Vaughan's five-stage normalisation of deviance model. See text for explanations of each stage of the process.

in NASA's procedures and risk criteria. However, as far as the rocket engineers at Thiokol were concerned, the solid rocket boosters were very much still in a developmental stage.

Over the next few years, 14 more launches took place. One in 1983 showed evidence of hot gasses from the rockets touching but not eroding the 'O Rings', Another, in 1984 showed unprecedented 'O Ring' damage. Both cases were interpreted as lying within the 'experience base' of observed damage that could be considered 'acceptable' and indeed expected to occur again in future launches. Vaughan's model clearly shows how the initial concern and alarm over a potentially dangerous phenomenon gradually became more and more accepted and normalised over an extended period of time, until it simply became one of the anticipated and expected consequences of launching the Space Shuttle. In other words, the cultural worldview that emerged from these processes contributed to the normalisation of deviance that preceded the *Challenger* launch. Vaughan labels this element 'the production of culture'.

Of course, this normalisation did not occur in a vacuum but in the context of other organisational and cultural values and processes. In particular, Vaughan devotes a further portion of her analysis to two additional structural factors which she refers to as 'the culture of production' and 'structural secrecy'. The first of these refers to the 'institutionalized beliefs' of the engineering and aerospace industries and the organisations around them such as NASA and Thiokol. Over time these changed from a purely technical outlook to one that incorporated the essential components of a capitalist business culture concerned with production schedules, efficiency and cost-cutting. These cultural values become the taken-for-granted background and specific decisions such as the *Challenger* launch decision then become the foreground. Vaughan argues that these unquestioned cultural assumptions provide people with the necessary framework for making sense of ongoing events. In these terms, what might appear deviant when viewed from outside this frame of reference makes perfect sense and can seem normal and appropriate from within that same frame: 'In this manner the culture of production contributed to the normalization of deviance in the work group'.

Structural secrecy refers to a diverse set of organisational structures and processes that essentially clouded the significance of the early signals of danger that emerged from examination of the 'O Rings' in the solid rocket boosters. Both the impersonal technical language used to describe the catastrophic failure and the regular acceptance of the risks before each launch gradually diminished the impact of the initial concerns. In large technical organisations, both work specialisation (engineers and managers have their own distinctive vocabularies that make different aspects of the world more salient) and hierarchy make it difficult for those at upper levels to fully understand the work done at lower levels. Vaughan points to one of the great ironies of organisational life where 'efforts to communicate more can result in knowing less ... masses of information is not useful ... obfuscation parades as clarity'. Both the tendency to skim over the large amounts of information that are communicated daily via internal memos and nowadays by email and the impersonal technical vocabulary serve to distance those at the top of the organisation from the reality of the consequences of failure. Vaughan also includes the problems of regulation, outlined further below, into this category.

Vaughan's exhaustive analysis of years of technical and managerial decision making at NASA can be summarised in her own words:

> Launch decisions accepting more and more risk were products of the production of culture in the SRB work group, the culture of production, and structural secrecy. Each (of these elements) taken alone is insufficient as an explanation. Combined, they constitute a nascent theory of the normalization of deviance in organizations.

The elements of Vaughan's theory are shown graphically in Figure 7.5.

SAFETY OVERSIGHT AT NASA

There is no doubt of NASA's formal commitment to safety as an essential part of its mission. A significant difficulty for unique organisations operating at the very edge of known science and technology is that the necessary expertise to oversee the safety of their operations is necessarily contained within the organisation itself. For this reason, NASA relied largely on 'regulated self-regulation' as described in Chapter 1. There were two internal bodies set up to oversee operational safety. These were the 'Safety, Reliability and Quality Assurance Panel' (SRQA) and the 'Space Shuttle Crew Safety Panel' although the latter was disbanded several years prior to the *Challenger* disaster. It is notable that there was no representation from the SRQA panel at the eve-of-launch meeting. Following the Apollo launch pad fire that killed three astronauts in 1967, the US Congress set up an additional safety body for NASA known as the 'Aerospace Safety Advisory Panel', largely comprising experts from the aerospace sector.

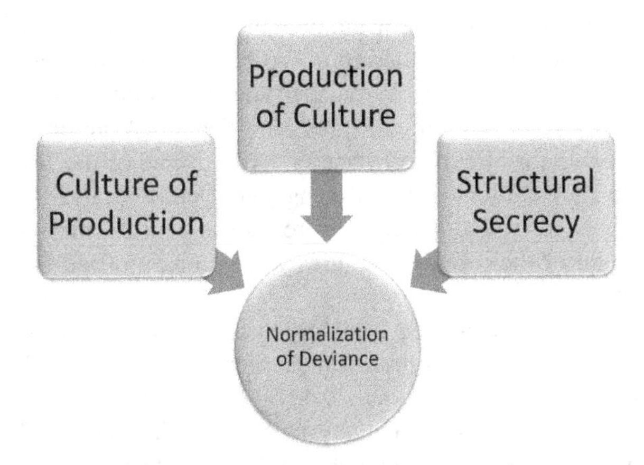

FIGURE 7.5 The three main elements of Vaughan's model of the environmental and organisational factors that lead to the normalisation of deviance. See text for explanations of each factor.

In theory, combining internal self-regulation with external regulation should have minimised the disadvantages of either approach. In practice, all three bodies adopted a compliance approach to safety monitoring where the operations of NASA and its sub-contractors were monitored to ensure compliance with standards and procedures. None of the bodies had any powers to levy sanctions or punishments for violations. All the 'regulators' were either internal to NASA or dependent on their relationships with NASA and contractor personnel. Because of the size and internal complexity of NASA, both the internal and external regulators were subject to significant limitations on their abilities to gather and understand information. Unfortunately, despite its commitment to safety, none of the bodies set up to monitor and guide the safety of operations at NASA were wholly successful. Diane Vaughan went as far as to say: 'just as caution was designed into the NASA system so was failure'.[10]

Chief amongst the failures was the 'O Ring' problem which had first come to light in 1977. This led to the component receiving a criticality rating of 'C1' meaning that failure of the component could lead directly to the failure of the entire assembly and loss of the shuttle. Unfortunately, this change from the earlier 'C1R' classification (meaning that redundancy should prevent total failure) was never picked up by the various safety monitoring bodies. The fact that NASA cut 71% of its funding for safety oversight between 1970 and 1986 may have had a bearing on that. A second critical failure was the lack of monitoring of trend data. Post-launch examinations of the rocket boosters revealed 'O ring' problems on a number of previous launches. Data on these trends were not gathered or utilised by the regulators. A third problem was blurred lines of reporting of safety-critical issues. Different (sometimes contradictory) rules were listed in different documents creating confusion and uncertainty. The marginalised status of the internal regulator (SRQA) was reflected in the fact that no safety specialist was invited to the eve-of-launch teleconference deliberations.

Similar issues have arisen in the aviation industry. With the increasing technical sophistication of modern aircraft design, it has become steadily more difficult for the government authority responsible for certifying an aircraft as safe to fly to exert appropriate oversight. In the United States, the Federal Aviation Administration (FAA) is responsible for aircraft certification. At the time of the Boeing 737 Max certification, the FAA allowed certain Boeing employees to exert regulatory authority on its behalf. An obvious conflict of interest arose which contributed to the lack of proper oversight of the technical failures which compromised the safety of the 737 Max design.

The failure of safety regulation at NASA was not down to incompetent individuals but to the inherent organisational problems associated with any form of regulation whether internal or external. Despite good intentions, the difficulties of self-regulation of cutting-edge technology in a high-risk operational environment are considerable. Vast amounts of complex technical information must be filtered and interpreted to determine what is of significance, and the best people to do this are the very people whose work is being regulated. These organisational problems are generally not reflected in technical risk estimates, which as Vaughan points out means that 'risk is always underestimated, creating unwarranted confidence in all risky technological systems'.

NOTHING FAILS LIKE SUCCESS

William Starbuck, a specialist in organisational behaviour and especially on how organisations learn from experience, has proposed a similar approach to the gradual development and acceptance of the 'O Ring' problems by NASA and Thiokol.[11] *Challenger* was the 25th launch of a space shuttle – the previous 24 had been successful with no losses. The likely effect of a long run of successes is to make future success seem even more assured. Clearly, the technology has lived up to specifications, and the people and organisation must be highly competent to have delivered such outstanding performance over a long period of time. Starbuck suggests that this simple idea provides an excellent characterisation of the behaviour of the managers at NASA and Thiokol: 'As successful launches accumulated, these managers appear gradually to have lost their fear of design problems and grown more confident of success'. For example, Thiokol gradually 'tweaked' the design of the rockets to make them lighter and more powerful. These changes exacerbated the problems of the 'O Rings' failing to seal and led to the change in criticality rating mentioned above.

Over time, the repeated successes of the Shuttle missions seem to have resulted in a gradual acceptance of the ability to live with the 'O Ring' problems. As problems continued to occur, NASA adopted a pressure testing procedure designed to force the 'O Rings' into place. Starbuck reports that although aware of the problems, NASA and Thiokol both began to view these as 'allowable erosion' and 'acceptable risk'. Because of the problems, a 'launch constraint' was imposed. There were a large number (more than 800 on the shuttle alone) of these that had to be considered prior to launch. To permit the launch to go ahead a 'flight waiver' had to be issued. This became normal practice in the run-up to each Shuttle launch. Inevitably, each successful launch and recovery led to diminishing concerns about the significance of the 'O Ring' problem.

In hindsight, it becomes obvious that the 'O Ring' problem should have been properly corrected or else the Shuttle launch programme suspended. The problem 'signals' stand out clearly against other factors now known not to have been involved. However, experience can all too easily teach us the wrong lessons as it did in this case. It has been argued that NASA did not draw the appropriate conclusions from the *Challenger* disaster, leading inevitably to another Space Shuttle calamity in 2003 when *Columbia* was destroyed on re-entry just 16 minutes before landing.

BREAK-UP OF THE SPACE SHUTTLE *COLUMBIA*, 1 FEBRUARY 2003

'The physical cause of the loss of *Columbia* and its crew was a breach of the Thermal Protection System on the leading edge of the left wing, caused by a piece of insulating foam which separated from the left bipod ramp section of the External Tank at 81.7 seconds after launch and struck the wing in the vicinity of the lower half of Reinforced Carbon panel number 8. During re-entry this breach in the Thermal Protection System allowed superheated air to penetrate through the leading-edge insulation and progressively melt the aluminium structure of the left wing, resulting in a weakening of the structure until increasing aerodynamic forces caused loss of control, failure of the wing and breakup of the Orbiter.

The Board recognized early on that the accident was probably not an anomalous, random event, but rather likely rooted to some degree in NASA's history and the human space flight program's culture … with the result that this report, its findings, conclusions and recommendations, places as much weight on these causal factors as on the more easily understood and corrected physical cause of the accident'.[12]

Echoing Starbuck's analysis of the *Challenger* case, the Board investigating the *Columbia* disaster found that 'reliance on past success as a substitute for sound engineering practices' was one of the organisational factors responsible for this event. As with *Challenger*, there was a clear technical issue that led to the destruction of the shuttle. Bits of foam had broken off on launch previously and many shuttles had returned with numerous impacts on the thermal insulating tiles. As with *Challenger*, this repeated experience of successfully launching and returning the affected shuttles became 'normalized to the point where they were simply a "maintenance" issue'. Over time, managers and engineers began to lose their fear of the possible dangers of such damage to the integrity of the shuttle. **Scott Sagan**, a professor of political science at Stanford University, captured the essence of this in his memorable remark that NASA turned 'the experience of failure into the memory of success'.

The initial evidence that a large chunk of foam had broken away and impacted the Shuttle came from the examination of launch videos and high-speed cameras stationed around the launch pad. None of these clearly showed the actual impact with one key camera producing out-of-focus images from a poorly maintained lens. Had NASA been sufficiently concerned about the potential effects of debris strikes on the integrity of the Shuttle's Thermal Protection System then camera evidence might have been accorded higher priority than appeared to be the case. The successful launch and return of the 87 missions between *Challenger* and *Columbia* contributed to the sense of normalcy about debris strikes just as the successes preceding *Challenger* led to complacency regarding the 'O Ring' problems. The sporting cliché that a team is only as good as its last performance seems not to have found currency amongst the decision-makers at NASA where its applicability was just as relevant.

Of course, there were many other organisational issues connected to the *Columbia* failure and the *Columbia* Accident Investigation Board addressed a number of these. As is fairly commonplace in relation to complex disasters, issues of communication and leadership inevitably crop up. As with the HROs described in Chapter 6, NASA had a steeply hierarchical formal structure but also had decentralised decision-making groups such as the Debris Assessment Team that were responsible for assessing the foam damage. Connections to, and therefore communication with, the Mission Management Team were missing or ineffective. The *Columbia* Board noted that when the issue of photographic evidence was raised there was resistance from the formal hierarchy: 'Program leaders spent at least as much time making sure hierarchical rules and procedures were followed as they did trying to establish why anyone would want a picture of the Orbiter'.

As well as comparing NASA with the record of organisations previously considered as HROs, the *Columbia* Board added several comparisons with additional organisations that it considered to have successfully operated high-risk technology over an extended period of time. Chief amongst these was the US Navy's Submarine and Reactor Safety Program, which had accumulated over 5,500 'reactor years' of accident-free operating experience at the time. The 70 nuclear-powered submarines of the US Navy have now exceeded 6,200 years.[13] Amongst the organisational practices, the Board considered pivotal to this success were effective communication practices that ensured that technical personnel were kept fully informed by detailed briefings rather than slide-ware presentations and peer-reviewed formal correspondence. Following his criticisms of the data presentations used in the *Challenger* launch decision, Edward Tufte's also criticised the abbreviated PowerPoint™ presentations used by the Debris Assessment Team. The *Columbia* Board investigation concluded that 'the endemic use of PowerPoint™ briefing slides instead of technical papers (is) an illustration of the problematic methods of technical communication at NASA'.

Both the Space Shuttle disasters illustrate the importance of detecting and responding to occurrences of 'normalization of deviance'. In both cases, better trend monitoring would have provided essential warning of what was happening. Ideally, this would be one function of an independent safety monitoring group and the findings communicated with the technical personnel in proper technical briefings rather than bullet-pointed slideware. For a system as complex as the Space Shuttle, this is indeed a challenging task as the shuttle has over 4,000 critical components, any one of whose failure can result in the destruction of the Orbiter and crew.

To avoid some of the problems of decision making in cohesive groups referred to previously as 'groupthink' the *Columbia* Board also commended the US Navy nuclear reactor safety programme for encouraging the expression of minority or unpopular opinions as well as specific concern with worst-case scenarios. Janis's case studies frequently highlighted the role of the suppression of divergent thinking and the fostering of an air of invulnerability and superiority amongst the group members. It is plausible to suggest that both of these may have played some role at NASA with its reputation as an elite workforce, pulling together in the face of ever tightening fiscal constraints to achieve the previously unimaginable. The *Columbia* Board specifically noted that the safety personnel at NASA 'were present but passive and did not serve as a channel for the voicing of concerns or dissenting views'.

APPLICATIONS OF THE 'NORMALISATION OF DEVIANCE' CONCEPT

Since all workplace behaviour is governed by both explicit rules and procedures and implicit norms of behaviour, it would be expected that there would be a great many examples where the two do not precisely align. Part of becoming a proficient practitioner in almost any domain involves learning the 'tricks of the trade' which often consist of ways of working within and around the explicit rules and procedures so that productivity and efficiency can be maximised. Learning what is done in practice, or what is considered 'normal' in a workplace setting is an important part of developing any skill or profession.

The same is true of any social or cultural situation where it is necessary to acquire a sense of what is normal or expected behaviour. One of the most disturbing applications of the 'normalisation of deviance' concept was the rapid acceptance of the deliberate bombing of civilian populations in the Second World War. Laws and principles governing warfare were established by The Hague Convention (1899) and a series of Geneva conventions beginning in 1864 and updated after the Second World War. The bombing of the town of Guernica in Spain (commemorated in a famous Picasso painting) in 1937 was followed by global outrage and condemnation. However, by 1945 such acts had become a relatively 'normal' part of the war culminating in the 1945 atomic bombing of the civilian population of Hiroshima in Japan.[14]

The 'normalisation of deviance' concept has mainly been brought to bear on issues in healthcare and process safety. In some cases, the concept has been equated with 'complacency' and the development of working practices that deviate from prescribed processes and procedures. As noted above, work done in practice invariably differs in some respects from the officially prescribed version but this is often a necessary adaptation and is not the same process as that described by Vaughan in terms of incremental changes in official monitoring and acceptance of emergent signs of risk that occurred at NASA. Similarly, individuals and organisations might become *complacent* about the hazards and risks involved in their activity or technology through a simple process of habituation – planes don't crash every day and power plants don't blow up very often – so the hazardous can easily seem commonplace. In everyday terms, safe behaviours and unsafe behaviours produce the same outcomes so there is little to discriminate between them or to reinforce the safe option.[15] Complacency can certainly appear as part of the final outcome of Vaughan's five-stage model but is not itself strictly synonymous with the 'normalization of deviance' concept.

Healthcare provides plentiful examples of adaptation to sub-optimal working practices (e.g. failing to follow hygiene standards) or to follow 'best-practice' procedures. Most of the healthcare references to 'normalization of deviance' in fact refer to the more general processes of accommodation and adaptation to working practices through observation of fellow workers as well as through more explicit advice and guidance.[16] Medical practitioners quickly learn which procedures are strictly adhered to and which are regarded more flexibly. Reports of workplace practices that deviate from standard practice in surgery and anaesthesia, for example, are widespread. Examples include 'Turning off alarms because 'we are right there'; Not wiping IV ports with alcohol for 15 secs'; 'No double checks for medication administration' and so on.[17]

One healthcare example that comes closer to Vaughan's model was the recurrent problems with the universal connector (known as a 'luer' connector) for medical tubing. A patient might have several different 'lines' set up to deliver drugs intravenously as well as food into an intra-gastric tube for example. Should the two tubes be mixed up and the feeding solution delivered into the patient's veins, the consequences could be extremely serious or even fatal. One recent study collated 116 cases from US hospitals where fluids had been wrongly directed between IV (intravenous) and enteral (gastric) lines leading to 21 deaths and many other serious complications.[18] Common human factors/ergonomics issues were regularly associated with these events such as busy, acute care settings, multiple lines on a single pole, poor lighting, similar appearance of the liquids, as well as the tubing connector design.

The authors cite this as another example of 'normalization of deviance' and certainly clear signs of danger were officially recognised and ad hoc efforts made to overcome them through the usual routes of encouraging 'greater vigilance' and the like. Unlike the NASA examples, however, in this case, healthcare providers did not conclude that the danger had been adequately examined and dealt with and that operations could safely continue. As of late 2016, however, a set of luer connector standards designed to address these inter-connection problems has been drawn up by the International Organization for Standardization(ISO).[19] These may go some way to reducing the incidence of such problems in the future.

Vaughan's concepts are also regularly mentioned in the context of organisational failures, particularly where dubious business practices have become established. **Blake Ashforth** and **Vikas Anand** have described the process by which legally and ethically dubious actions can easily become embedded in the structure and processes of an organisation.[20] Employees become socialised into believing that these actions are an integral part of the way in which this organisation goes about its business. Over time, actions that might by themselves seem slightly questionable come to seem more and more 'normal'. This model is close to Vaughan's in that an initial signal becomes rationalised and explained and thereby incorporated into corporate memory. Other employees are socialised into this way of doing things which henceforth becomes routinised, unremarked and, therefore, normalised. The journey from initial deviant actions to routinised, normal practice does not involve a sudden leap but typically takes place gradually and incrementally. Applied to businesses and technologies operating in risky environments, this process can be described as a 'drift into failure'.

An interesting and unusual application of Vaughan's *normalization of deviance* concept has been made to a persistent and troublesome phenomenon in science which is the misquoting of previous scientific research. Studies have typically shown quite high rates of quotation error in fields ranging from medicine to social work. These are typically around the 20–25% range. Some of these errors may be quite minor involving typographical mistakes, but others involving misrepresentation of the ideas or evidence contained in a publication may have more serious consequences and are easily perpetuated through subsequent (accurate) quotations of the erroneous material. A recent study found that nearly 13% of the material in the scientific literature referring to Diane Vaughan's work on the *Challenger* disaster were inaccurate.[21] The authors refer to the production pressures faced by academics along with norms that foster multiple referencing and the avoidance of direct quotation, as amongst the cultural factors that increase the likelihood of quotation error in science.

DRIFTING INTO FAILURE

First coined by the Danish electrical engineer and systems safety expert, Jens Rasmussen, the term has since been popularised by Sidney Dekker currently Professor in the School of Humanities, Languages and Social Science at Griffith University in Australia.[22] Sidney Dekker has a relatively unusual background for a university professor, being also a qualified airline pilot with ratings to fly the B737. For several decades he has been a frequent conference speaker as well as a prolific book author. As a European well versed in European philosophy as well as the more European field of cognitive engineering, Dekker has been a fierce critic of the philosophical approach underlying the subject of human error, which Dekker views as reductionist, simplistic and naïve. In most legal and 'common-sense' or folk understanding, 'human error' forms one of the preceding steps in a chain that ends in an undesirable outcome, e.g. an accident. This is in the traditional scientific approach that culminated in Newton's universal laws. Error is the cause and the accident the effect. To understand failures, we need to find these 'causes' which are most often located in human performance and occasionally in mechanical failure. In the third section of the book, we will examine alternative frameworks that take a 'systems' view of performance and failure.

CRASH OF ALASKA AIRLINES FLIGHT 261 OFF THE COAST OF CALIFORNIA, 31 JANUARY 2000

Flight 261, a McDonnell Douglas MD-83, crashed into the Pacific Ocean. All 88 people on board were killed and the aircraft was destroyed on impact. The scheduled flight had departed from Puerto Vallarta in Mexico and was heading for San Francisco. The National Transportation Safety Board (NTSB) determined that the probable cause of the crash was a loss of airplane pitch control resulting from the in-flight failure of the horizontal stabiliser trim system jackscrew assembly's acme nut threads. The thread failure was caused by excessive wear resulting from Alaska Airlines' insufficient lubrication of the jackscrew assembly.[23]

Modern airliners are miracles of electronic and mechanical complexity. Even the most complex systems, however, can involve some pretty basic componentry. One of the most important control functions on any aircraft is pitch control – basically the ability to point the aircraft upwards or downwards. This is accomplished by the elevators – resembling a smaller set of wings at the rear of the fuselage. The elevator itself moves up and down in a similar manner to a basic mechanical jack for raising and lowering an automobile. A large threaded bolt turns inside a nut and thus raises and lowers the elevator. This simple mechanical component requires only regular lubrication to function safely. The manufacturer generally determines the lubrication intervals which initially were set at once every 600 flight hours. All aircraft components have additional inspections and checks to determine if any deterioration is taking place.

Over a period of about a decade, an incremental series of changes were made to the lubrication schedule for this component. Each increment was itself only a few hundred hours – for example, from 1,000 hours in 1988 to 1,200 hours in 1991. Between 1987 and 1994, the interval incrementally stepped up by more than 250% and then effectively quadrupled to become once every 8 months which, for Alaska Airlines equated to around 2,550 hours of flight time. Every one of these small changes was approved by the regulator – the FAA. Something similar seems to have happened to the inspection check intervals. The investigators noted wide variation amongst the airlines flying this aircraft with some lubricating the component at the original manufacturer's recommended 600-hour intervals, with others including Alaska Airlines at 2,550-hour intervals. Inspection check intervals varied from every 2,000 hours to once every 10,500 hours. The official accident investigation report identified the lack of lubrication of the elevator jackscrew assembly and the extension of the check interval as the two key factors in bringing about this catastrophic accident. At the same time, the lack of provision of a fail-safe mechanism for this critical component by the aircraft manufacturers was also noted.

It is important to appreciate the power of small changes. Small, incremental changes in personal habits – say exercise and diet – are much easier to live with than large-scale dramatic changes. Before you know it, the smaller food portion or the longer run become the 'new normal'. Salespeople know that it is much easier to shift people towards their desired goal in small incremental steps rather than in one large hit. The odds are that a relatively small change in lubrication intervals will save the company some money and have absolutely no observable negative effect. This will quickly seem the sensible, normal standard as will another small change a couple of years later. And so on. As Dekker notes, this makes identifying the 'drift into failure' path very difficult, as it is paved with innocuous-looking small changes that appear entirely reasonable and normal at the time.

Dekker considers the typical hindsight search for causal factors in accident investigation to be misplaced and to convey an illusory sense of understanding. As noted in the *Challenger* and *Columbia* disasters, the unfolding events of a typical accident reveal that people are generally going about their normal jobs following routines and practices that they normally follow. It is only in hindsight that certain behaviours stand out and appear faulty or ill-judged. Dekker's view is that identifying and counting up 'errors' in hindsight conveys an entirely illusory sense of understanding. Echoing William Wagenaar's observations about maritime accidents (see Chapter 2), Dekker argues that most accidents involve normal people carrying out normal work in their normal organisational environment. Vaughan refers to the *Challenger* and other disasters as 'embedded in the banality of organisational life'.

Typically, there are no major large-scale shifts in organisational practices that immediately precipitate disaster. Rather there are smaller, incremental adjustments to maintenance schedules, codes of practice, regulatory requirements and so forth. These constitute what Dekker labels 'the drift to failure'. Diane Vaughan made precisely the same point about the incremental development of the working culture at NASA: 'If we can distill a single important insight … it is about the potential of small precedents established early to have disproportionately larger consequences later'.

Dekker argues against the traditional safety science view that incidents are the precursors to accidents: 'Incidents do not precede accidents: Normal work does'. Dekker spends quite a bit of time dissecting traditional notions of cause and effect and their role in the development of mechanistic models of accident causation. Whilst it is possible in hindsight to structure the events of the *Challenger*, *Columbia* and Alaska Airlines disasters in a way that shows these causal links, this line of reasoning is divorced from the reality of the events as they appeared at the time to the people involved. This insensitivity to the complexities and subtleties of human behaviour in organisations means that the resulting simplistic mechanistic understandings of complex disasters are inadequate if we wish to understand how to better manage the operations of risky activities and technologies. Dekker suggests that if we are to make real progress in understanding – and ultimately better controlling – safety in complex human-engineered systems, then we need to adopt a radically different scientific conceptualisation and different vocabulary. This is the view that will be explored more fully in the third section of this book as we explore systems theory and its application to the modern science of safety.

NOTES

1. Aronson, E. (1984). *The Social Animal, 4th Ed.* New York: W.H. Freeman & Co. The book has now reached 12 editions!
2. The first appearance of Solomon Asch's work was in a book chapter entitled: Effects of group pressure upon the modification and distortion of judgment. In M.H. Guetzkow (Ed.), *Groups, Leadership and Men* (pp. 117–190), published in 1951. Follow-up work

by Morton Deutsch and Harold Gerard showed that the socially conforming expressed judgments were greatly reduced when made anonymously. Their work appeared in an article published in 1955: A study of normative and informational social influence upon individual judgment. *Journal of Abnormal and Social Psychology, 51*, 629–636.
3. Coch, L., & French, J.R.P. (1948). Overcoming resistance to change. *Human Relations, 1*, 512–532.
4. *Report to the President by the Presidential Commission on the Space Shuttle Challenger Accident*, June 6th, 1986. Washington, DC: Government Printing Office. Retrieved from: https://spaceflight.nasa.gov/outreach/SignificantIncidents/assets/rogers_commission_report.pdf. The quotations on pages 157–158 are taken directly from the report.
5. The House Committee on Transportation and Infrastructure. (2020). *Final Committee Report: The Design, Development and Certification of the Boeing 737 Max*. Downloaded from: https://transportation.house.gov/imo/media/doc/2020.09.15%20FINAL%20737%20MAX%20Report%20for%20Public%20Release.pdf
6. Tufte, E.R. (1997). *Visual Explanations*. Cheshire, CT: Graphics Press. A sumptuously illustrated large-format book.
7. Janis, I. (1972). *Victims of Groupthink*. Boston, MA: Houghton Mifflin. The abortive US attempt to invade Cuba in 1961 was so bungled that Janis labels it a 'Perfect Failure'.
8. Esser, J.K., & Lindoerfer, J.S. (1989). Groupthink and the space shuttle challenger accident: Towards a quantitative case analysis. *Journal of Behavioral Decision Making, 2*, 167–177.
9. Vaughan, D. (1996). *The Challenger Launch Decision: Risky Technology, Culture and Deviance at NASA*. Chicago, IL: The University of Chicago Press.
10. Vaughan, D. (1990). Autonomy, interdependence, and social control: NASA and the space shuttle challenger. *Administrative Science Quarterly, 35*, 225–257.
11. Starbuck, W.H., & Milliken, F.J. (1988). Challenger: Fine-tuning the odds until something breaks. *Journal of Management Studies, 25*, 319–340.
12. Columbia Accident Investigation Board. (2003). *Report Volume 1*. Washington, DC: Government Printing Office. Downloaded from: https://www.nasa.gov/columbia/home/CAIB_Vol1.html
13. According to Wikipedia: https://en.wikipedia.org/wiki/United_States_Navy_Nuclear_Propulsion
14. Kramer, R.C. (2009). Resisting the bombing of civilians: Challenges from a public criminology of state crime. *Social Justice, 36*(3), 78–97.
15. Bogard, K., Ludwig, T.D., Staats, C., & Kretschmer, D. (2015). An industry's call to understand the contingencies involved in process safety: Normalization of deviance. *Journal of Organizational Behavior Management, 35*, 70–80.
16. Banja, J. (2010). The normalization of deviance in healthcare delivery. *Business Horizons, 53*, 139–148.
17. These examples are from: Odem-Forren, J. (2011). The normalization of deviance: A threat to patient safety. *Journal of Perianesthesia Nursing, 26*, 216–219. Other authors have suggested that these issues are widespread in healthcare. For example: Price, M.R., & Williams, T.C. (2018). When doing wrong feels to right: Normalization of deviance. *Journal of Patient Safety, 14*(1), 1–2.
18. Simmons, D., Symes, L., Guenter, P., & Graves, K. (2011). Tubing misconnections: Normalization of deviance. *Nutrition in Clinical Practice, 26*(3), 286–293.
19. A news release describing the new ISO standards for luer connectors can be found here: https://www.iso.org/news/2016/10/Ref2131.html
20. Ashforth, B., & Anand, V. (2003). The normalization of corruption in organizations. *Research in Organizational Behaviour, 25*, 1–52.

21. Lock, J., & Bearman, C. (2018). Normalization of deviation: Quotation error in human factors. *Human Factors, 60*(3), 293–304.
22. Dekker, S. (2011). *Drift into Failure: From Hunting Broken Components to Understanding Complex Systems.* Farnham, UK: Ashgate.
23. National Transportation Safety Board. (2002). *Aircraft Accident Report: Loss of Control and Impact with Pacific Ocean Alaska Airlines Flight 261 McDonnell Douglas MD-83, N963AS about 2.7 Miles North of Anacapa Island, California January 31, 2000. NTSB/AAR-02/01.* Washington, DC: National Transportation Safety Board.

Section 3

Systems

8 Cognitive Engineering
Constraints and Boundaries

In one sense, we live in a world of limitless possibilities. An almost infinite number of lives and lifestyles are on offer. However, our life course is much more constrained by the circumstances of our birth and by a host of other factors – mental, physical and environmental – that will shape the course of our lives. Even at the level of individual action, all human activities – at home, at work or during travelling in between, are subject to a wide variety of constraints.

CONSTRAINTS

A constraint is a limitation that restricts the ability to do something. The most familiar examples are economic constraints. My desire to purchase a Bugatti Veyron is constrained by my lack of the necessary financial resources! Economic constraints are amongst the most powerful forces acting on individuals, organisations and governments alike. Physical constraints, such as lack of height, constrain my ability to become an NBA basketballer (average height around 2 m). Some physical constraints (such as height) are relatively fixed whilst others (such as muscle strength) are more malleable. Similarly, cognitive constraints such as memory capacity, sensory ability and know-how can, in some cases, be relatively enduring and in other cases highly changeable through training and practice.

The environment constrains behaviour on many levels. Temperature (see Chapter 4) constrains human performance at both higher and lower extremes. It also constrains how materials behave, as in the effects of the colder temperatures on the flexibility of the O-Ring seals on the *Challenger* space shuttle (see Chapter 7). Nuclear power plants are constrained by the laws of physics. Weather is a major constraint on aviation and shipping and certain meteorological conditions (e.g. fog, snow and ice) can also constrain driving, cycling and even walking. Social constraints – religious, cultural, gender and other social constructions affect our every action. Almost every activity in and out of the workplace is constrained by applicable rules and regulations, sometimes enshrined as legal constraints.

Finally, the machines and tools that we use constrain our behaviour in numerous ways. I can only interact with my personal computer in highly restricted ways – chiefly through a keyboard and mouse – and not in others such as by voice or gesture. My interactions with the digital world are constrained to viewing a relatively small, flat two-dimensional surface. Written documents such as books constrain our experience to follow a linear progression of words that transmit information visually (or aurally in the case of audiobooks) in a way that is quite different from our other interactions with the world.

DOI: 10.1201/9781003038443-11

Any human activity, whether it be work, recreation or travel is constrained in one or more of these ways. Professor **Jens Rasmussen** was trained as an electrical engineer and began his career in control room design for nuclear power plants in his home country of Denmark. Over a lengthy career spanning six decades, Rasmussen became a leading figure in the fields of human-machine interaction and human error, becoming one of the founders of what has since become known as *cognitive engineering*. In addition to his work with the Atomic Energy Commission, he was a research professor at Riso National Laboratory and the Technical University of Copenhagen until his death in 2018.

Understanding any human activity requires understanding the various constraints that guide and limit what we can do. Much of Rasmussen's work was directed at identifying the constraints that shape performance in the workplace, especially in complex sociotechnical systems such as nuclear power plants. Workplace activity involves three components – actors or 'agents', which mostly refers to humans but increasingly extends to other intelligently acting components, the environment or external *world*, and the tools or 'artefacts' that we interact with. The various possible interactions between agents, artefacts and the world create enormous degrees of complexity and possibility.

These three interacting components are depicted in Figure 8.1. The interactions between humans and machines (especially computers) have been studied for several decades and are usually known as *human–computer interaction*, or HCI. The interactions between machine artefacts and the external world are largely covered by engineering, especially the design and operation of sensors. Interactions between

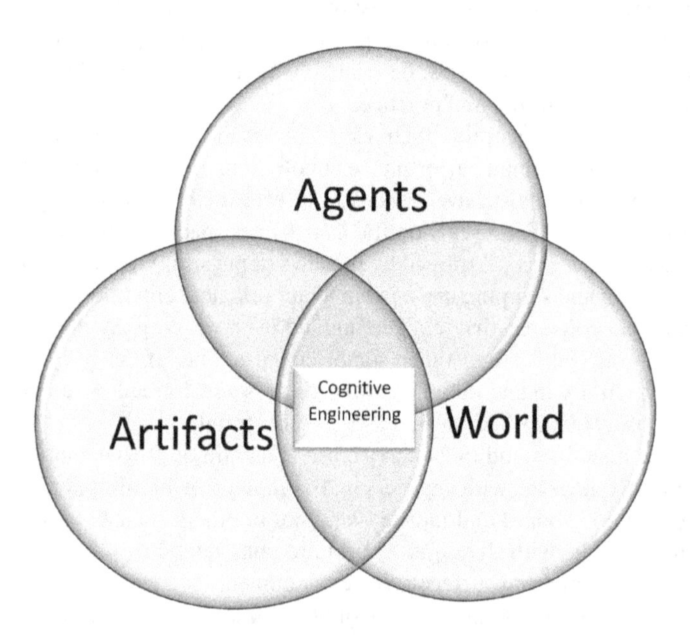

FIGURE 8.1 Cognitive engineering is concerned with the performance of intelligent agents in conjunction with technical artefacts in real-world settings.

human agents and the world are covered generally in psychology and, more specifically, in human factors and ergonomics. The intersection of all three (agents, artefacts and the world) defines the subject matter of cognitive engineering.

The interdisciplinary subject of cognitive engineering is concerned with the nature of the constraints on human performance in settings such as power plant control rooms. As a young electrical engineer given the task of working on control room design, Rasmussen turned to the psychology literature for guidance. At this time (in the late 1950s and early 1960s), apart from some of the early post-war work in human factors described in Chapter 4, the orientation of psychology was still largely behaviouristic focusing on overt behaviour but not on the mental activity or environmental conditions behind that activity. This led Rasmussen to the conclusion that 'it appeared to be necessary to start our own selective research program to find models useful for engineering design'.[1]

The initial aims of the Rasmussen programme were to discover the constraints on human performance in engineering tasks such as fault-finding and troubleshooting as well as finding ways of describing the problem environment itself. Later a third aim was developed which was to find ways of making the constraints on performance visible to the human operator through appropriately designed display technology. Later in his career, Rasmussen explored broader themes of controlling safety in sociotechnical systems and understanding the accidents that sometimes result. These aims, along with some of their corresponding developments within cognitive engineering, are outlined in Table 8.1.

If you were to suddenly find yourself at the control desk of an atomic reactor, after say one day's training, how would you perform? According to Rasmussen, you would operate at a *knowledge-based* cognitive level. Armed with your very limited knowledge, you would try and interpret the display indications in front of you and make the appropriate control adjustments accordingly. You would probably spend a good deal of time consulting the manual beside you (the one that begins: 'Congratulations

TABLE 8.1
Aspects of Cognitive Engineering Developed from the Work of Jens Rasmussen.

Aim	Resulting Tools and Concepts
Understanding the constraints on human cognition	Skill/Rule/Knowledge (SRK) Taxonomy Signal/Sign/Symbol Decision Ladder
Describing and analysing the constraints from the problem environment	Abstraction Hierarchy Cognitive Work Analysis
Making the constraints visible to the operator in a useful way	Ecological Interface Design (EID)
Design safe working systems	Safety Space Safety Boundaries
Understanding accidents	AcciMap

on your purchase of a *Westinghouse Atom Smasher 3000*') for relevant rules (e.g. 'if water in-flow exceeds 20 m^3 per minute then close the relief valve'). After gaining a little experience you would have internalised a good number of these rules and these would enable you to control the plant under normal operating conditions. Unsurprisingly, this is referred to as the *rule-based* cognitive level. Meanwhile, the very experienced operator on an adjacent panel is making necessary control movements in response to the information in front of her without needing to consult the manual or think from first principles. This is known as *skill-based* cognitive control. This simple taxonomy (Knowledge-Rule-Skill) for levels of cognitive control has become enormously popular and widely used. As Rasmussen acknowledges, it is very similar to other tri-partite schemes developed earlier to describe the stages of perceptual-motor skill acquisition.[2]

Traditional HF/E work on display design has concentrated on ensuring that displays are as legible and readable as possible. One of Rasmussen's insights was that the same display could work in very different ways depending on the cognitive processes (Skill, Rule or Knowledge) employed by the observer. For example, a simple meter showing a moving pointer at some value – say 20 m^3 as in the example above – can be read as a *signal* (just make control response to adjust the flow up or down to some set point), a *sign* (find an associated rule for action to be taken when this value is reached) or *symbol* (figure out what this value might mean by reasoning through system knowledge or from first principles). A summary of Rasmussen's basic concepts is shown in Table 8.2.

These simple qualitative distinctions have provided a terminology and conceptual framework for discussing the cognitive activities involved in real-world tasks such as operating a control panel, electronic fault-finding, diagnosis and so on. The idea that operators can move up and down the levels as situations dictate, interpreting display indicators in qualitatively different ways, has provided an appealing way of looking at operator behaviour in a wide variety of situations. By the usual scientific measure of influence (number of citations by other authors) the Skill/Rule/Knowledge (SRK) taxonomy has a huge number of citations, many of them relatively recently in the healthcare literature.[3]

THE DECISION LADDER

A logical analysis of most human action above the merely reflexive responses (such as the jerk of the knee to a hammer tap or the closing of the eyelid to a puff of air)

TABLE 8.2
Rasmussen's Models of Information Processing in Work Activities.

Level of Control	Information from Environment	Characteristics
Knowledge-based	Symbol	Causal, functional understanding
Rule-based	Sign	Normal professional activity
Skill-based	Signal	Continuous control tasks

suggests that at minimum there must be three components. There must be a state or situation requiring a decision and this must be apprehended in some way. A decision involves a choice between two or more courses of action so some kind of weighing up must be achieved. This is normally thought of as involving deliberative thought, but as we saw in Chapter 2, purely affective or emotional factors can also weigh heavily. Finally, there must be an execution or action component where the person or animal moves to a new state or situation.

Rasmussen's Decision Ladder is a somewhat more elaborate version of this logical framework.[4] In total, Rasmussen identifies eight logically required information processing activities and associated states of knowledge that might be present in complex problem-solving and decision-making contexts such as control of a power plant or the electronic troubleshooting activities he studied. This sequence is set out in Figure 8.2. As shown, it more closely resembles a snake than a ladder! The ladder visualisation comes from Rasmussen's depiction showing an ascending sequence of steps up the left-hand side with an inflexion point at the 'interpret' step and then a descending sequence down to execution. Like a ladder, the left- and right-hand sides are connected by a series of 'shortcuts' which map onto the SRK classification. For example, an experienced operator detecting a change in some state or variable might execute the appropriate corrective response without the need for further thought in the same way as a driver 'automatically' executes a series of corrective steering, accelerator and brake adjustments to maintain lane position and following distance on a highway. This is skill-based control in an environment generating the appropriate signals that have been used to develop an internal model of the driving dynamics.

A particularly commonly used 'shortcut' occurs between identification ('I know what that is') of some state or condition and knowing what needs to be done about it. This occurs in tasks as varied as electronic troubleshooting and medical diagnosis. Such 'condition-action' pairings can be instantiated as 'IF-THEN' statements in computer code. This activity corresponds to the rule-based level of processing. Knowledge-based processing is represented by the innermost loop which may involve multiple iterations between interpretive and evaluative processing until an outcome that meets some criteria of satisfactoriness is reached. The reason that skill-based processing frees up so much in the way of cognitive capacity can be easily seen from the figure as none of the intervening stages of processing need to occur. By comparison, rule-based processing requires more cognitive resources than skill-based responding but rather less than knowledge-based processing.

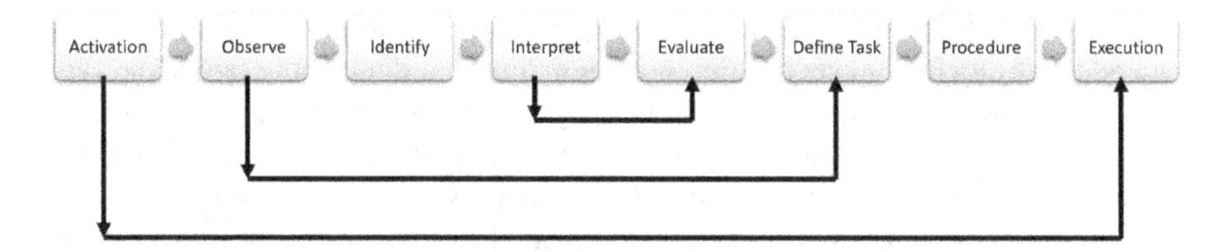

FIGURE 8.2 Rasmussen's 'decision ladder' illustrated in snake-like form. The three modes of cognitive control (Skill, Rule and Knowledge) are illustrated by the outermost, middle and innermost connecting lines, respectively.

The decision ladder provides a handy template for looking at examples of operator behaviour – either in simulations or in 'real-life'. In the latter case, performance can be analysed from accident or incident scenarios. When this was done for a set of aircraft accident reports there was a clear association between the severity of the outcome and the type of activity taking place. In cases where the investigators reported that inappropriate procedures or inadequate execution were involved, the outcomes tended to be much less severe than when investigators found evidence for missing or compromised functions such as detection or accurate diagnosis of the situation. In these cases, severe or fatal outcomes were much more likely. It seems clear that trying to deal with the wrong problem is apt to be much more dangerous than incorrectly dealing with the right problem.[5]

This has important implications for the design of tools and information displays to support work performance, especially in high-risk industries. Neville Stanton's team at Southampton University have been using the decision ladder to generate new design ideas for a range of problems from more economical driving to aircraft fuel management.[6] The critical importance of gaining an accurate understanding of a situation is reflected in the extensive academic literature on the topic of what has become known as 'situational awareness' (often ungrammatically referred to as 'situation awareness'). Some situations declare themselves relatively unambiguously (such as an iceberg ahead for a vessel (see Chapter 1) or most engine failures on take-off in a commercial airliner) whilst others do not. Power plant operators at Three Mile Island (Chapter 5), for example, struggled to understand what was happening at the plant. An important outcome of Rasmussen's analytical work is therefore its contribution to display design. We will return to this topic after a brief outline of cognitive work analysis and the abstraction hierarchy.

THE ABSTRACTION HIERARCHY

A key property of systems is that they are organised hierarchically. Human-designed systems are developed to serve some purpose such as to transport people or goods from one place to another or to generate power and so forth. Any given purpose can be achieved in many different ways and by many different means. Transportation can be achieved by land, sea or air. Power can be achieved by burning coal, running water or splitting atoms. Equally, any given means can serve multiple purposes. A vehicle can transport people or goods but could also serve as shelter. Atoms can be split to power domestic heaters or as an instrument of warfare. In studying the behaviour of maintenance and plant control personnel, Rasmussen noticed that they were adept at moving between different system levels in diagnosing faults and failures. Sometimes, they would focus on the physical components and at other times they would reason downwards from more abstract ideas about the system's functions.

The abstraction hierarchy was developed to provide a useful way of understanding the constraints in a work environment that come from the environment rather than the worker. **Kim Vicente**, the youngest ever (at age 34) full professor of mechanical engineering and founder of the cognitive engineering laboratory (CEL) at the University of Toronto in Canada, was one of Rasmussen's most frequent collaborators and has written extensively on most aspects of Rasmussen's work. He has provided one of the more useful guides to the practical applications of ideas such as the abstraction hierarchy to understanding how people work in complex systems and designing tools to support them in that work.[7] This is generally referred to as *cognitive work analysis* (CWA). A key tool in CWA is the *abstraction hierarchy* which maps out a series of levels between the system's overall purpose (labelled 'functional purpose') and the chosen means of achieving that purpose ('physical form'). Between these are three other levels of abstraction as shown in Figure 8.3.

The level of 'physical functions' refers simply to the designated functions that the specific components fulfil. For example, there is componentry in a computer that has the physical function of long-term storage of information. This can be achieved by various different configurations such as a hard disc drive or a solid-state drive which take quite different physical forms. The top three levels represent the functions of the system in a more general way without reference to any specific physical processes or materials. For example, an air traffic management (ATM) system is designed to move aircraft safely and efficiently through a sector of airspace. This is its functional purpose. The achievement of this purpose or goal involves the transmission and reception of information between ground and aircraft and an understanding of aircraft control and energy management. These are the abstract functions necessary to achieving the system purpose. The generalised functions required to do so, in this case, refer to the various strategies and procedures available to controllers. Some examples of these were referred to previously (Chapter 6) in discussing the ways in which controllers at San Francisco (SFO) were able to rapidly adapt to extremes of working conditions.

Human factors consultants are often asked to work solely with the physical form chosen by the manufacturers – more often than not, tasked with overcoming already built-in defects! However, as Rasmussen noted, workers may be trying to

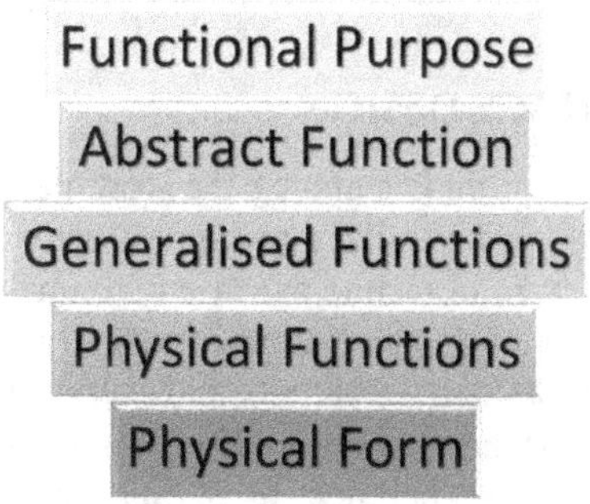

Functional Purpose
Abstract Function
Generalised Functions
Physical Functions
Physical Form

FIGURE 8.3 Basic outline of Rasmussen's abstraction hierarchy. The most abstract level ('functional purpose') represents the over-arching system objectives; the next level down ('abstract function') describes the necessary functions to achieve that objective in abstract terms such as through the flow of energy or information. At the bottom of the hierarchy are the least abstract, most concrete descriptions of the physical and tangible aspects of the system.

comprehend system functioning on several levels simultaneously, so a focus on 'fixing' a problem at the lowest level may have a limited impact on their overall ability to perform the task successfully. Carrying out a thorough cognitive work analysis provides much greater scope for understanding what operators are trying to do and how best to support that work on a number of levels.

Vicente gives an example, also from air traffic management, based on a previous study of air traffic controllers' behaviour. When the workload was very light, controllers would take on some of the more abstract functions of devising optimal flight paths for each aircraft based on principles of energy management and knowledge of the control abilities of each individual aircraft. Under higher levels of workload, controllers would gradually revert to more generalised functions involving standard procedures for routing aircraft, such as putting some in 'holding patterns' to relieve pressure on managing the active runway. This kind of detailed understanding of operator behaviour provides valuable information for a range of activities such as selection, training and information design to best support the variety of cognitive strategies employed in complex tasks.

This example also highlights the limitations of conventional task analysis methods that focus on decomposing a given 'task' as performed by a worker into its constituent elements. The aim of the exercise is generally to find a more efficient way of performing the elements to accomplish the task. Cognitive task analysis (CTA) expanded the method to include less easily observable mental operations. CWA goes further still by analysing the way in which environmental constraints shape the tasks that may be performed, as well as the way in which these affect the way in which the work can be conducted. This has implications for both the design of information presentation to best support work, especially in unusual or abnormal situations, and understanding the knowledge, skills and abilities required to accomplish the task. CWA has been very widely applied in areas as diverse as nuclear power, aviation, healthcare, rail transportation, product design, criminal network analysis and environmental flood protection.[8]

ECOLOGICAL INTERFACE DESIGN

Both the SRK framework for understanding the work performed by operators and the abstraction hierarchy can be utilised to suggest better ways of supporting the information needs of workers in technological environments. Taken together, they suggest the need for information presentations such as visual displays, to support knowledge-based activity, skill-based activity and everything in between. They also point to the need for such displays to both support reasoning at higher, abstract levels and facilitate speedy perceptual apprehension of important information. Vicente has provided detailed worked examples of the process of designing information displays according to such principles. Some examples of ecological interface design (EID) display design were described earlier in Chapter 4. This approach stands in contrast to the old-fashioned engineering approach of providing a single display indication for every variable that could be measured (e.g. a temperature gauge, a pressure gauge, etc.). The problems with the resulting collections of overloaded panels have been highlighted in Chapter 4. In addition to taking up unnecessary amounts of space, such displays fail to support operator behaviour at any of the three levels. Operators must do everything in their heads whilst relying on limited memory and scanning. Inefficient at the best of times and downright dangerous in adversity.

The approach developed by Rasmussen has been labelled 'ecological' as it is based on a far more comprehensive understanding of the problem environment itself. The term 'ecological' also refers to earlier, seminal work in visual perception by **J.J. Gibson**, whose work focused on the information directly available from the environment. Gibson was born at the start of the 20th century and spent most of his career at Ivy League Cornell University in New York State. Like many other psychologists at the time of the Second World War, Gibson found himself working for the military and, in particular, aircrew selection and testing. This led Gibson to consider how a pilot is able to land so precisely on a narrow strip of ground. His insight was that this skill was largely controlled by information generated by optical information generated by the relative motion of the pilot over the ground. For example, the rate of flow of the visual field in our periphery is directly interpreted by the brain as showing how fast we are moving. Crafty road designers can thus utilise this feature to 'fool' our brains into believing we are travelling too fast by painting road markings seen in our periphery increasingly close together. The brain 'sees' this directly, without needing to be analysed or interpreted, and responds directly by directing the body to slow itself down. An example was given in Chapter 4 of railway crossing design where this principle was used to enhance safety at railway crossings.

The principles of 'ecological interface design' were laid out by Rasmussen and Vicente in papers published in 1989 and 1992.[9] The problem of how to communicate essential information about a work domain is inextricably linked to the problem of how to understand the constraints that characterise people's interactions with the work environment. The abstraction hierarchy and the SRK framework, therefore, provide the essential theoretical concepts underlying EID. These provide two key insights underlying the EID approach. The first of these is that people may approach the interface with different needs and goals in mind depending on circumstances. In normal operations, the focus is lower down the abstraction hierarchy where display information can be treated largely as signals linked to well-practised routines and responses. Conventional display interfaces, as described in Chapter 4, are generally designed with this in mind.

Problems typically arise in unusual situations – either where issues arise that have never been previously anticipated by the designers or previous operators or where problems arise that exceed the knowledge and capacity of the current operators. In either case, reading the displays as signals quickly becomes confusing and even recourse to rule-based processing yields contradictions and puzzlement. In this case, the operators must have recourse to knowledge-based processing relying on the display information as symbolic. Conventionally designed displays are generally lacking in their ability to support this higher-level problem-solving as the information supplied by the array of displays is typically low level, designed for normal operating situations.

Vicente and Rasmussen consider that normal functioning using skill or rule-based processing is best facilitated by displays that allow rapid perceptual processing. Much has been made recently of the idea that human cognition can be divided into two distinct 'systems' – a perceptually based, rapid processing system and an analytically based, slower more demanding system. Nobel prize-winning psychologist **Daniel Kahneman** refers to these as 'system 1' and 'system 2', popularised in his best-selling book *Thinking, Fast and Slow.*[10] Vicente and Rasmussen's recommendation is for display interfaces to be designed so as to make maximum use of the quick and effective perceptual processing system. They point out that even where analytical processing would appear most effective – as in making complex judgements about people (e.g. medicine) or situations (e.g. firefighting) – research has shown a tendency for experienced decision makers to draw on perceptual-recognitional processing rather than the slow, analytical mode.[11]

The EID approach suggests that display interfaces should primarily facilitate rapid perceptual processing but also be designed to provide effective support for knowledge-based processing. Whilst rarely employed, this is most likely to occur in more critical, unforeseen and unrehearsed situations. Vicente provides detailed examples of how the interface design for a process control task was accomplished. Vicente summarises the aim of the design process as a question of 'how the information in the abstraction-decomposition space should be displayed in terms of signals, signs and symbols'. In other words, the interface should make visible the constraints that govern the work process.

For the thermal process control task, the main work constraints are the rate at which mass is flowing into the system, the rate that mass is flowing out of the system and the current mass inventory. The proposed interface display that supports both rapid-fire perceptual processing and higher-level analysis of the system's functioning is shown in Figure 8.4. The relative length of the top horizontal bar showing input and the lower bar showing output indicates whether there should be increasing or decreasing volume as shown by the shaded rectangle. The state of all three variables can be apprehended at a glance rather than being inferred from separate meters as would normally be the case in conventional display design.

This 'mass balance' display was one component of a much larger display developed using the same principles to provide an interface for operators of this process control task (see Figure 8.5). Vicente and his research student carried out an experiment to compare performance with the 'EID' display and a more conventional ('Piping and Instrumentation Diagram') display design.[12] A dozen engineering students worked with one of the interfaces on a variety of tasks simulating both normal and 'off-normal' operations involving fault management. As expected, the EID interface led to better fault diagnosis although under normal operating conditions performance did not differ significantly between the two interface groups. There was a difference in how the groups performed, however, with the EID group depending more heavily on spatial processing and the conventional interface leading to more reliance on verbal resources. Vicente suggests that it might be beneficial to rely more on spatial resources thus freeing up verbal resources for other tasks such as communication or reading manuals.

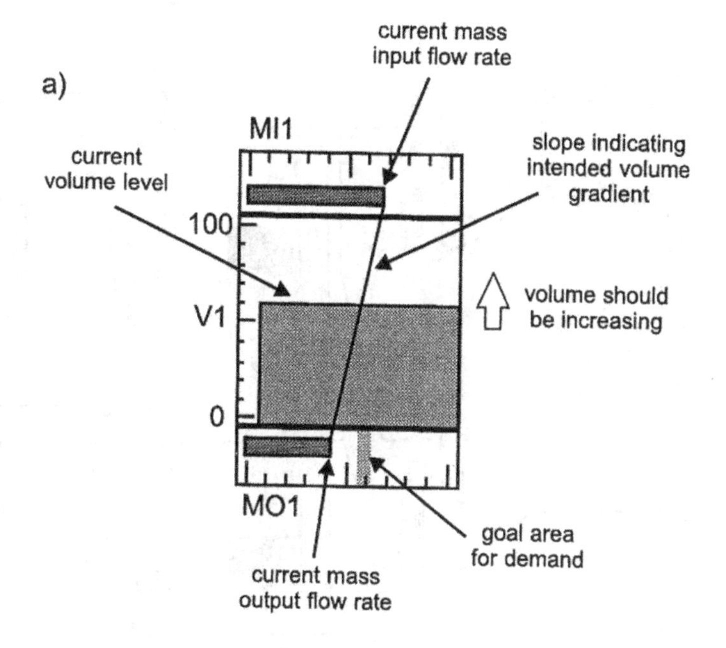

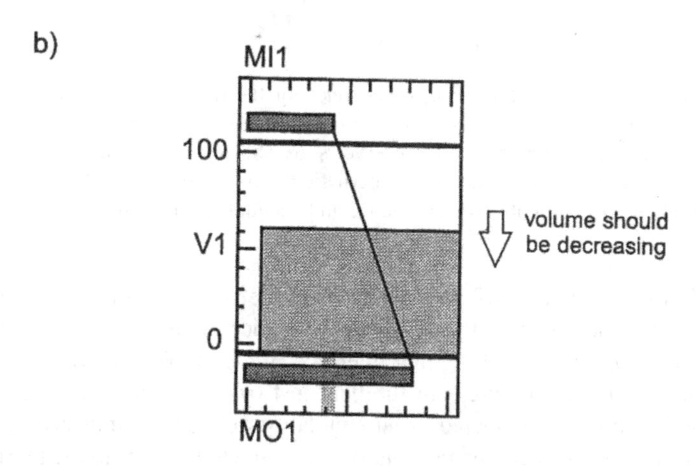

FIGURE 8.4 A display for a thermal process control task designed according to ecological interface design (EID) principles. The rectangle represents the current volume of water in a reservoir tank. The upper horizontal bar represents the inflow rate and the lower bar represents the outflow rate. If the upper bar is longer than the lower, then the volume in the tank should be increasing and vice versa. From Pawlak, W., & Vicente, K.J. (1996). International Journal of Human-Computer Studies. Reproduced by permission of Elsevier Science and Technology Journals.

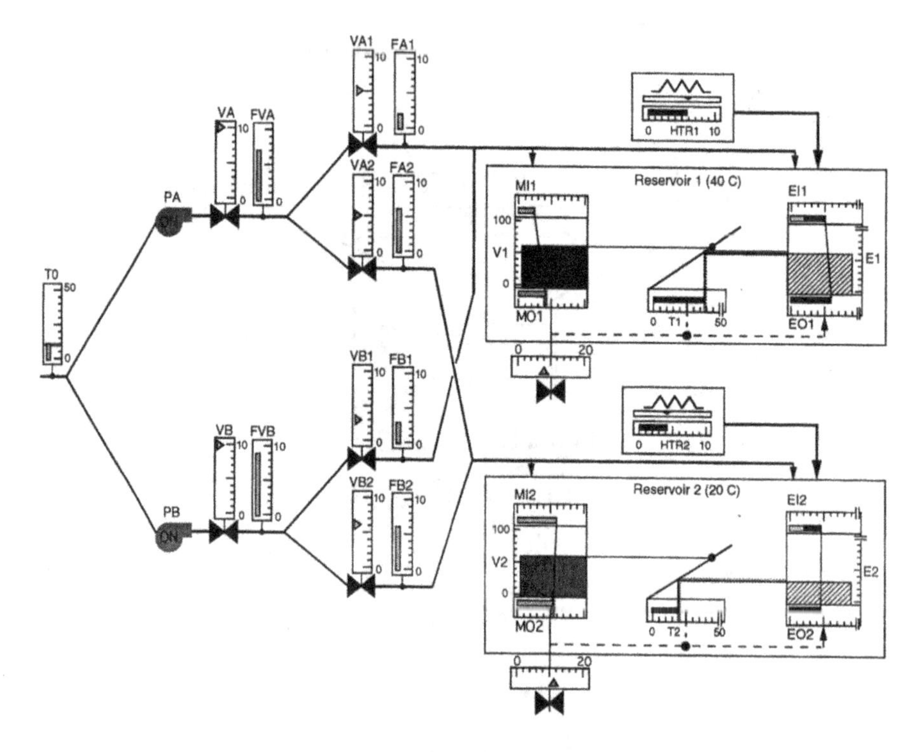

FIGURE 8.5 The complete EID operator panel for the thermal process control system. The vertical bars on the left represent flows through individual valves (VA1, VA2, etc.) and the results on the accumulation in the reservoirs are shown in the rectangles on the right. From Pawlak, W., & Vicente, K.J. (1996). International Journal of Human-Computer Studies. Reproduced by permission of Elsevier Science and Technology Journals.

How has EID panned out? Since 1989 when the principles of EID were first set out in the scientific literature there has been considerable interest in this new, theoretically based approach to designing interfaces for technological systems. Vicente was the first to review progress in the field just over a decade later.[13] Almost all the research had been conducted using simulations of various industrial processes (cement milling, nuclear power, conventional power, etc.) similar to the thermal processes simulation used in the initial studies by Vicente's Toronto research group. Vicente's main conclusion echoed that of the earlier research study described above: interfaces designed according to EID principles provided better support for operators under off-normal operating conditions but not under normal, expected conditions.

Vicente noted a variety of other projects that utilised cognitive work design principles to uncover information not currently utilised in interface design. These projects included systems management for the Lockheed Hercules C-130 aircraft, computer network management, software specifications for the TCASII (Traffic Alert and Collision Avoidance System described in Chapter 4), military command and control and others. However, comprehensive evaluations involving experienced operators had not been conducted on any of these systems. The fact that evidence for the effectiveness of EID largely comes from novice operators on simulated process control systems means that the question of the effectiveness of EID designs for large-scale industrial systems with experienced operators has not been demonstrated.

The most recent evaluation of the contribution of EID was carried out by Neville Stanton and the Transportation Research Group at the University of Southampton in England.[14] An extensive search for relevant research articles produced 75 published examples of EID applications between 1993 and 2014. Half of these applications came from the three areas of aviation, medicine and power generation. Two major textbooks on EID have also been published.[15] Stanton's review does not cover the effectiveness of the EIDs compared with conventional designs but looks at how the ecological designs were developed. In many cases, there was a substantial deviation from the core theoretical foundations of combining a cognitive work analysis involving some form of abstraction hierarchy, with an analysis based on the SRK framework. In fact, 40% of the applications did not cite the SRK framework at all. As Stanton points out, this creates a problem – how can the value of the 'method' be evaluated if many of the applications do not actually follow the method at all?

Kevin Bennett, a psychology professor and head of the Cognitive Systems Engineering Laboratory at Wright State University in Dayton, Ohio, and a long-standing expert on EID, has provided the most recent review of the field.[16] Bennett emphasises the importance of using the abstraction hierarchy in a CWA to establish the core constraints inherent in the work tasks and using this information to design interfaces that make these constraints directly visible to the operator. As an example, he cites recent work on military command and control funded by the US Army. The key constraint that was uncovered by the CWA was the 'force ratio' which refers to the relative amount of combat power between the two forces. A specific display representation for this was designed and included in an overall EID to support the decision-making performance of land-based tactical operations controllers at skill, rule and knowledge-based levels of cognitive control.

There is no doubt that Rasmussen's creation of analytical tools for revealing the constraints that govern the working task itself as well as the constraints that arise from the human's cognitive abilities have resulted in a major shift in focus for display interface design. While traditional HF/E concerns (Chapter 4) remain important, Rasmussen's work provided the tools to design interfaces that can both take advantage of the human capacity to very quickly and accurately detect patterns in visual information but at the same time provide information at more abstract levels to facilitate problem-solving in unusual circumstances. The academic literature fails to reveal the extent to which scaled-up applications have been actually put into practice although Vicente mentions a prototype EID development for a German conventional power plant. Whether consciously based on EID principles or not, modern interface display design has certainly moved away from the 'single sensor, single meter' approach to providing a broader range of system information, often in more abstract, schematic form. Such displays are now routinely incorporated in modern aircraft flight decks and in process control. The idea of providing information in a more integrated form that relieves the operator of the need for detailed mental computation, instead of placing that information out in the world in a form that can be directly perceived, is wholly consistent with the principles of EID.

THE BOUNDARIES OF SAFE WORK

Whilst Rasmussen's earlier work was grounded in his involvement with the Danish nuclear power industry, and in particular, his close observations of troubleshooting and diagnostic problem-solving, his later work broadened out to a wide-ranging consideration of the nature of safety and risk management in all technological activities. The development of tools such as the abstraction hierarchy (AH) and cognitive control taxonomy (SRK) was essential to identifying the constraints on real-world work. To more broadly understand the nature of safety and the risks that arise in various activities, Rasmussen introduced two more conceptual tools. The first of these is a basic notion of systems theory which is that any complex system can be described as a hierarchy of levels each with its own properties.

The universe is composed of systems. We have solar systems, geological systems, biological systems and ecosystems. We are used to thinking about our bodies as a system of systems – our overall physiology is built up of respiratory, circulatory, digestive, reproductive and other bodily systems. Outside of our biology, we have social and cultural systems – we often refer to the socio-economic system for instance. The kind of systems that Rasmussen was concerned with and which are the subject of the present book are *sociotechnical systems*. Examples include transportation, healthcare, manufacturing and so forth. The general structure of such systems can be laid out along the lines shown in Figure 8.6.

Governments exert some control over safety through the legal system which governs the government agencies tasked with overseeing safety as well as other interested bodies such as worker unions. As we saw in Chapter 1, some countries have more extensive legal and regulatory frameworks than others. The requirements of the regulatory bodies are passed on to companies and organisations that make the decisions about resource allocation to meet these requirements. These decisions affect the day-to-day managerial processes of managing the business of the organisation and reconciling that with the equipment and personnel provided by the company. At the lower levels, the staff carry out the daily operations of the organisation in the context of the company's policies and procedures, disciplinary ethos, the standard of equipment, etc. Different disciplines traditionally focus on just one or two of these levels. Management and legal scholars focus on the top levels, organisational theorists on the middle levels, whilst human factors/ergonomics researchers typically focus on the lower levels. The aim of systems science is to study the interrelationships and connections between all the levels. In this sense, systems science can be considered a *metascience*.[17]

As Rasmussen notes: 'The usual approach to modelling sociotechnical systems is by decomposition into elements that are modelled separately. This has some peculiar effects'.[18] In other words, this follows the traditional reductionist approach to science where the complex is separated into its constituent elements which are then studied separately. The 'peculiar effects' Rasmussen is referring to come from the tendency

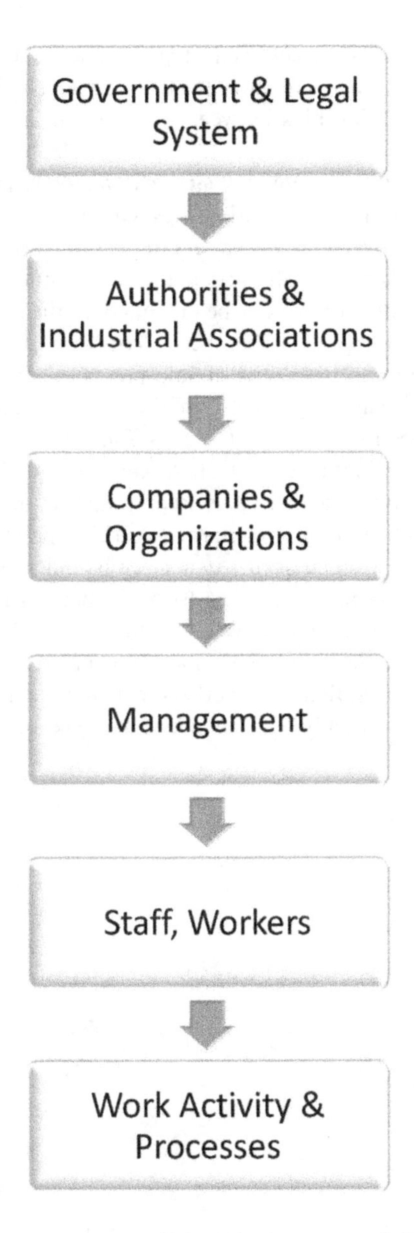

FIGURE 8.6 An outline of the general hierarchical structure of all sociotechnical systems.

to view each of the levels in Figure 8.6 separately without reference to the lower levels. So, management studies compare across organisations without reference to the process being managed. This has led to the trend for management to be considered a profession independent of what it is that is being managed. So, managers switch between managing a factory and managing a hospital for example. Not everyone believes this is a good thing. Rasmussen suggests that we need to study each level of the hierarchical structure in relation to the other levels. What matters are the interconnections and relationships between the levels in a healthcare system or a transportation system, etc.

Rasmussen proposes the concept of *safety boundaries* as a means of representing some of these constraints and interrelationships. Organisations are continually seeking efficiencies and improvements which require working around existing, normally accepted practices. Workers naturally seek to minimise unnecessary effort and avoid excessive workload. Ordinary work can be thought of as the exercise of some choices that achieve the operational goals within these constraints. At some point, this brings with it 'the risk of crossing the limits of safe practices'. Examples abound in the disaster literature. For example, various small changes in practices chipped away at the safety boundaries for the *Herald of Free Enterprise* (Chapter 2) and for the operators of the Number Four reactor at Chernobyl (Chapter 5).

As usual, Rasmussen provides a visual representation of these boundaries (see Figure 8.7) to represent the reality of normal working practices in sociotechnical systems where 'normal' practice occurs in a space bounded by these economic and working constraints. Continual pressures to work more efficiently and to reduce costs result in constant pressure to move closer to the boundary of safe and acceptable performance, as does the need to economise on effort and workload. In a well-run sensibly designed operation, the workers will remain far enough away from these boundaries to comfortably avoid unnecessary risk and damage. Under other

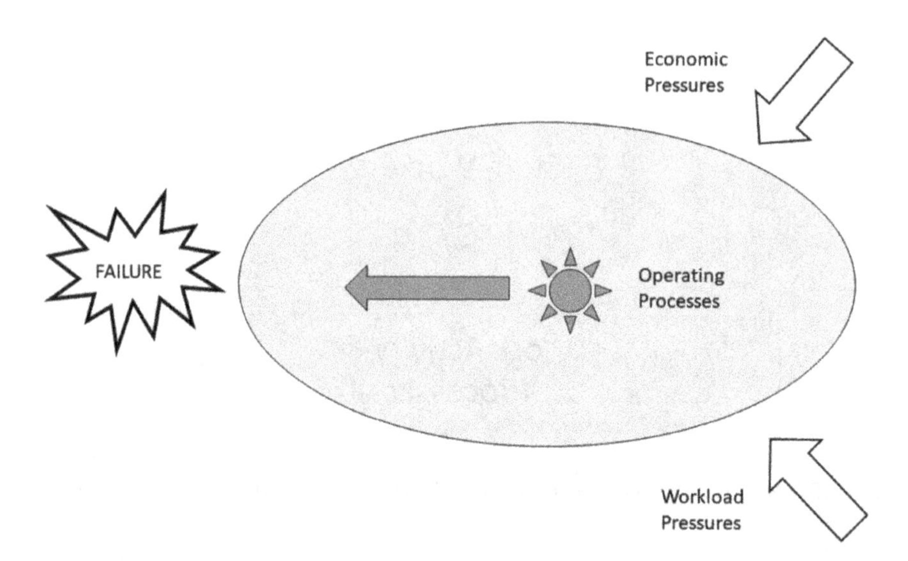

FIGURE 8.7 The boundaries of safe work. Work is conceptualised as operating within a space constantly squeezed by economic and workload, pressures. Work practices vary dynamically in response to these pressures. The closer work practice comes to the performance boundary, the greater the risk of ultimate failure and potential disaster. Knowing where these boundaries are located forms an essential part of risk management.

conditions of excessive commercial pressures, for example, work can be brought close to, and occasionally across, the boundary of safe performance. In Rasmussen's view, numerous accident reports show that far from being due to independent 'human errors', accidents are generally attributable to a 'systematic migration of organisational behaviour'.

CLAPHAM JUNCTION RAIL CRASH, 1988

'On the morning of 12th Dec 1988 at 0810 a crowded passenger train crashed into the rear of another train that had stopped at a signal just south of Clapham Junction railway station in London, and subsequently sideswiped an empty train travelling in the opposite direction. A total of 35 people died in the collision while 484 were injured.

The collision was the result of a signal failure caused by a wiring fault. New wiring had been installed but the old wiring had been left in place and not adequately secured. An independent inquiry chaired by Anthony Holden, QC found that the signalling technician responsible had not been told that his working practices were wrong and his work had not been inspected by an independent person. British Rail was fined £ 250,000 ($ 350,000) for violations of health and safety law in connection with the accident'.[19]

The Clapham Junction railway disaster provides a good illustration of Rasmussen's ideas about the tendency of working practices to 'drift' as workers adapt to the practical (i.e. economic and workload) constraints of the tasks they are asked to perform. Workers are inevitably under pressure to do their work as economically and efficiently as possible and thus there is a continual process at work of varying work practices to satisfy these goals. Rasmussen explains how this led to the signal failure in the Clapham Junction case:

> safety checks following modifications of signal system wiring were planned to be independently performed by three different persons, the technician, his supervisor, and the system engineer. Work force constraints and tight work schedules, however, led to a more 'efficient' division of work. The supervisor took part in the actual physical work and the independent check by him as well as by the engineer was abandoned. In addition the technician integrated the check (i.e. a 'wire count') into the modification task although it was intended to be his final separate check.[20]

Diane Vaughan's analysis of the *Challenger* disaster at NASA, which was discussed in detail in Chapter 7, provides another illustration of these processes at work in complex sociotechnical systems.

Rasmussen's dynamic theory of risk and the genesis of accidents was the first comprehensive attempt to formulate a truly systems-based approach to safety. Previous work in safety science had followed the traditional reductionist scientific approach of breaking down complex systems into elements, such as the individual or the organisation, and developing analyses and theories focusing on that one element.

Much useful information has been generated by this approach but the larger question of how safety can be managed in large-scale complex systems cannot be adequately dealt with in this way. Rasmussen's 'safety boundary' metaphor illustrates another fundamental notion of systems science which is that phenomena can dynamically 'emerge' from their background conditions rather than being directly 'caused' by a deterministic sequence of preceding events.

Biology provides numerous examples of 'emergence' from the emergence of life itself out of inorganic compounds (abiogenesis), to the emergence of language and consciousness in some forms of life (and potentially in artificial forms of life in the not-so-distant future).[21] Systems inherently tend to become more complex. Simple inorganic molecules when stimulated by appropriate amounts of energy can become more complex molecules, such as amino acids, the building blocks of life. Single-cell organisms become multi-celled organisms and so on. Sociotechnical systems of work and transport work in the same way as other biological and cultural systems. The network of connections across different levels of their hierarchical structure tends towards greater levels of complexity as new, and often unexpected, connections form.

These are the systems properties that lead to the emergence of unexpected properties such as accidents. Rasmussen's insights provide the building blocks for better understanding systems failures. Traditionally these have been seen as the unexpected results of isolated individual technical or human failures. Safeguards, defences, training and motivation have been the traditional, invariably ineffective, approaches to dealing with this problem. Rasmussen has pointed out that systems can find numerous ways of carrying out any particular function, growing in complexity as they do so. Workers' flexibility in responding to the dynamic constraints (e.g. economy, performance, etc.) is essential to system functioning but at the same time can push the boundaries of safe performance. Accidents and disasters occasionally emerge as a consequence of natural work practices in complex systems. Unsurprisingly, Rasmussen later developed a representational tool known as 'AcciMap' for depicting and analysing the vertical interrelationships between the levels of a sociotechnical system as illustrated in Figure 8.6. This tool has become widely used in accident analysis and investigation and will be described in more detail in Chapter 11.

MAD COWS AND ENGLISHMEN

THE UK BSE OUTBREAK

From the mid-1980s, cattle in the UK became infected with bovine spongiform encephalopathy (BSE) known as 'mad cow disease'. BSE is a neurodegenerative disease related to scrapie found in sheep. In 1994 the first fatal case occurred of Creutzfeldt-Jakob Disease (CJD) caused by eating BSE infected beef. Over four million cattle were slaughtered in an effort to eliminate the disease and it is estimated that 177 people died of CJD as a result of becoming infected. The subsequent parliamentary inquiry concluded that the epidemic developed as a result of recycling animal protein in the food fed to cattle. The inquiry added: 'At the heart of the BSE story lie questions of how to handle hazard – a known hazard to animals and an unknown hazard to humans'.[22]

Kim Vicente and colleagues have put Rasmussen's risk management framework to use in analysing several public health outbreaks including the BSE outbreak in the UK.[23] Several predictions were derived from the structural mapping of the different levels of the health management system as shown on an AcciMap version of Figure 8.6 and from the dynamic model of safety boundaries shown in Figure 8.7. For example, the models predict that accidents and safety threats result from 'a lack of vertical integration (i.e. mismatches) across levels of a complex sociotechnical system'. In accordance with Rasmussen's concepts of migration of work practices towards the safety boundaries, it was also predicted that evidence of this would be found in the BSE outbreak. Data to test these predictions were obtained from the detailed (16 volume, 5,000 pages) BSE inquiry report and several other sources.

Described as a 'qualitative case study methodology', Vicente suggests that the evidence scrutinised from these sources, and independently reviewed by several other experts, is strongly consistent with the predictions derived from the Rasmussen framework. For example, there was a clear lack of coordination and feedback between the actors and agencies at the different levels of the system rendering the government agency responsible (MAFF) relatively ineffective. There was considerable evidence of economic pressures driving work practices at different levels of the system from the regulatory agencies to the local abattoirs and feed mills. The absence of feedback loops within and between the various levels of the system was clearly evident here and in many other cases of systems failure. The importance of feedback in safety is taken up in detail in Chapter 9.

Whilst one of the first to systematically explore the usefulness of Rasmussen's framework and its associated tools (e.g. structural mapping/AcciMap), there have been many other examples appearing over the past decade. Some of the accident investigation and analysis examples will be discussed in Chapter 11. Undoubtedly, the overall impact of the suite of systems analysis tools provided by Rasmussen from the abstraction hierarchy and SRK taxonomy onwards on safety science has been significant.

'WORKAROUNDS' IN HEALTHCARE

Rasmussen has had an enormous influence on the development of safety science and his articles have been widely cited by other researchers and writers and his influence charted in several books and special editions of journals.[24] The SRK framework, whilst not original in describing the qualitative differences that occur in human cognitive control as a function of task familiarity and experience, has provided the theoretical background to numerous studies of operator performance, training design, human error analysis and display design for nearly 40 years. Recent examples have used the framework to describe and analyse errors in aircraft maintenance in Australia, adverse events in healthcare in the Netherlands and the United States, and the performance of nuclear power plant controllers in Taiwan. Researchers in the United States used the framework to guide training programme development for aircrews to make appropriate use of flight automation.[25]

Rasmussen's risk management framework simultaneously encompassing both the static hierarchical organisation of systems and the dynamics of worker performance within those systems has also been highly influential. The healthcare system provides a good example of a multilevel system engaged in high-risk operations under ever-changing conditions. The previously unsuspected levels of unintended harm associated with healthcare delivery have been discussed in Chapter 6. **David Gaba**, a Professor of Anaesthesiology at Stanford University, has written extensively about patient safety in healthcare with particular reference to organisational perspectives such as normal accident theory (NAT), high-reliability organisations (HROs) and normalisation of deviance discussed in Section 2 of this book.[26] Gaba makes the point that 'systems thinking' has yet to take hold in the medical community, mired as it is in an emphasis on individual performance and a culture of blame.

Another American healthcare expert **Richard Cook** collaborated with Rasmussen on a paper describing the application of systems thinking in general and Rasmussen's risk management framework in particular to healthcare.[27] They suggest that whilst healthcare, and specifically hospitals, is inherently 'loosely coupled' (see Chapter 5), the constant drive to greater operational efficiency (especially in an entirely profit-driven healthcare system such as the United States) can lead to tighter coupling becoming a regular occurrence as, for example, bed occupancy is maximised. This can lead to severe bottlenecks – for example, an elective surgery gets cancelled when the expected ICU (intensive care) bed is needed for another patient whose condition worsens. This has effects throughout the system – not only directly on patients but on surgical and nursing rosters and the like.

The effects of this can be conceptualised in terms of Rasmussen's safety boundaries. The 'squeeze' at the economic efficiency boundary effectively pushes worker performance closer to the boundary of safe operations. They also point out that certain high-risk operations (such as the US Naval aircraft carriers discussed in Chapter 6) operate stably at the very boundary of safe operations by containing performance deviations to the absolute minimum. Once chaos envelops a system (as when an out-of-control pandemic occurs), then in their natural desire to operate effectively, workers' performance can start to vary widely, tipping the system across the edge.

A key concern in this discussion of metaphorical boundaries is how the location of the boundaries is made apparent to workers and management and also how variations in performance become known in real-time or shortly thereafter. Rather than focus on 'safety culture' and the like, Cook and Rasmussen suggest reconceptualising the debate in terms of the dynamic safety model. In other words – how to establish where the safety boundary is which means effectively establishing where current performance is located. Much more research is still needed to move Rasmussen's model from purely conceptual to one that can be fully operationalised in practice.

There is good evidence on one point – the adaptability and mobility of worker performance to systemic pressures resulting from the drive for greater economy and efficiency and the need to manage workload, often in the face of dwindling resources. Healthcare again provides good examples. In response to the original revelations by Lucian Leape and others of the unexpectedly widespread adverse effects of healthcare, which showed that drug-dispensing errors were one of the most prevalent areas of concern, technical innovations such *as barcoded medication administration systems* (BCMA) have since been introduced. These systems involve scanning a barcode worn by the patient as well as scanning barcodes on the medication. The results are matched with *electronic medical administration records* (eMARs). In theory, such a process should ensure that the right patient gets the right dose of the right medication at the right time.

In theory. In practice, hospital personnel have been found to operate a variety of 'workarounds' to administer the process. A study of five hospitals in the United States involving over 300,000 medication administrations found more than 30,000 cases where aspects of the system were 'overridden' in some way.[28] These 'workarounds' or violations of procedures come into being as a result of the pressures on drug administration staff to get the job done in the time required. One set of problems were due to the design of the trolleys used to move around the medications and the computerised devices. Referred to as COWs (computers-on-wheels) these devices can be too large to enter patient rooms or else require tethering to electrical outlets in hallways. These system design problems make it difficult or impossible to administer the medications according to protocols. For example, the procedure involves scanning user, medication and patient barcodes (in that order) and then documenting administration. If the COW can't be positioned correctly, the nurse then takes the medication into the room and scans the patient's barcode but may be unable to see any warning displayed on the device indicating wrong patient delivery, for example. Alternatively, although the medication has been scanned (indicating delivery), the patient may not ingest it for some reason.

Altogether, these 'workarounds' were found to occur in approximately 10% of all medication administrations. It is much too simplistic to view them simply as 'violations' since they may very well have positive as well as negative effects on patient safety. In fact, as the study authors point out, some of them may 'have become institutionalised as "best practices"'. They arise from the pressures to accomplish complex tasks in the required amount of time in the face of numerous technological (e.g. COW design, software problems, etc.), organisational (e.g. missing barcodes on patients and medications) and other (e.g. lack of wireless connectivity, location of medications) constraints. One of the most ironic examples is 'medication barcodes covered by a label reminding users to scan barcodes'. This healthcare example provides another good illustration of how the dynamics of workplace pressures can result in work practices migrating towards the safety boundary described in Rasmussen's dynamic model of risk.

SUMMARY AND CONCLUSIONS

For more than four decades, Danish electrical engineer Jens Rasmussen's work has laid the foundations for a true systems approach to safety and inspired the work of a significant number of researchers and academics. Rasmussen's investigations and empirical studies were always based on actual work as done in practice and not on stripped-down experimental analogues. This reflected a conceptual approach that sought to deal with systems as a whole rather than selected subsystems piece by piece. Systems as a whole have certain properties that are not contained in individual subsystems. As mentioned previously, hierarchical interconnectedness arises from the individual componentry but exists on a systems level. As a result, various properties are *emergent* from these relationships. For Rasmussen, safety is one of these emergent properties. This means that safety cannot be fully understood at a subsystem level – for example, neither as the outcome of individual risk-taking or human error nor as the property of particular organisational structures or processes. Safety *emerges* from the interconnections and information flows between these various components.

Safety Science ultimately becomes a science of safety in sociotechnical systems. Whilst understanding safety from both an individual and an organisational perspective remains essential, further advances in understanding involve tackling safety directly at a systems level. Whilst this may seem to have the most direct practical benefits when analysing large-scale disasters, all work and travel are socially constructed activities that take place within elaborate systems of social relationships involving governments, regulators, organisations and workforces in combination with hardware and software engineering practices that develop, build, operate and maintain the technologies involved.

The complex interactions between these systems components primarily involve the exchange of information and control. The next steps in safety science, therefore, involve further analysis and understanding of the nature of information and control, and the part these play in managing safety. Chapter 9 introduces cybernetics, information theory and the development of a systems safety approach built directly upon some of the key ideas and concepts developed by Rasmussen.

NOTES

1. This statement is in the Preface of an early book: *Information Processing and Human-Machine Interaction* published in 1986 in New York by Elsevier. A few years earlier, Rasmussen published one of his most highly cited papers: Rasmussen, J. (1983). Skills, rules, and knowledge: Signals signs and symbols and other distinctions in human performance models. *IEEE Transactions on Systems, Man and Cybernetics, SMC-13*(3), 257–266.
2. See Fitts, P.M., & Posner, M.I. (1967). *Human Performance*. Belmont, CA: Brooks/Cole.
3. Wears, R.L. (2017). Rasmussen number greater than one. *Applied Ergonomics, 50*, 592–597.
4. A clear outline of the decision ladder can be found in Chapter Two of: Rasmussen, J. (1986). *Information Processing and Human-Machine Interaction: An Approach to Cognitive Engineering*. New York: Elsevier. Various modifications to the original ladder have been proposed. For example: Lintern, G. (2010). A comparison of the decision ladder and the recognition-primed decision model. *Journal of Cognitive Engineering and Decision Making, 4*(4), 304–327.

5. O'Hare, D., Wiggins, M., Batt, R., & Morrison, D. (1994). Cognitive failure analysis for aircraft accident investigation. *Ergonomics, 37*(11), 1855–1869.
6. Banks, V.A., Plant, K.L., & Stanton, N.A. (2020). Leaps and shunts: Designing pilot decision aids on the flight deck using Rasmussen's ladders. In R. Charles & D. Golightly (Eds.), *Contemporary Ergonomics and Human Factors.* Boca Raton, FL: CRC Press.
7. Vicente, K.J. (1999). *Cognitive Work Analysis.* Mahwah, NJ: Lawrence Erlbaum.
8. Stanton, N.A., Salmon, P.M., Walker, G.H., & Jenkins, D.P. (Eds.). (2018). *Cognitive Work Analysis: Applications, Extensions and Future Directions.* Boca Raton, FL: CRC Press.
9. Vicente, K.J., & Rasmussen, J. (1992). Ecological interface design: Theoretical foundations. *IEEE Transactions on Systems, Man, and Cybernetics, 22*(4), 589–606.
10. Kahneman, D. (2011). *Thinking, Fast and Slow.* London: Allen Lane.
11. Gary Klein was one of the first researchers to draw attention to the tendency for expert decision makers to rely primarily on perceptual-recognitional processes rather than deep, analytical ones. See: Klein, G.A. (1989). Recognition primed decision making. *Advances in Man-Machine Systems Research, 5*, 47–92.
12. Pawlak, W.S., & Vicente, K.J. (1996). Inducing effective operator control through ecological interface design. *International Journal of Human-Computer Studies, 44*, 653–688.
13. Vicente, K.J. (2002). Ecological interface design: Progress and challenges. *Human Factors, 44*(1), 62–78.
14. McIlroy, R.C., & Stanton, N.A. (2015). Ecological interface design two decades on: Whatever happened to the SRK taxonomy*? IEEE Transactions on Human-Machine Systems, 45*(2), 145–163.
15. Burns, C.M., & Hajdukiewicz, J.R. (2004). *Ecological Interface Design.* Boca Raton, FL: CRC Press. The more recent book is: Bennett, K.B., & Flach, J.M. (2011). *Display and Interface Design.* Boca Raton, FL: CRC Press.
16. Bennett, K. (2017). Ecological interface design and system safety: One facet of Rasmussen's legacy. *Applied Ergonomics, 59*, 625–636.
17. An excellent recent guide to systems theory can be found in: Mobus, G.E., & Kalton, M.C. (2015). *Principles of Systems Science.* New York: Springer.
18. Rasmussen, J. (1997). Risk management in a dynamic society: A modelling problem. *Safety Science, 27*, 183–213.
19. https://en.wikipedia.org/wiki/Clapham_Junction_rail_crash. The report of the official inquiry can be downloaded from: https://www.railwaysarchive.co.uk/docsummary.php?docID=36
20. Rasmussen, J. (1994). Risk management, adaptation, and design for safety. In Sahlin, N.E., & Brehmer, B. (Eds.), *Future Risks and Risk Management.* Dordrecht: Springer Netherlands.
21. Macklem, P.T. (2008). Emergent phenomena and the secrets of life. *Journal of Applied Physiology, 104*, 1844–1846.
22. Report of the Parliamentary Inquiry into BSE. Downloaded from the UK national archives at nationalarchives.gov.uk.
23. Cassano-Piche, A.L., Vicente, K.J., & Jamieson, G.A. (2009). A test of Rasmussen's risk management framework in the food safety domain: BSE in the UK. *Theoretical Issues in Ergonomics Science, 10*(4), 283–304.
24. A specially edited collection of articles to celebrate Rasmussen's 60th birthday was published in 1988 by Goodstein, L.P., Anderson, H.B., & Olsen, S.E. (1988). *Tasks, Errors, and Mental Models.* London: Taylor & Francis. A whole issue (Vol 59, Part B) of the journal *Applied Ergonomics* was dedicated to articles on Rasmussen's work in 2017.

25. Some of the many recent examples of work in the Rasmussen tradition include: Hobbs, A., & Williamson, A. (2002). Skills, rules and knowledge in aircraft maintenance: Errors in context. *Ergonomics, 45*, 290–308; Smits, M., et al. (2010). Exploring the causes of adverse events in hospitals and potential prevention strategies. *BMJ Quality and Safety, 19*(5), 1–7; Lin, C.J., Shiang, W.-J., Chuang, C.-Y., & Liou, J.-L. (2014). Applying the skill-rule-knowledge framework to understanding operators' behaviours and workload in advanced main control rooms. *Nuclear Engineering and Design, 270*, 176–184.
26. Gaba, D.M. (2000). Structural and organizational issues in patient safety. *California Management Review, 43*(1), 83–102.
27. Cook, R., & Rasmussen, J. (2005). 'Going solid': A model of system dynamics and consequences for patient safety. *BMJ: Quality and Safety in Healthcare, 14*(2), 130–134.
28. Koppel, R., Wetterneck, T., Telles, J.L., & Karsh, B.-T. (2008). Workarounds to barcode medication administration systems: Their occurrences, causes, and threats to patient safety. *Journal of the American Informatics Association, 15*, 408–423.

9 The Cybernetics of Safety
Information and Control

It would be difficult to find a geographical place-name more strongly associated with academic and intellectual excellence than that of Cambridge. The University of Cambridge founded in 1209 and based in the English town of the same name not only is nearly a thousand years old but has produced more Nobel Prize winners than any other institution on earth. The almost identically sized town of Cambridge in Massachusetts, just outside Boston, also has one of the top-ranked universities in the world. Not quite as ancient, at just over 150 years old, the Massachusetts Institute of Technology (MIT) based in Cambridge, MA, has produced nearly as many Nobel laureates (96 versus 121) as the more venerable Cambridge across the Atlantic.[1] Although a relatively small institution, MIT, specialising in science and technology, has been associated with the careers of almost every one of the major figures in the fields of information, control, cybernetics and their modern applications to safety science that populate this chapter.

FEEDBACK AND CONTROL

For many centuries of warfare, the principal concern was to ensure that weapons (arrows, artillery shells, etc.) found their target. Ensuring artillery shells land where they are supposed to was, at one time, largely a matter of trial-and-error. After firing some shells, observation of the results, either by the gun operators themselves or by external observers (e.g. pilots flying above the area) enabled the operators' to make adjustments to the gun's elevation and horizontal positioning to bring the shells incrementally closer to their target. The same processes apply in non-military examples of projectile aiming such as throwing darts at a dartboard or firing arrows at an archery target. It is important to note that the maxim 'practice makes perfect' omits one vital ingredient – feedback. A blindfolded dart player or artillery gunner will not improve no matter how many darts or shells are discharged. Feedback in the form of information concerning the outcome of each attempt is vital to the development of any skill or ability.[2]

More specifically it is information about the difference between the desired state (e.g. hitting the bullseye) and the actual state (e.g. dart landed 2 cm to the right of the bullseye) that enables the appropriate adjustment (e.g. throw slightly to the left) to be made. Information about both the target or desired state and the obtained or current state are needed to provide an accurate estimate of the difference that can then be used to modify future actions. This basic notion of feedback is shown in Figure 9.1. Many mechanical and biological processes are guided by feedback loops of this kind. The most often-quoted example of the former is a household thermostat

DOI: 10.1201/9781003038443-12

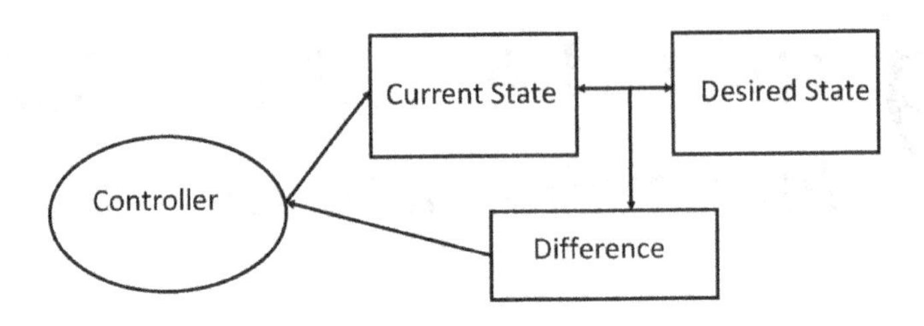

FIGURE 9.1 The process of feedback whereby the difference between the current and desired state is used to guide future action.

used to control a heating source, such as a boiler or furnace. The desired temperature state is set by the user, and the sensors measure the actual temperature state and the difference determines the actions of the thermostat – if the difference is positive then instructions are sent to an actuator to shut down the boiler and if negative, to turn up the boiler.

Similar processes are widely used by living organisms to keep internal conditions such as body temperature and blood alkalinity within very narrow ranges. These are known as 'homeostatic' mechanisms. Such a mechanism has also been proposed to account for human responsiveness to risks, as described in Chapter 2. Homeostatic mechanisms are perfect examples of the general principle of *feedback control* which is at the core of what was named *cybernetics* by **Norbert Wiener** in 1947. Wiener was a child prodigy who graduated from Tufts University at the age of 14 and received his PhD from Harvard at the age of 19! Unable to join the military in the First World War because of poor eyesight, Wiener somewhat ironically found himself working instead in the area of ballistics. After the war, Wiener was hired by MIT and spent his career there as a professor of mathematics. With the onset of the Second World War, Wiener again became involved with ballistics, particularly that related to anti-aircraft artillery. Unlike the simple artillery problems mentioned above, this problem becomes much more complex as the target is rapidly moving, so simply aiming at the target will be useless as the target will be somewhere else by the time the projectile arrives! Working on these problems led Wiener to focus on the issues of human-machine systems and particularly on the role of feedback in the control of complex systems.

Wiener decided to unite the diverse literature on biological organisms with those on the control in mechanical systems under a single label – *cybernetics* from the Greek word for 'steersman' as in the person who controls the direction of a ship. Cybernetics was defined by Wiener as 'the science of control and communication in the animal and the machine'.[3] The examples mentioned so far illustrate *negative feedback* which reduces the variability in control output thus maintaining stability. Whenever room temperature or blood sugar levels start to vary from the set level then output signals are sent to the boiler, in the first case, or pancreas in the second, to initiate actions to stabilise the system back to its set point (e.g. the pancreas releases either insulin to lower blood sugar or glycogen which acts on the liver to release glucose to elevate blood sugar since you ask).

Positive feedback in contrast amplifies the variability of the system's output. To take another biological example, platelets are produced by the blood in response to bleeding. The appearance of platelets stimulates more platelets to appear thus amplifying the original action. Various other physiological processes operate along similar lines. All complex systems – biological, mechanical or social involve some system of control based on feedback. Such systems are referred to as *closed-loop* systems whereby feedback from the difference between desired and actual state is delivered to a controller, which may be biological, mechanical or human, that then issues instructions that are designed to alter the current state of the system generating further feedback *ad infinitum*. Organisations attempt to influence employees' behaviour by disseminating policies and procedures and then monitoring their effects, making subsequent adjustments and so forth. In this way 'Closed loop control is the basic form of operational control. It is found in every complex dynamic system'.[4] Cybernetics was based on the fact that this process of control is common to all systems from the biological to the organisational.

Pathologies of control: Given this basic structure of control there are several ways in which things can go wrong. On the control side, commands might be too strong, too weak, mistimed or missing altogether. Numerous physiological examples illustrate this from an over-active thyroid to an under-active pancreas. In the organisational realm, lack of policy or strategic guidance, or ambiguous or misleading management actions can result in failures to influence the system as intended. On the feedback side, lack of measurement, or inaccurate measurement of the current state of the system, missing or poor channels of communication, or ambiguous signals can all result in missing or degraded feedback with consequent effects on the controller's ability to manage the system.

CRASH OF AIR NEW ZEALAND DC-10, MOUNT EREBUS, ANTARCTICA, 28 NOVEMBER 1979

'On 28 Nov 1979 an Air New Zealand DC-10 airliner crashed into the northern flanks of Mt Erebus while on a scenic flight over Antarctica. All 257 people aboard died instantly'. The official air accident investigation report found that the crash was due to the Captain's decision to continue 'the flight at low level toward an area of poor surface and horizon definition when the crew were not certain of their position'. A subsequent Royal Commission of Inquiry beginning in mid-1980 reached a rather different conclusion:

> The single effective cause of the crash was the act of personnel in the (airline's) Flight Operations Division in altering the latitude and longitude of the destination waypoint without the knowledge of the aircrew, and in omitting to notify the aircrew … of the fact that an alteration had been made.

The government regulator, the Civil Aviation Division (CAD) was also accountable for failures of oversight.[5]

As systems become more complex so control structures become more hierarchical. Every complex system has lower levels of control intimately related to the system's operation. For example, stock management and personnel scheduling systems are closely related to the day-to-day operations of many businesses and organisations. Above this immediate operational level, in all but the smallest organisations are levels concerned with more logistical and tactical matters such as payroll and human resources, and above that upper management, concerned primarily with strategic and long-term goals. In addition to these inner layers, within an organisation are yet more levels of regulatory and governmental control, each with its own upper and lower levels.

The case of New Zealand's worst-ever air disaster in Antarctica provides an illustration of multiple failures amongst and between a number of these levels of control. Subsequent investigations revealed an extensive list of control action failures. Those between the regulator (the CAD which was a sub-division of the Ministry of Transport) and the airline included waiving regulatory requirements for the Captain of a flight to have flown the route previously and the lack of any regulatory requirements for specialised training for crews operating in polar regions where unusual atmospheric and meteorological conditions can occur. The operating crew were misled by one such phenomenon known as 'sector whiteout' whereby reflected light from underneath an overcast sky combined with the bland white texture of the ground to remove the normal textural clues that would indicate the gradient of the land. In this case, under exactly such conditions, the slopes of Mt Erebus ahead became effectively invisible to the flight deck crew. This phenomenon and its safety implications were known to military crews who operated in polar regions, but the CAD neither utilised this experience nor required knowledge of it for Air New Zealand crews making their maiden flights to the polar ice.

The airline's flight planning department mistakenly shifted the destination waypoint from overhead Mt Erebus to the middle of the ice shelf and back again without making this information known to the pilots and crew. The audio-visual briefing that took place for the crew showed a flight tracking down the ice shelf of McMurdo Sound. The airline also produced publications disseminated to passengers and to the general public showing the route over the ice shelf and describing 'low-level' flight around Mt Erebus when the airline's Flight Operations Division mandated minimum height restrictions that would not have allowed this to occur. Feedback from crew and passengers of their experiences on earlier flights appear not to have even been collected much less attended to.

If control of complex systems depends on feedback cycles of the type outlined earlier, then the opportunities for things to go wrong in complex sociotechnical systems are legion. The tragic Air New Zealand disaster illustrates this all too clearly. The concepts of hierarchical control systems and feedback cycles are at the core of a systems science approach to understanding safety. A more comprehensive perspective on the systems approach to safety (developed by another MIT scientist) will be outlined in subsequent sections. Before doing so we should consider another key concept in systems science – that of information.

INFORMATION: THE DIFFERENCE THAT MAKES A DIFFERENCE

Claude Shannon was a mid-Westerner who undertook his graduate (Masters and PhD) studies at MIT. In 1940 as the United States became involved in the War, Shannon began work at Bell Laboratories, an important centre of research and innovation in advanced engineering and telecommunications. Part of his work was in the area of cryptography, which played a vital role in the conduct of the war. This work brought him into contact with **Alan Turing**, the famous British mathematician whose work was pivotal to the decoding of the German 'Enigma' machines.

However, it was also his work with Norbert Wiener on anti-aircraft artillery that led Shannon to formulate his 'mathematical theory of communication' first published in 1948. Along with Wiener's emphasis on feedback loops, Shannon's conceptualisation of information provided two of the core concepts of cybernetics and thereby of modern safety science.

When we think about communication it seems natural to focus on the content of the communication and the means by which it is to be communicated, i.e. the medium and the message. Numerous rules and guidelines can be found for preparing and structuring communication and selecting the appropriate media. Despite this, failures of communication are all-too-frequent. To take a recent example, in May 2020, the UK government chose to send out a new Covid-19 communication (see Figure 9.2) to the British public which replaced the previous message of 'Stay Home, Protect the NHS, Save Lives'. In cybernetic terms, the messaging was a control action

FIGURE 9.2 The UK government's new Covid-19 messaging introduced in May 2020.

designed to engender the appropriate actions (e.g. staying home, social distancing, hand washing, etc.) by members of the British public. Instead of being informative, the new communication engendered much confusion due to the vagueness and ambiguity of the message.[6] The new message actually increased uncertainty about what to do and not to do. Many health and safety messages suffer from similar problems (e.g. as in the standard warnings to 'Be Careful') and are similarly ineffective in controlling safety in the manner intended.

Traditional definitions of 'information' have focused on the content of a message itself. Information is usually equated with knowledge or facts. Recent world affairs have brought into question just what does count as information with terms such as 'misinformation' and 'disinformation' in widespread use. Shannon's insight was the realisation that the informativeness of a message actually lay, not in what it contained or added but rather in its ability to reduce uncertainty in the mind of the recipient.[7] Uncertainty depends on the number of possible alternatives and the likelihood of their occurrence. In the simplest case, such as tossing a fair coin, there are only two equally likely possibilities (Head or Tail). A communication that 'Tails' occurred eliminates any uncertainty and is said to provide one 'bit' of information. If there were four equally likely outcomes then a message that identifies which one of them has occurred provides a greater reduction in uncertainty than in the coin toss and contains two bits of information and so on. This ability to precisely quantify information has had significant impacts on many areas from telecommunications to human performance.

Gregory Bateson, a British anthropologist educated (need you ask) at Cambridge University in the UK, later moved to the United States, where he became famous for several things including marrying the extremely famous anthropologist Margaret Mead as well as proposing the *double-bind* theory of communication. Made famous by 'Catch-22', a double-bind is a paradoxical communication where one part of the communication is in direct contradiction to another part. In the novel 'Catch-22', the double-bind is that to be eligible for discharge from active service you have to be insane but to want to be discharged proves that you must, in fact, be sane. Bateson was part of the group that debated and developed cybernetic ideas and applied the ideas of homeostasis and feedback loops to ideas in anthropology. He coined the memorable definition of information as 'a difference which makes a difference'.

Communication can serve many functions in organisations from the aspirational to the motivational to the informational. Any communication that is intended to influence how the system functions in some way through peoples' behaviour must be informative. As we have seen, this means that it must reduce uncertainty about what to do to make a difference. In the UK government's Covid-19 messaging, 'Stay Home' has informational value as it directly answers the question 'can I go out or not?' In contrast, their subsequent messages such as 'Stay Alert' contain no information at all, as people's prior uncertainty regarding working, shopping, socialising and other matters remains exactly as it was before receiving the message. Looking at public health and safety campaigns in particular, as well as many examples of organisational messaging in general, through the lens of information theory provides a useful tool to assess whether organisations are supplying the necessary information to control the system safely and effectively. As we saw in the Air New Zealand disaster, mistimed, ambiguous or missing control actions, and missing, incomplete or ineffective feedback are frequently found when systems fail to function safely. This has led to a comprehensive new theory of safety based on some of the principles of system science that we have described so far, namely hierarchical organisation, information and closed-loop feedback control.

THE STAMP MODEL OF SYSTEM SAFETY

Nancy Leveson is Professor of Aeronautics and Astronautics and Engineering Systems at MIT in Cambridge, MA. Trained in computer science and mathematics, she initially focused on systems engineering and software safety. For the past couple of decades, she has broadened her focus and published extensively on a systems science approach to safety in a wide variety of contexts. She has developed a conceptual approach to system safety that can be used both proactively to design safer systems and retroactively to investigate system failures.

Reliability ≠ Safety: It might seem fairly self-evident that the more reliably elements in a system perform then the safer that system must be. So pervasive is this belief that Leveson starts her book about systems thinking and safety by addressing this very issue.[8] When designing technological systems engineers are very much concerned with the reliability of the components they are installing. Reliability refers to the probability that the component will perform as expected and this is sometimes measured as the *mean time between failures* (MTBF). Where the components are safety-critical, such as in the case of aircraft flight-control computers, then systems depend on redundancy to ensure that the probability of a single component failure leading to an adverse outcome is extremely low.

In simple mechanical or electronic systems, a component failure can lead to disaster as when a leak in a vehicle's hydraulic system leads to fluid loss and subsequent brake failure. However, as systems have grown more complex, particularly with the growth in software control of systems from cars to spacecraft, so the possibility of failures arising from the interactions between components becomes more important. Leveson provides a number of examples from the control of a chemical process plant to the loss of NASA's Mars Polar Lander in 1999. Whilst descending towards the Martian surface, the deployment of the three landing legs generated signals to on-board sensors that the software interpreted as indicating that the lander was on the surface, and not 40 m above it as was, in fact, the case. The software was designed to cut off the engine thrust in response to a touchdown indication and this sudden cessation of thrust resulted in the lander impacting heavily from an altitude of 40 m above the surface. Leveson argues that the landing legs, sensors and software all performed as designed and were, therefore, reliable, but the operation was not safe. She describes this as an example of a *component interaction accident* where the individual components performed reliably but not safely.

Healthcare provides many examples of unexpected component interaction events. The response to the revelation of significant effects of 'human error' on patient outcomes has been to introduce increased levels of automation into the system.

Drug-dispensing errors have again been the focus of some of these efforts (see Chapter 8 for further examples). Richard Cook and colleagues have described a case study at a large urban hospital where the automated dispensing unit (ADU) for medications became unresponsive as a nurse from the emergency department attempted to urgently obtain critical medications for a patient in severe respiratory distress. Eventually, a pharmacist arranged for the necessary medications to be brought by hand from the pharmacy department. As Cook says: 'The crux of this incident was the unanticipated interaction of two nominally separate computer systems'.[9]

The ADU system was meant to interact with the hospital information system (HIS). Following several 'upgrades', the HIS experienced both hardware and software failures. A procedure to enable the ADU system to override the need to obtain authorisation from the HIS was initiated but created a torrent of messages between each ADU computer and the central computer which snarled up the system completely rendering it unresponsive. None of this had been anticipated when the ADU system was introduced. Cook makes some valuable points about the vulnerability of healthcare to systems failures. One reason is the severe financial constraints in this sector. Another is the fragmentation of the workforce into different groups whose interests may compete with one another. In a memorable phrase, Cook and colleagues sum up the situation: 'healthcare organisations are in a sense barely organizations at all but rather tense social webs of sometimes competing, sometimes cooperating groups, whose governance seems best modelled by the feudal system'.

Leveson goes on to argue that efforts to increase reliability (such as by building stronger tanks for chemical processing) do not necessarily increase safety as this may result in any eventual failure becoming more severe. Organisations can also seem to be performing reliably, in the sense that individuals and other components at each level of the organisation are carrying out their designated responsibilities exactly as prescribed. Even when this is the case, threats to safety may manifest through unexpected interactions between the different levels of the organisation and its surrounding system. The Erebus disaster (discussed earlier) as well as many others, such as the sinking of the Korean ferry, *Sewol* (discussed later), illustrate the complex nature of systems failures. The ability of high-reliability organisations (HROs) discussed in Chapter 6 to operate safely depends more on their abilities to respond flexibly and adaptively to rapidly changing conditions than on their reliability in the sense described here.

A Control-Theory Model of System Safety: At the core of Leveson's model of safety is the cybernetic feedback loop shown in Figure 9.1. A notable feature in Leveson's control model shown in Figure 9.3 is the addition of some kind of system model to the controller. The controller, whether human or mechanical, must have some kind of internalised 'model' or representation of the system being modelled. In the case of a human controller, the term 'mental model' is commonly used. At its simplest, a heating thermostat's model need only represent a couple of variables (set point temperature and sensed temperature) whereas a controller for an air traffic management system or hospital surgery scheduling would need a considerably greater number of variables. The closed feedback loop model at the core of Leveson's systems approach is shown in Figure 9.3.

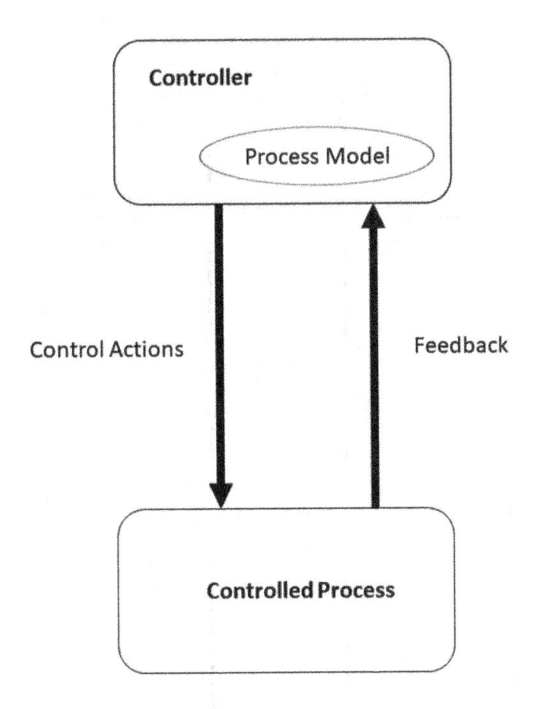

FIGURE 9.3 The basic closed feedback loop involved in the control of many biological, technological and social systems. For a mechanical controller, the process model would be instantiated mechanically or electronically. For a human controller, a mental model of the process is required to execute appropriate control actions and monitor the results.

Leveson builds directly on Rasmussen's hierarchical model of sociotechnical systems outlined in Chapter 8 (see Figure 8.6) by emphasising the feedback control loops between the different levels of the hierarchy. As an example, the hierarchical structure of a civil aviation system showing some of the possible feedback loops is shown in Figure 9.4. The government sets performance and safety targets for the regulator (control actions) and also monitors the regulator's progress in meeting those targets (feedback). The regulator may be responsible for controlling and regulating various organisations within the sector (e.g. airlines, flight training, sightseeing, etc.) as well as controlling entry into the sector (e.g. pilot licencing) and setting standards for the equipment used (e.g. aircraft maintenance and certification, etc.). These all require necessary control actions (regulations, auditing, inspections, etc.) and monitoring activities to fully achieve the desired control of the sector. Control actions are the means of influence or effect by which each level of the hierarchy constrains the behaviour of the level below through the provision of information. The nature of constraints was described in Chapter 8. Similar control structures to the one illustrated in Figure 9.4 are found throughout nature as well as in almost every engineered and organisational system.

Ross Ashby, another noted developer of cybernetics, proposed the *Law of Requisite Variety* (sometimes referred to as the *First Law of Cybernetics*) which

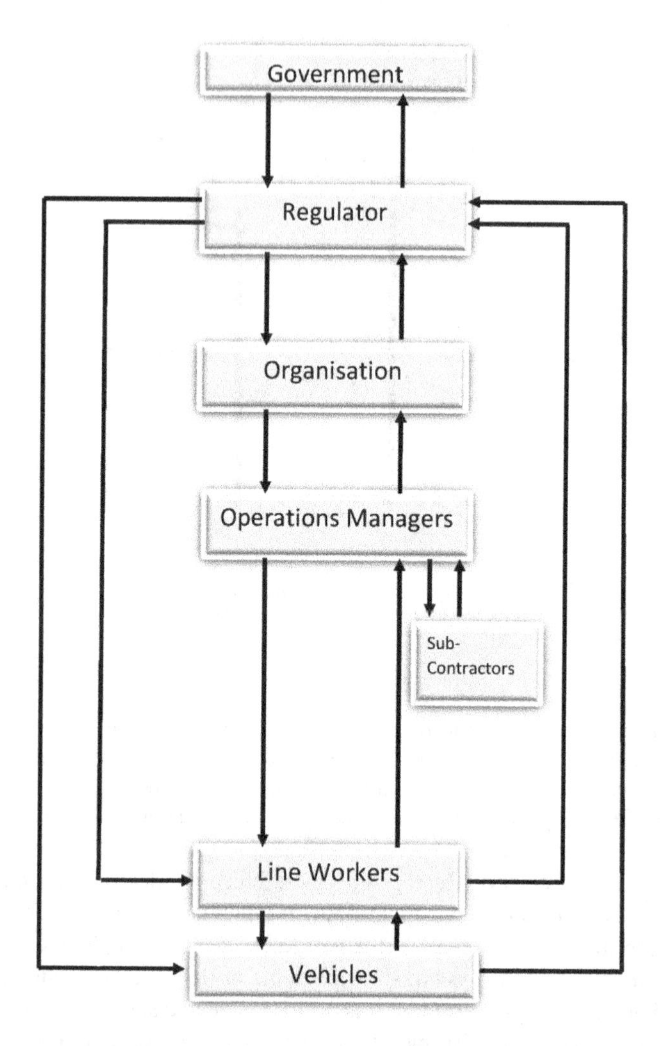

FIGURE 9.4 An outline of the hierarchical structure of a civil aviation system from the government and its regulator, down through several levels of individual organisations in the sector and their equipment.

states that to manage a stable system the number or variety of possible states of the controller must be greater than, or equal to the number or variety of possible states of the process being controlled.[10] In effect, this means that the controller's model of the system (mental or software) must be as complex (or have as much variety) as the system itself. A simple thermostat has two possible responses (turn heat on; turn heat off) to two states of the environment (temperature rising; temperature falling) and therefore has the requisite variety to control the process. The same system would lack the requisite variety to control a more sophisticated system such as an aircraft flight management system. Numerous examples can be found where the human controller's mental model of an aircraft's flight management system lacked the requisite variety to match the number of possible states that the automatic system could attain and was, therefore, much too simple and incomplete for the crew to fully understand what the aircraft's system was really doing.

ASIANA AIRLINES FLIGHT 214 CRASH ON APPROACH AND LANDING AT SAN FRANCISCO INTERNATIONAL (SFO)

The Boeing 777-200ER had earlier departed from Incheon Airport, Seoul and was making its final approach to runway 28L at SFO in fine weather at around 11.30 a.m. local time on 6 July 2013. For the last few miles of the approach, the aircraft was flying too low and too slow. The aircraft's landing gear impacted the seawall at the end of the runway, followed by separation of the left engine and tail section as the aircraft impacted the runway heavily. Three of the passengers died in the crash. The aircraft was destroyed.

The crash was investigated by the US National Transportation Safety Board (NTSB). The aircraft had been operating reliably and there were no component failures in any of the aircraft's systems. The investigators focused on the crew's interactions with the automated flight management systems which controlled the aircraft's flight path and throttle settings. The NTSB found that 'Mismanagement of approach and inadequate monitoring of airspeed led to the crash of Asiana 214'.[11] The crew's faulty mental model of the automated systems led to the inadvertent deactivation of the automatic airspeed control.

Whilst the automated systems that controlled the Asiana B777 were highly complex with multiple operating modes, the pilots' understanding of these systems was less sophisticated. In cybernetic terms, this meant that the automatic system's 'variety' of possible states exceeded that of the controller's model leading to instability of outcomes. Similar issues have arisen in other airliner crashes as well as in other automated systems in marine transportation and in industrial and process control. A good part of the difficulty arises from the necessity for the human controller of an automated system to have a mental model of both what is being controlled (e.g. flight path management) and how the automated system functions to control that process.

Leveson refers to her new systems theory model by the acronym *STAMP (Systems-Theoretic Accident Model and Processes)*. The approach is based on the concept of constraints, introduced in Chapter 8: 'Safety is reformulated as a control problem rather than a reliability problem'. In this view, lack of safety comes from problems in the control of safety constraints at every level from regulation to design to operation. Central to safety then is the appropriate communication of information downwards through the levels of the hierarchy as well as the appropriate return of information upwards. Just as the human body's homeostatic regulating mechanisms can be disrupted by problems on either side of the loop, so sociotechnical systems are vulnerable to these communication and information-flow problems.

The STAMP framework has been used to develop two slightly different techniques. One, known as *STPA* (*Systems-Theoretic Process Analysis*) is designed as a hazard analysis technique to be utilised in place of the traditional techniques such as fault trees and event trees (see Chapter 2). Like these techniques, STPA is designed to be used in designing safety systems in advance rather than for the investigation of accidents and incidents after the event, for which *CAST* (*Causal Analysis Based on STAMP*) was designed. Most applications simply refer to the techniques developed by Leveson under the umbrella term 'STAMP'. We will look at the STAMP/CAST accident analysis technique in more detail in the final chapter (Chapter 11).

Hazard Analysis with STAMP/STPA. How might STAMP be integrated into a conventional Safety Management System (SMS) as described in Chapter 2? The first step is to map out the overall hierarchical control structure from government and regulators to the individual workers as illustrated in Figure 9.4. The key concept is that each level of the hierarchy imposes some kind of constraints on the operation of the level below. As described above, the concept of safety constraints is central to the STAMP framework. As previously described, constraints can be entirely physical, as when the design of a handle on a surface limits the possible actions to pushing or pulling rather than twisting or turning or a physical interlock on a machine that prevents operation if both hands are not on the handle.

Constraints may take other forms such as those prescribed in policies and procedures. In its strictest form, the standard operating procedures (SOPs) that are mandatory in airline operations, for example, are very clear constraints on exactly what must be done and when. The most critical phases of flight are governed by the most restrictive SOPs. Other forms of specification, requirements, reporting procedures and policies all set out constraints designed to limit behaviour to those that are judged likely to keep the system operating safely. The problem with 'soft' constraints such as these is that it is never possible for system designers or managers to predict in advance every possible safe and unsafe behaviour that could occur. 'Hard' constraints imposed through design can work well with simpler, more predictable, systems where undesirable behaviours (e.g. placing a hand too close to a cutting surface) can be predicted with some certainty.

In general, constraints are those variables that must be controlled to keep the system operating within safe limits. Problems may arise due to inadequate control of these constraints at any level of the system – not just at the lower levels. Traditional hazard analysis techniques are likely to uncover some of these lower-level constraints as part of the normal exercises in hazard identification and risk matrix construction. Higher levels of control are often neglected or given cursory treatment in traditional hazard assessment approaches. Traditional techniques are best suited to identifying more static threats to safety, whereas constraints are inherently more dynamic processes dependent on appropriate and timely control actions and sensitivity to information feedback and are more likely to be revealed by a systems-based approach such as STAMP/STPA.

The additional information provided by a STAMP/STPA analysis involves the identification of those control actions and feedback loops that are necessary to prevent constraints from being violated – in other words, to prevent the system from moving closer to the boundaries of safe operation, as described in Chapter 8, with the increased possibility of adverse outcomes. In sum, the three key steps are (1) identify the safety constraints or the variables that need to be controlled; (2) outline the control structure showing what actions need to be taken and what feedback needs to be received; and (3) consider potential unsafe control actions and identify potentially hazardous control actions in advance.

Communication Gaps: Two kinds of risks frequently arise in complex sociotechnical systems. One of them involves 'gaps' between the safety requirements and the roles and responsibilities of individuals. Job descriptions may leave out essential safety constraint responsibilities so that no individual has clear responsibility, leading to gaps in the management of safety constraints. The reverse problem occurs when several individuals have overlapping responsibilities for similar control actions. Overlaps can lead to the assumption that someone else has taken some action or followed up on some feedback when in fact no one has. Gaps mean that no one realises the need to take action or follow up on the results of a previous action. Both kinds of failures are very commonly found when adverse outcomes have occurred: 'One of the basic causes of accidents ... is multiple controllers with poorly defined or overlapping responsibilities'.

Leveson describes an application of STAMP/STPA to the identification of risks involved in NASA's new safety oversight programme known as the *Independent Technical Authority (ITA)*. This resulted in a much wider range of identified safety risks (more than twice as many as in the conventional analysis) covering the entire safety control structure from the government downwards. A number of other studies carried out by Leveson's students and collaborators have also been documented. More recently, reports of the application of STAMP/STPA to a variety of problems have been reported by investigators not associated with Leveson's group at MIT. Applications include safety management systems for cruise ship and ferry operators, aviation safety and a number of applications to aspects of road safety and advanced vehicle design such as adaptive cruise control, collision avoidance and lane-keeping-assist systems.[12]

To date, most of these applications have been illustrative, designed to demonstrate how the method can be used and the kind of results that can be obtained. Most of the studies take some pains to distinguish the systems-based approach of STPA from the traditional hazard analysis techniques based on component failure and accident chain models, such as fault tree analysis, failure mode and effects analysis (FMEA) and hazard and operability analysis (HAZOP). The STPA technique focuses more broadly across all the elements that contribute to system functioning rather than solely on a narrower subset of potential faults or failure events. All the techniques require a considerable amount of subject-matter expertise and the number of identified *unsafe control actions* (UCAs) found in an STPA analysis can therefore vary widely depending on the individual experts' perspective and time involvement.

The Transportation Research Group at the University of Southampton has used STAMP/STPA to explore potential safety improvements for a class of events that occur in commercial aviation involving loss of the on-board oxygen supply due to problems with either the aircraft pressurisation system or some kind of structural failure. One of the most notorious examples involved a British Airways BAC1-11 where the cockpit windshield became detached soon after take-off with the alarming result that the Captain became partially sucked out of the cockpit window.[13] In another example, a Boeing 737 flying from Cyprus to Athens suffered depressurisation leading to the incapacitation of all on board. The aircraft eventually ran out of fuel and crashed to the ground.[14]

The Southampton researchers developed the STAMP/STPA analysis over three workshops involving one experienced airline pilot each. The overall hierarchical control structure for an airline decompression event was generated and then a list of (78) potential UCAs was developed. This large number was reduced to 21 by considering just 4 key control loops (e.g. 'Ensure adequate oxygen levels for crew' and 'Descend to 10,000 ft') between crew and aircraft. Some new safety constraints (e.g. provision of new warnings) were identified as part of this process. The authors conclude: 'STAMP/STPA can be used to identify potential safety constraints that can inform … the development of new technology and software that could improve the overall safety of future flight operations'.

More recently the same group have applied the STAMP/STPA methodology to analyse the hazards and risks involved in extreme low-level Royal Navy jet missile simulation exercises.[15] Following the same methodology as in the previous study, 88 UCAs were identified. Safety constraints were then identified for all 88 UCAs. Examples included 'Fail to provide procedures for risk assessment' and 'Failure to maintain safe control of the aircraft below 250 ft'. In the first case, the UCA occurs between the regulator (in this case, the *UK Military Aviation Authority* or MAA) and the organisation (i.e. the Royal Navy) and the second UCA is between the pilot and the aircraft. The authors conclude that the STAMP/STPA approach appears to yield a much broader range of hazards and risks than conventional techniques but note that the exercise is extremely time-consuming and would benefit from the development of software support tools.

Other analysts have obtained similar results with varying numbers of identified UCAs and possible new constraints that can be used to drive changes in design and operation. For example, researchers from Beihang University, Beijing, examined the case of a 2008 electrical fire in a US Air Force Minuteman missile launch facility in Wyoming, finding 41 UCAs.[16] Rather than focusing on component failures and human errors, the UCAs covered a much broader range of systems control issues throughout the control structure from upper management to the personnel directly involved in maintenance and operations.

The process of using STAMP/STPA is resource-intensive and demanding on subject-matter expertise. The main advantage of the STAMP/STPA technique is that it is truly system focused rather than on individual behaviour or organisational structure. It provides outputs that can be used to strengthen the system control structure, reducing the chances of future recurrences. The UK's Military Aviation Authority (MAA), a part of the *Defence Safety Authority* (DSA), was established in 2010 to oversee all military aviation activity in the UK. All military organisations must be able to show that risk assessments have been carried out and that all risks are as low as reasonably practicable (ALARP) as described in Chapter 2. STAMP/STPA is listed as an appropriate technique to be used in the recognition and mitigation of hazards.

SINKING OF *MV SEWOL*, 16 APRIL 2014

The 18-year-old ferry had been purchased from a company in Japan in 2012 and modified with the addition of two extra passenger decks and expanded cargo space, adding 239 tons to her overall tonnage. The passenger capacity was 956 people. On 16 April, the *Sewol* left the port of Incheon bound for Jeju with 476 people on board. In calm conditions and good visibility, the ship entered the Maenggol Channel travelling at 19 knots and made several course alterations. At the speed she was travelling, these produced an outward heel leading to a pronounced list to port. Cargo shifted suddenly and water entered through the cargo-loading door and the rear car door increasing the list to 60 degrees. The ship sank over two hours later with the loss of over 300 lives. Shortly after the sinking, four of the crew, including the Captain, were arrested and charged with murder. The CEO of the ferry company was also arrested and charged with causing death by negligence. The sinking had widespread social and political repercussions throughout South Korea.[17]

The unfortunate *Sewol* tragedy provides a good example of the intricate interconnectedness of every level of the shipping system from the government of South Korea to the individuals on the bridge of the *Sewol* itself. Ultimately, the tragedy led to the demise of the government and the resignation of the Prime Minister. Analysts have used STAMP to provide detailed insight into the connections within the system and to show where the principal failures in control and communication occurred. Regulation of the shipping system in South Korea involved four separate agencies. Overall control was provided by the Ministry of Oceans and Fisheries (MOF) who, in turn, had oversight of three independent agencies, the Korean Shipping Association (KSA), the Korean Register of Shipping (KR) and the Korean Coast Guard (KCG).

The KR was responsible for the structural integrity of Korean ships whilst the KSA and the KCG had oversight of operational matters. Mapping out the hierarchical control structure and the necessary control actions and feedback loops using STAMP/CAST provides insight into the numerous unsafe control actions and faulty feedback loops between these four agencies. These dynamic systems problems meant that the assumed safety constraints that were intended to ensure the structural and operational safety of vessels in Korea were either missing or ineffective. For example, approval of the modifications to the *Sewol* that increased its cargo capacity were meant to be contingent on strict observation of operational limitations requiring additional ballast to be carried. These limitations were not made known to either the Korean Shipping Association (KSA) or the Korean Coast Guard (KCG). In practice, the *Sewol* was operating overloaded and under-ballasted compromising the stability of the vessel. Some of the control problems between the four official bodies involved in shipping in Korea revealed by the STAMP-based analysis are illustrated in Figure 9.5.

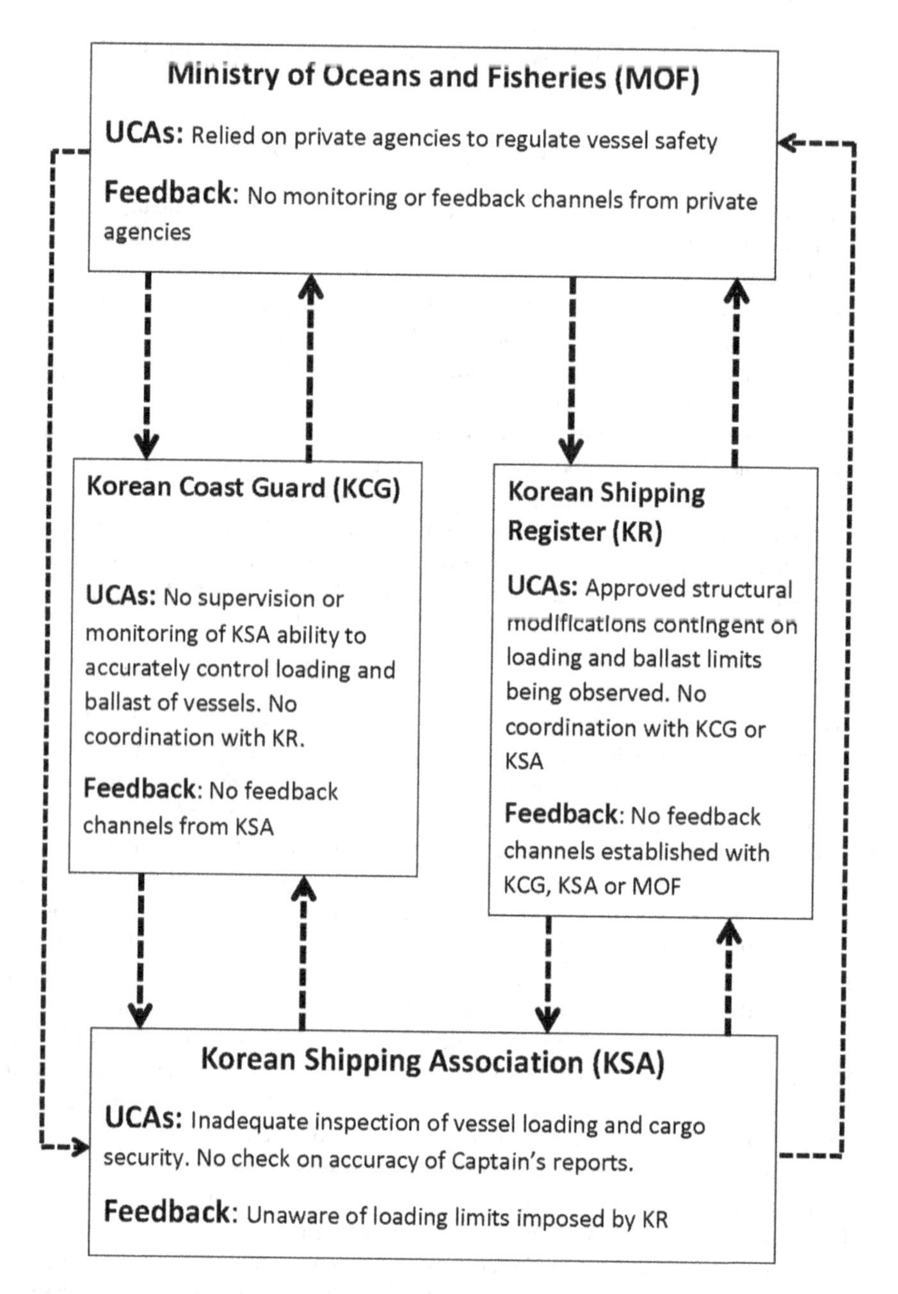

FIGURE 9.5 Part of the hierarchical control structure for Korean shipping showing the four principal agencies involved in relation to the sinking of the *Sewol*. The broken lines represent failures in control actions and feedback channels between each of the agencies involved.

The *Sewol* tragedy generated intense public outcry in South Korea leading to the downfall of a government and serious criminal prosecutions against the crew that were operating the vessel that day. The value of the STAMP systems-based analysis is in exposing a wide variety of deficient control actions and feedback loops at every level of the system within which the *Sewol* operated. These defects suggest corresponding corrective remedies that can be used to improve the safety of the whole system. These apply at every level from the government agencies that were charged with regulating the safety of maritime operations in South Korea, to the specific actions taken by individuals in the shipping company, as well as of those on board the vessel that day. In the systemic approach, illustrated here with the STAMP model, no one of these deficiencies can be considered the 'cause' of the tragedy. Instead, failures located in the system of controls and feedback that were designed to constrain operations to a safe standard allowed the system to vary beyond the desired limits. Disasters such as the *Sewol* emerge from the failure of these constraints rather than being caused by any single one of them. This is what is meant by systemic failure and as the *Sewol* case illustrates, this lies at variance with both most legal frameworks and the commonly expressed desire for simple solutions (i.e. someone to blame) for complex phenomena. In this sense, legal and moral accountability need to be distinguished from systemic safety. Satisfying the first two may do little to advance the third and, in fact, may run counter to it.

MANAGEMENT CYBERNETICS: THE VIABLE SYSTEM MODEL

Stafford Beer was a colourful English character whose career combined wartime service in India with the Gurkha Rifles, post-war work with the operations research branch at the War Office, posts in the industry with United Steel and academic posts at Manchester Business School. He spent some time in South America, most notably in Chile at the request of Salvador Allende's government, at the time under siege from the CIA. Beer was recruited in 1970 to advise the Chilean government on how to apply his management principles to improve the efficiency of the manufacturing sector of the economy. Project 'Cybersyn' involved creating a novel network of computers and telex machines to provide real-time feedback about manufacturing performance from around the country to management in Santiago.[18] Although the project was never fully completed due to the 1973 CIA-backed coup and installation of the Pinochet dictatorship, Beer was later involved in similar projects in Mexico and Venezuela.

Beer had become interested in the then newly developing field of cybernetics and had seen an opportunity to apply the principles of cybernetics to the business of organisational management. Beer also proposed a hierarchical conceptualisation of systems consisting of five levels from the lowest, operational levels that carry out the basic work of the organisation, through middle levels including line management that organise and coordinate work in the here and now (Level 2) and middle management that organises the necessary resources (Level 3). The upper levels (Levels 4 and 5) are concerned with longer-range tactical and strategic control. All complex adaptive systems, from simple biological entities to complex sociotechnical systems, have similar patterns of hierarchical organisation. The ability to adjust to external circumstances as well as the ability to regulate internal processes depends entirely on the efficient and timely dissemination of information within and between these levels. Information plays a key role in cybernetics and therefore in any systems science including safety science.[19]

Beer conceived of the operations of each of these levels in cybernetic terms. For example, as described earlier, negative feedback is used to reduce or attenuate variety, and organisations achieve this by monitoring the environment in which they operate for new technologies and updated government regulations so that the organisation's outputs remain matched with societal trends and competitor organisations. Positive feedback, on the other hand, operates to amplify the effects of internal (e.g. policies and procedures) and external (e.g. market research and advertising) communications. Attenuation and amplification both serve to bring the variety of management in line with the variety of operations, for example, in accordance with Ashby's Law of Requisite Variety. Effective feedback and control require the rapid transmission of information which led Beer to be an early advancer of distributed information technology in organisations decades before the advent of the World Wide Web made this commonplace.[20]

Beer was an applied cybernetician who adopted and adapted ideas from the biological world to the business of organisational management. He defined cybernetics as 'the science of effective organization'. Management was, therefore, 'the profession of regulation and therefore of effective organization'. The management of any organisation needed to function like the higher levels of control in any biological organism by being constantly aware of its environment in order to control and regulate its behaviour to adapt to changing circumstances. This is what Beer meant by a viable organisation which is one that could constantly adapt to its circumstances and thrive. The constant scanning of the environment and the provision of real-time information to the higher levels (i.e. Levels 4 and 5) and the translation of the results into immediate and effective action was the heart of the *viable systems model* (VSM). The explosive growth of modern businesses such as Amazon depends entirely on gathering as much information about their environment as possible and using machine learning algorithms to translate the results into immediate adjustments to further expand their business. This was exactly the process set out by Beer and enabled with technology not yet developed in Beer's time.

By conceiving of organisations in cybernetic terms, Beer brought into prominence the notion of adaptability to their environment that all complex systems have to achieve to remain viable. This is true for biological, mechanical and social organisations. Again, the cybernetic notions of *attenuation* (through negative feedback) and *amplification* (through positive feedback) are central. Beer's model of organisational structure and control would appear to be highly compatible with Leveson's STAMP control model of safety. Researchers have recently combined the two to provide detailed insight into the safety management of a road tunnel system.[21]

Longish road tunnels such as the 11.6 km tunnel under Mont Blanc between France and Italy require elaborate ventilation systems so that there is breathable air more than 5 km away from the entrances. An unavoidable risk is that of fire.

The last such event in the Mont Blanc tunnel killed 39 people in 1999. An effective safety management system is therefore essential. STAMP/STPA was used to identify the constraints that need to be enforced and to define the necessary safety control structure. Drawing on the five-level structure of the VSM model the researchers elaborated on potential organisational sources of control flaws. Examples at Level 5 include ambiguous safety policies; at Level 3 an inappropriate balance between centralisation and decentralisation; and at Level 1 many of the frequently cited problems such as lack of reporting (feedback) systems, failure to learn from experience, gaps and overlaps in responsibilities, lack of leading safety indicators, etc., may occur. Each of these is discussed in greater depth.

The VSM model has not been extensively adopted in management studies or in safety science, but it provides a natural complement to the well-structured STAMP framework and its associated hazard control (STPA) and accident analysis (CAST) techniques. With its emphasis on the real-time acquisition of information and the translation of the results into immediate action, it was far ahead of its time as the same strategies underlie the current success of modern mega-businesses such as Facebook and Amazon. The usual reference point for the systems-based approaches described in this third part of our 'safety journey' is large-scale technological systems such as air traffic management, power generation, heavy manufacturing, etc. However, the majority of workers are situated in much smaller-scale enterprises where there are unlikely to be the full five levels of organisational structure as described by Beer. Nevertheless, all enterprises, at least in those countries with developed health and safety systems as described in Chapter 1, are embedded within a hierarchy of social structures (e.g. government, health and safety authority, enterprise associations, etc.) that exert control over the relevant safety constraints in the ways described by STAMP and VSM.

For smaller enterprises and organisations, the core of an effective safety management programme lies in the identification of the constraints that need to be controlled and the implementation of effective control strategies. A simple example illustrates the process. A small firm engaged in the business of applying chemical dyes to garments might identify a critical constraint as that of ensuring the workers don't get the chemical in their eyes or inhale the fumes. One appropriate control strategy would be to require workers to wear a mask and goggles. In many cases, this strategy would be implemented by notices and posters in the workplace explaining the need to wear the protection. Job done! Not quite. Does the manager receive timely and accurate feedback about the extent to which the workers are wearing the assigned *personal protection equipment* (PPE)? It is not uncommon for workers not to wear PPE for various reasons usually related to comfort or impeding their ability to perform the work.

In this example, even though an appropriate constraint has been identified, the chosen control action is ineffective. Poor feedback further compounds the problem. Gap and communication analysis may show that nobody has clear responsibility for auditing PPE behaviour or that it is assumed to be the responsibility of someone else. Recriminations may follow and workplace safety meanwhile continues to drift further towards the boundaries of safe operation (Chapter 8). Whilst the language of constraints and control actions may be unfamiliar, the basic elements of this scenario are likely to be recognised by anyone involved in workplace safety.

CONCLUSIONS

All systems, living and non-living are based on a capacity for self-regulation. Processes occurring within the system are subject to internal regulation through the mechanisms of control and feedback. The human body has numerous such systems governing everything from body temperature to heart rate and digestion. Interactions with the outside world are governed by similar processes of action and feedback. A recognition of the universality and communality of these processes led to the development of the field of cybernetics in the 1940s. The applicability of these ideas to the operation of complex sociotechnical systems and their relevance to understanding and controlling safety has been developed through the work of engineers such as Jens Rasmussen (Chapter 8) and Nancy Leveson (this chapter).

Building on Rasmussen's explication of the hierarchical nature of the sociotechnical systems that govern much work, transportation, healthcare and other activities, Leveson's STAMP control theory provides a simple but powerful framework for thoroughly understanding how complex systems function. STAMP has spawned approaches to both accident analysis (CAST) and hazard identification and risk management (STPA). There is a growing scientific literature comparing the STAMP/CAST accident analysis approach with other systems-based techniques such as Rasmussen's AcciMap and FRAM based on the work of **Eric Hollnagel** to be described in the next chapters. Scientific literature is also developing on the STAMP/STPA approach to hazard identification but direct comparisons with other approaches (such as fault tree analysis and HAZOP) have yet to appear so it is too early to provide definitive evidence-based guidelines on the effectiveness of these new approaches.

There has been a recent reawakening of awareness in the importance of the early work in cybernetics for fields such as human factors and ergonomics and safety science. Much cross-fertilisation between the two developing fields took place in the mid-20th century.[22] As more safety professionals begin to understand the importance of moving away from viewing safety as a problem of individual behaviour or organisational structure and towards a view of safety as a property of systems themselves, so ways of thinking about the real-time dynamics of system behaviour become more important. This requires thinking about how complex sociotechnical systems develop and control the activities that take place within them and with their external environment. The cybernetic systems view is that safety is an emergent property of systems which are designed or developed to exert constraints on system behaviour to maintain stability and control interactions with whatever variations occur in the operating environment. This represents a significant departure from the traditional view of safety as failures of mechanical components or human behaviours or other forms of unreliability.

Domains such as healthcare with their highly complex structures and diverse workforces are ripe for the application of the system-based approaches outlined in this chapter. The last couple of years with the Covid-19 pandemic have created huge environmental pressures that have challenged the viability of healthcare in many parts of the world. Some systems have adapted and met these challenges more successfully than others. Investigations using the core concepts of cybernetics (such as the Law of Requisite Variety) and their more recent developments, in terms of the Viable System Model and STAMP framework, could generate valuable new insights into the nature and functioning of successful healthcare systems.

Safety science can usefully draw upon earlier work in cybernetics and systems science to develop tools and techniques specifically designed to address the safe functioning of modern workplaces and transportation systems. In Chapter 10, we extend the systems approach another step further to look in more detail at how systems can exhibit adaptability and resilience in workplace environments. These concepts are important components of a proposed 'new wave' of safety science known as 'Safety II' proposed by Eric Hollnagel which has gained considerable traction in certain areas of safety management, notably within the healthcare domain. We will critically examine the nature of Hollnagel's proposals next.

NOTES

1. Both Cambridges have a population of around 120,000 as does Dunedin, New Zealand, home of the country's oldest university. The numbers of Nobel laureates credited to each institution varies according to whether they were full time employees, former graduands and so forth. Regardless, the numbers are impressive. For comparison, only three New Zealanders have been awarded Nobel prizes and all three (you guessed it) either worked at, or received degrees from Cambridge University, England.
2. This is not strictly correct as there are many sources of feedback. A key one is an extrinsic one: 'knowledge of results' which tells us where the dart or shell landed in relation to the target. However, there are other intrinsic sources of feedback such as the feel of the dart or bow tension as well as associated sounds. These would enable our blindfolded darts player to gradually improve, however this is not recommended as a training method!
3. Wiener, N. (1948). *Cybernetics or Control and Communication in the Animal and the Machine.* New York: Wiley.
4. From: Mobus, G.E., & Kalton, M.C. (2015). *Principles of Systems Science.* New York: Springer. As noted previously, this provides an excellent recent guide to all the basic elements of systems science from information and communication to cybernetics.
5. The quoted text comes from Vette, G. (1983). *Impact Erebus.* Auckland: Hodder & Stoughton. Captain Vette was a senior Air New Zealand pilot whose research led to many new insights into the tragedy. He dedicated the rest of his life to improving aviation safety before passing away in 2015.
6. See: PR pros lambast new Government 'Stay alert' slogan as 'unclear' and 'unhelpful' | PR Week
7. Shannon originally published his 'Mathematical theory of communication' in *The Bell System Technical Journal, 27*(3), 379–423. A subsequent book (co-authored with Warren Weaver) provided a more accessible version of the theory. See: Shannon, C., & Weaver, W. (1949). *The Mathematical Theory of Communication.* Urbana, IL: University of Illinois Press.

8. Leveson's theory is described in her 2011 book: *Engineering a Safer World: Systems Thinking Applied to Safety.* Cambridge, MA: MIT Press. It was also laid out in a journal publication: Leveson, N. (2004). A new accident model for engineering safer systems. *Safety Science, 42*(4), 237–270.
9. Wears, R.L., Cook, R.I., & Perry, S.J. (2006). Automation, interaction, complexity, and failure: A case study. *Reliability Engineering and System Safety, 91*, 1494–1501.
10. Ashby's ideas are set out in his major work: Ashby, W.R. (1956). *An Introduction to Cybernetics.* New York: Wiley. The Law of Requisite Variety is set out in: Ashby, W.R. (1958). Requisite variety and its implications for the control of complex systems. *Cybernetica, 1*(2), 83–99.
11. National Transportation Safety Board. (2014). *Aircraft Accident Report: Descent Below Visual Glidepath and Impact with Seawall. Asiana Airlines Flight 214 Boeing 777-200ER, HL7742 San Francisco, California July 6, 2013. NTSB/AAR-14/01.* Washington, DC: NTSB.
12. Examples of recent applications of STPA include: Salmon, P.M., Read, G.J., & Stevens, N.J. (2016). Who is in control of road safety? A STAMP control structure analysis of the road transport system in Queensland, Australia. *Accident Analysis and Prevention, 96,* 140–151; Mahajan, H.S., Bradley, T., & Pasricha, S. (2017). Application of systems theoretic process analysis to a lane keeping assist system. *Reliability Engineering and System Safety, 167,* 177–183; Allison, C.K., Revell, K.M., Sears, R., & Stanton, N.A. (2017). Systems theoretic accident model and process (STAMP) safety modelling applied to an aircraft rapid decompression event. *Safety Science, 98*, 159–166.
13. The official report of the BAC1-11 windshield departure can be downloaded from: www.gov.uk/aaib-reports/1-1992-bac-one-eleven-g-bjrt-10-june-1990. A vivid account of the incident can be found in James Reason's 2008 book: *The Human Contribution.* Farnham: Ashgate.
14. The official Greek report, published in 2006 can be downloaded at: https://www.webcitation.org/5zayqkbkF?url=http://www.aaiasb.gr/imagies/stories/documents/11_2006_EN.pdf
15. Stanton, N.A., Harvey, C., & Allison, C.K. (2019). Systems theoretic accident model and process (STAMP) applied to a Royal Navy hawk jet missile simulation exercise. *Safety Science, 113*, 461–471.
16. Rong, H., & Tian, J. (2015). STAMP-based HRA considering causality within a sociotechnical system: A case of Minuteman III missile accident. *Human Factors, 57*(3), 375–396.
17. Some of the information comes from the Wikipedia article ('Sinking of MV Sewol') and also from: Kim, T.-E., Nazir, S., & Overgard, K.I. (2016). A STAMP-based causal analysis of the Korean Sewol ferry accident. *Safety Science, 83,* 93–101.
18. A telex network was a telephone-like system of interconnected teleprinters developed in the 1930s. Unlike fax machines, only text messages could be sent and received. Beer's work in Chile is described in: Medina, E. (2006). Designing freedom, regulating a nation: Socialist cybernetics in Allende's Chile. *Journal of Latin American Studies, 38*(3), 571–606.
19. These points are explained in much greater detail in the excellent '*Principles of Systems Science*' by George Mobus and Michael Kalton (New York: Springer Science, 2015).
20. Beer, S. (1985). *Diagnosing the System for Organizations.* New York: Wiley. A good short summary of the VSM model can be found in: Umpleby, S.A. (2007). The viable system model. In S. Clegg & J.R. Bailey (Eds.), *International Encyclopedia of Organization Studies.* New York: Sage.

21. Kazaras, K., Kontogiannis, T., & Kirytopoulos, K. (2014). Proactive assessment of breaches of safety constraints and causal organizational breakdowns in complex systems: A joint STAMP-VSM framework for safety assessment. *Safety Science, 62*, 233–247.
22. Baber, C., Golightly, D., & Waterson, P. (2019). Editorial: The cybernetic return in human factors and ergonomics. *Applied Ergonomics, 79*, 86–90.

10 Resilience, Adaptability and System Safety

Close to a billion people, mostly in sub-Saharan Africa, lack access to one of the most basic, and most essential, elements of a modern society – electricity. With a few exceptions of dedicated off-grid pioneers, the remaining 90% of the world's population depend on the reliable distribution of electrical energy to power their homes and businesses. Electrical power is mostly generated in large plants powered by fossil fuels, renewable energy or nuclear power and distributed across large-scale networks. With the rate of climate change accelerating, more extreme climatic events are occurring more frequently than in the past. Even prosperous and technologically advanced economies, such as the United States, are starting to see supplies of electrical power more regularly interrupted by severe weather events such as droughts and storms.

In 2020, the Californian grid was subject to numerous shutdowns as extreme heat and fire risk led to overloads and preventative shutdowns. In 2021, Texas suffered major problems when severe winter storms ramped up demand from consumers and triggered blackouts. At the same time, plants were closed due to extreme icy conditions which caused pipes and valves to freeze over, as did wind turbine blades. A winter storm a decade previously had led to similar results. Similar effects on critical infrastructure from previous events such as Hurricane Katrina in 2005 and 'Super Storm Sandy' in 2012 that affected the power distribution and energy infrastructure in 21 US states triggered a surge of interest in studying the ability of critical infrastructures that underlie electricity, water, transport, telecommunications, etc., to recover from increasingly frequent severe disruptions from environmental adversity to other potential challenges such as deliberate attacks from individuals or states.[1]

In the previous chapter, we noted Nancy Leveson's important distinction between reliability and safety. Similarly, in the context of designing and operating critical infrastructure systems the ability to perform reliably under a well-defined, relatively constant, set of conditions is not related to the capacity of a system to recover from novel, unexpected or severe disturbances and continue to function. The term most widely used to refer to this ability is *resilience*. All complex systems whether they be biological, mechanical or organisational need some degree of resilience to recover from these more extreme disturbances. Systems that lack resilience are *brittle* and may be highly compromised or destroyed by relatively minor fluctuations in their environment.

The US power grid failures show that systems that function reliably under normal circumstances may be more brittle and less resilient than expected and may not possess the ability to adapt to more extreme conditions. Any successful system must be

DOI: 10.1201/9781003038443-13

adapted to whatever environment it operates within. If that environment remains relatively constant and unvarying, then sudden change may exceed the inherent ability of the system to recover from an extreme event. The meteor impact that led to the end of the dinosaurs was an example of a severe challenge that exceeded the capacity of those animals to adapt whilst other life forms were able to survive and thrive. Similarly, organisations may be faced with challenges that exceed their ability to adapt leading to business failure.

Indeed, resilience is the hallmark of all successful systems that can improvise and adapt to changing circumstances. We could rely on luck and good fortune to winnow out the brittle and increase the prevalence of resilience or we could explore ways of building and designing resilience into people, organisations and systems. Becoming a more resilient individual is the theme of numerous self-help and personal development manuals and it will be best to leave that aspect of resilience to those experts. For the present, we will focus on the question of the role that resilience plays in systems and in system safety.

RESILIENCE ENGINEERING

Eric Hollnagel has been a leading figure in modern safety science for several decades. He is currently a Professor at the Jönköping University in Sweden where his main interest is now centred around healthcare safety and performance. He began as a traditional experimental psychologist studying information processing and cognition before becoming interested in safety and control in complex technological systems. For nearly 40 years Hollnagel has been writing and consulting on human performance in these systems becoming one of the early developers of cognitive engineering (see Chapter 8) occupying a variety of academic positions in Denmark, France and Sweden.

The idea that safety in technological and other workplace systems is a facet of the system's ability to adapt to environmental and operational challenges, was first formalised at a conference organised by Hollnagel, Nancy Leveson and **David Woods** (now an Emeritus Professor of Systems Engineering at Ohio State University) held in Söderköping, Sweden, in 2004. A selection of the presentations at that meeting was published under the title 'Resilience Engineering: Concepts and Precepts' a year later.[2] Hollnagel and Woods had been collaborators since the 1980s on what was then known as cognitive systems engineering, and both had come to reject previous accounts of systems failures as the result of faulty individual cognition or decision making. Both had emphasised the importance of viewing individual performance *in context* with reference to environmental and operational constraints and pressures. A natural consequence of this viewpoint is to emphasise that success occurs far more frequently than failure, and that the roots of failure are, therefore, more likely to be bound up with these same success-making processes than as resulting from some special failure-prone mechanisms. The recent development of what Hollnagel has labelled 'Safety II' revolves around this notion that variability in performance is a desirable quality, leading more often to success than failure, and that safety science needs to concern itself more with ensuring that 'things go right' rather than preventing 'things going wrong'. We will explore 'Safety II' in more detail in the subsequent sections.

The term resilience has been used in a variety of contexts – initially as a property of materials such as wood or metal and later as a property of all complex adaptive systems to recover from environmental challenges. In both cases, resilience has been seen to be essentially reactive – the potential to return to shape following some sort of disturbance. Woods and Rasmussen's application of the term to sociotechnical systems is more broad-ranging covering all that a system does to ensure success in an unpredictable and changing environment. To do so requires that the system or organisation act more proactively to manage anticipated disturbances as well as to 'bounce back' from actual, experienced challenges. Diligent readers will recall that this is somewhat reminiscent of the earlier studies of high-reliability organisations (HROs) covered in Chapter 6.

Numerous definitions of resilience have been suggested and Hollnagel has proposed several different versions over the years. All the definitions focus on how an organisation or system performs (i.e. what it does) rather than its structural or other properties (i.e. what it has). The common themes of resilience are that it enables systems to 'survive and thrive' by adjusting to conditions ahead of time, coping with both the anticipated and the unexpected. If we return to Rasmussen's concept of a safety space, described in Chapter 8, then resilience involves the ability of the system to remain safely within these boundaries and to quickly recover if the boundaries are transgressed. David Woods' review of NASA's organisational culture suggested that resilience engineering could provide the necessary tools to 'manage risk proactively'.[3] All of which, of course, is somewhat abstract. To make it more practical we need to have objective ways of measuring or assessing resilience and then we need suitable tools to develop and enhance resilience within an organisation or system.

Indicators of Resilience: Hollnagel proposed that resilient performance in organisations and systems depends on certain potentials.[4] Most obviously, it depends on the *potential to respond* which means knowing how to adjust their functioning to meet the disturbances and challenges. This is related to the *potential to monitor* which involves knowing what to keep an eye on in terms of own performance as well as what is happening in the immediate environment. Third, Hollnagel suggests is a *potential to learn,* which means drawing the right conclusions from previous experiences. Finally, there is the *potential to anticipate*, which means having a well-developed sense of likely future developments, technologically, operationally and environmentally. There is some clear overlap between these terms used to characterise resilience and the terms used previously to characterise high-reliability organisations (HROs) as outlined in Chapter 6. Indeed, in that literature the 'potential to anticipate' was characterised as a 'commitment to resilience. The 'potential to monitor' is quite similar to the HRO characteristic of 'sensitivity to operations' and so forth. A number of the participants at the first Resilience Engineering conference in 2004, in fact, drew direct comparisons between the principles previously suggested to underlie high-reliability organisations with those currently suggested to underlie resilience. The 'flexible, mindful organization' of HRO research was directly equated, by some, with the resilient organisation of resilience engineering.[5]

The 'potential to monitor' necessarily involves attending to selected aspects of the organisational and operational environment and obtaining relevant information about them. The measures that do this are referred to as indicators and these largely fall into three groups: those indicators that reflect the past, called lagging indicators; those that provide real-time information or current indicators; and those that provide information about the future, referred to as leading indicators. Accident and incident statistics are themselves examples of lagging indicators and are often used to justify safety interventions. Unfortunately, the presence or absence of accidents or incidents does not necessarily indicate the presence or absence of resilience.

It has been argued that road transportation, whilst the source of millions of deaths and injuries worldwide annually is actually a resilient system.[6] This is based on a calculation of the likelihood of any individual traffic encounter leading to a crash which is exceptionally low, perhaps as low as 1.5×10^{-9} which is a very low probability indeed. By inference, drivers must successfully manage the overwhelming majority of their driving and encounters with other road traffic thus implying a high degree of resilience in the system. This conclusion is based entirely on a statistical calculation rather than on an analysis of the road transport system as a whole, but it does emphasise that the relationship between resilience and safety may not be a straightforward one. Much like the relationship between reliability and safety.

Hollnagel provided a framework for conducting assessments of resilience which he labelled the *Resilience Assessment Grid* (RAG). This consists of a set of questions, relating to each of the four potentials (respond, monitor, learn and anticipate), that are intended to guide individually tailored sets of questions for any particular system or domain. For example, the potential to learn involves issues such as the selection of events to learn from; procedures for collecting data about these events; and implementation of the results of this learning process. Hollnagel suggests specific examples of the kind of questions that could be used to assess each of these issues. In this case, some of the appropriate questions could include 'It is clearly established what should be reported'; 'Reports are being investigated sufficiently'; and 'We meet with other units/departments to learn from each other'. Hollnagel suggests that the questions should be answered on a five-point Likert scale. The questions are intended to be used repeatedly over time within a single organisation rather than for comparative purposes across multiple organisations. This is in accordance with Hollnagel's emphasis on measuring the processes involved in resilience and assessing the progress or development of an organisation over time to become more resilient. This can be done by comparing the resulting profiles at different points in time.

The RAG method has not been widely utilised to date although there have been some demonstrations of its relevance and ease of use in several contexts. For example, as in the United States, there have been increasing challenges to infrastructure resilience in the UK brought about by increasingly frequent extreme climatic events. A very clear application of the RAG method as outlined by Hollnagel to the water supply network in the UK has been reported recently.[7] The researchers found that changes in regulation introduced in 2014 had led to significant improvements in water network resilience that were observable in 2019. The authors praised the method's 'simplicity, flexibility, and easily-interpretable graphical results'. The ability to provide a profile of resilience over time is also compatible with the *Safety Culture Maturity Model* approach which proposes that organisations mature through a series of stages over time from a reactive, compliance-centred approach to the stage of mindful engagement and awareness of a high-reliability organisation (HRO).[8] The RAG technique could provide one diagnostic tool for assessing an organisation's maturity level.

FUKUSHIMA REACTOR CORE MELT, 11 MARCH 2011

A major earthquake (9 on the Richter scale) off the Pacific Coast of Japan occurred on the afternoon of 11 March triggering a powerful tsunami which caused major damage up to 10 km inland from the affected coastal areas. One of these was the area around Fukushima where the Daiichi Nuclear Power Plant operated six reactors. The automatic systems responded to the earthquake by shutting down the reactors. This led to loss of electrical power, triggering the emergency back-up diesel generators into action to provide power to continue the circulation of coolant through the reactor cores. The subsequent tsunami triggered by the earthquakes overwhelmed the seawall and flooded parts of four of the reactors, causing the diesel generators to fail. The resulting lack of circulating coolant led to meltdowns, explosions and release of radioactivity from these three reactors. The subsequent release of radioactivity led to the evacuation of 154,000 inhabitants of the surrounding area. The disaster was classified as Level 7 on the International Nuclear Event Scale, the highest possible classification making it the worst nuclear disaster since Chernobyl.[9]

Hollnagel has published an analysis of the Fukushima disaster in terms of the four resilience potentials described above (i.e. Responding, Monitoring, Learning and Anticipating) and more particularly the relationships between them.[10] In the case of Fukushima, the anticipating and responding potentials were of principal interest. The traditional risk analysis methods such as Probabilistic Risk Assessment (PRA) which are widely used in the nuclear industry as their formalised approach to anticipation (see Chapter 2) fail to provide an adequate basis for dealing with unforeseen, and generally highly unlikely, events such as a massive tsunami. The apparent rigour of such approaches can also lead to a strong sense of complacency which is antithetical to the mindset that is continually aware of the possibility of surprises. The ability of the organisation to respond to the unexpected electrical failures, and consequent lack of circulation of the critical coolant water, was, in Hollnagel's view, compromised by the same rigid PRA process.

Other than a trenchant criticism of PRA methods, Hollnagel does not offer very much by way of specific recommendations other than the standard injunction in resilience engineering to focus on the paths to success and to view the natural variability of human performance as an asset to be utilised rather than a problem to be eliminated. In a distinct echo of the characteristics of high-reliability organisations,

Hollnagel suggests that 'Resilience engineering advocates a constant sense of unease, that we should be mindful of what we do, to counteract overconfidence' In contrast, a team of researchers in the nuclear industry have suggested a strongly data-driven approach to developing leading indicators for nuclear power operations.[11]

Using the stress–strain model of resilience proposed by David Wood, the research makes use of the extensive data captured by the industry's *Corrective Action Programme* (CAP), which utilises employee reporting to generate around 10,000 annual 'condition reports' for each plant. Wood draws on the materials science approach to resilience whereby the ability of a material to absorb increasing amounts of 'strain' can be measured and plotted as a curve showing a plot of the resulting 'stress' on the material. A typical material curve shows a region where increasing strain is easily absorbed and then a region where the increasing demand starts to stress the material resulting, eventually, in sudden failure. The researchers use this analogy to generate a stress–strain curve for a nuclear power plant where stress is measured by the number of corrective actions filed and the strain measured by work orders and CAP actions. There was a clear correlation between the level of organisational demand (i.e. strain) and the likelihood of a significant event occurring (i.e. stress).

Whilst efforts to develop leading indicators in highly hazardous industries such as aviation, healthcare and nuclear power are undoubtedly valuable, it must be noted that these are essentially based on static models, derived from observed patterns of association, that do not fully account for the dynamics of complex systems. As previously noted the numerous feedback loops which characterise all complex, hierarchically structured systems can lead to the sudden emergence of outcomes in a way that is not captured by traditional cause and effect, linear models. In a technical sense, as a function of their inherent properties, complex systems exhibit *non-linear dynamics* which simply means that the association between any set of conditions and the outcomes can seem far from straightforward. Rather, they are dynamic, interweaving and resonating against one another in a dance of seemingly unpredictable cause and effect. Such systems can rapidly diverge away from their starting points (as opposed to predictable, linear changes) so that the eventual outcome can be highly sensitive to the initial starting point. This was the key insight of *chaos theory* which began to attract attention in the mid-1970s through the work of **Edward Lorenz** on modelling weather systems.

Such effects have been explored in many areas such as materials science, weather and human physiology, and it seems more than likely that similar effects characterise safety science. The attempt to model nuclear system safety in terms of the stress–strain model of resilience does partially address such concerns in that the effects of organisational strains are shown to be non-linear, at first easily absorbed, and only becoming problematic past some threshold. Whilst providing some useful insight, this doesn't fully address the nature of the underlying processes that initially accommodate the strains and then eventually buckle under them.

Hollnagel's resilience approach has generated numerous reports and published papers since 2004. Several reviews of the field have been published, the most recent in 2018. Close to 500 documents focusing on resilience, published between 2004 and 2016 were found in the scientific literature although the majority (nearly 60%) remained uncited by other papers in the literature.[12] Whilst this might be an indication of their lack of relevance or quality, it may also be that there had not yet been sufficient time for the more recent publications to become cited by others. Analysis of the content of this scientific literature reveals a preponderance of advocacy of the resilience perspective over substantive advances in measuring and managing resilience.

The conceptual overlap between resilience engineering and the earlier field of high-reliability organisations has been remarked on several times in the course of this discussion. That this is more real than apparent has been recently confirmed by David Woods and Sidney Dekker who draw on both fields of research in their analysis of the Helios Airways aviation disaster mentioned in Chapter 9. Drawing on the findings of chaos theory as well as the non-equivalence between reliability and safety noted by Leveson, they reiterate the view that 'Resilience engineering represents the action agenda of HRO'.[13] One area where this action agenda has had, and continues to have, a significant influence is that of healthcare.

RESILIENT HEALTHCARE

The inherent complexity and risks of technologies such as nuclear power generation and aviation have been apparent since their inception. In contrast, the complexities and harms associated with healthcare only began to be appreciated after the pioneering research of Harvard Professor **Lucian Leape** and the publication of a seminal report from the Institute of Medicine (IOM) in 1999. The attempt to transfer the focus of attention from individual mistakes and errors to system-wide problems initially met some pushback from people fearing that problems of incompetence and negligence would simply be swept under the carpet.

A second report from the IOM in 2001 set out six simultaneous goals that would characterise the 21st-century-healthcare system.[14] First and foremost was 'patient safety' followed by goals relating to effectiveness, efficiency and equity. David Woods has pointed out that these goals, whilst admirable individually, often conflict and interact so that achieving one can undermine the achievement of another. This is the fundamental difficulty facing all complex systems – mutually incompatible goals often define system success. NASA's ill-fated mantra of 'faster, better, cheaper' was the classic example, where faster and better are unlikely to be cheaper and cheaper is unlikely to be better. Too often the resolution of such conflicts is left to the individuals at the front line to deal with as best they can, and this is especially true in healthcare. The results can be seen in the compromises, workarounds and improvised practices that people come up with in the face of an inherently impossible mandate.

Five years after the original IOM report Leape concluded that whilst there had been some progress in certain areas, overall progress had been frustratingly slow.[15] Nonetheless, Leape saw a positive effect on organisations beginning to seriously engage with the broader notion of systemic problems rather than the individual error/blame focus that had previously characterised the healthcare system. Leape's sentiments have been echoed by others including Dr **Mark Chassin**, a member of the team that produced the original IOM report. He has concluded that 'we haven't moved the quality and safety needles as much as we had hoped'. Part of the problem is that the healthcare system remains strongly focused on root cause analysis and linear chain-of-effect models as explanations of adverse effects. As a result, healthcare has been slow to 'buy-in' to the safety message promoted by resilience engineering and 'Safety II' approaches of safety as an emergent property in a complex dynamic system. Public and legal sentiments that there must always be someone responsible for an adverse effect are another factor slowing down the application of systems analysis to healthcare.[16]

During the first decade of the present century, healthcare organisations increasingly began to look towards other high-hazard industries such as aviation and nuclear power for approaches to managing safety founded in human factors and safety science. The ideas about resilience had particular relevance to organisations involved in the care of individuals whose personal resilience was often being critically challenged. The first formal connection between the resilience engineering approach and healthcare practice was established in 2013 with the publication of the first volume on 'Resilient Health Care' based on the work of participants in the *Resilient Health Care Network* (see resilienthealthcare.net) sponsored by Macquarie University based in Sydney, Australia. One of the leading figures in the resilient healthcare network has been the incredibly prolific Professor **Jeffrey Braithwaite**, Professor of Health Systems Research at Macquarie University in Sydney.

Several subsequent volumes in the series have been published, with *Delivering Resilient Health Care* appearing in 2019 and the somewhat enigmatically subtitled sixth volume *Resilient Health Care: Muddling Through with Purpose* appearing in 2021. In fact, the subtitle refers simply to the distinction frequently made by Hollnagel between the idealised work process as envisaged by upper managers and the reality of people struggling to balance conflicting objectives and find ways to work around the obstacles and difficulties of everyday working life. This is the distinction between 'work as imagined' (WAI) and 'work as done' (WAD) which lies at the core of Hollnagel's revitalised 'Safety II' agenda as described in the next section.

The scientific literature on resilient performance in healthcare has provided ample demonstration of the enormous flexibility in performance that allows systems to be stretched to the limits but continue to function. This can only be accomplished by designing work systems to allow sufficient variability in human performance rather than robotic adherence to prescribed rules and procedures. This raises inevitable questions over responsibility and accountability when outcomes fall below expectations. Unfortunately, the frameworks used by lawyers, the public and scientists to address such issues are fundamentally different from one another and lead to different, and often conflicting, conclusions. Until, and unless, the legal framework in particular starts to catch up with the scientific framework then people inevitably face legal sanction and even criminal prosecution for performing in ways that normally keep systems operating successfully.

To date, there has been rather more limited progress in translating the concepts of resilient performance into usable, measurable characteristics and objectives. It remains difficult for workers and managers to assess whether a system is functioning at a safe distance from the 'safety boundaries' (see Chapter 9) or for workers to dynamically assess the impacts of temporarily trading off one set of objectives for another. In the absence of such information, decisions under adversity have to be taken 'on the fly' based largely on previous experience and know-how. In the vast majority of cases, this leads to successful outcomes and the system performs resiliently. As noted above, however, in the rarer cases where outcomes are not as desired, these same individuals may find themselves in the invidious position of being held to account, often with serious consequences.

'SAFETY II'

Every workplace throughout recorded history has most likely echoed to some version of the same refrain of 'the bosses don't understand what really goes on around here'. As we have previously noted, all systems from the biological organism to the complex technological system exhibit some form of hierarchical organisation. At the lowest level of the hierarchy, the work of the system is actually being performed, with higher levels (e.g. middle and upper management) responsible for coordinating the resources needed for that work, managing the allocation of those resources, interacting with other organisations and planning the long-term future of the organisation. Work itself is invariably governed by a framework of rules, procedures and established processes. It is commonplace to find that the world defined by these requirements invariably does not fully match the world of day-to-day operations. As previously noted, Eric Hollnagel coined a couple of catchy phrases for this discrepancy, referring to the former as 'work-as-imagined' and the latter as 'work-as-done'.

Inevitably the difference between the two 'works' is cited by accident and incident investigators in the aftermath of an untoward event. It is invariably easy to find some point at which the frontline operators diverged from the prescribed processes and procedures and this has been frequently pinpointed as the 'cause' of the crash or other unwanted outcome. Remediation then often consists of one or more of the following: the punishment of the individuals concerned, recommending additional training, or the addition of new procedures or rules to cover the situation that has occurred.

In these investigations, the key question of how frequently the same divergences lead to successful outcomes is never even raised, let alone answered. However, from the 'Safety II' perspective, this is the crucial issue as these same variations in work performance lead to success far more often than they lead to failure. From this perspective, since things go right far more often than they go wrong, the focus of safety analysis should be on the processes that underlie success rather than the rare and unusual examples where things go wrong. From this perspective, variability of performance is characterised as an asset to be exploited rather than a problem to be eliminated. *Safety* emerges from a proactive process where key information is available to people to assess the current pressures on the system and to correctly balance the competing and often conflicting goals and objectives. This contrasts with the traditional approach which relies on retrospectively identifying failures and implementing barriers and defences.

Some of the contrasts between traditional perspectives on safety (now referred to as 'Safety I') and the new perspective ('Safety II) as described by Hollnagel are shown in Table 10.1. It is important to note that Hollnagel and others have never suggested a wholesale abandonment of the traditional views of safety represented by Safety I and always proposed a complementary relationship between Safety I and Safety II.

Hollnagel suggests that the traditional 'Safety I' approaches tend to work well in smaller, simpler work systems that can be characterised in terms of sequences of sub-tasks and readily understandable principles of operation.[17] Much of everyday manufacturing, assembly, distribution and repair work carried out by small- to medium-size organisations match these characteristics. There are plentiful opportunities for unwanted outcomes and injuries but these are more easily addressed through traditional methods of hazard identification and risk mitigation (see Chapter 2). The Safety II approach has grown along with the rise in increasingly large-scale complex technologies that cannot be easily broken down into independent sub-units and whose operations can never be completely described. Launching space vehicles, air transport systems, advanced manufacturing and healthcare all fall into this category. Inevitably there will be other examples that fall somewhere in between.

Hollnagel has emphasised the complementary nature of the two approaches as shown in Figure 10.1. For most purposes, a focus on the avoidance of failures will remain appropriate in managing industrial and occupational safety matters. When a worker trips and falls into a hole, philosophical discussions of cause and effect take second place to pragmatic attempts to remediate and avoid the hazards that

TABLE 10.1
'Safety I' and 'Safety II' Contrasted.

	Safety I	Safety II
Definition of safety	As few things go wrong as possible	As many things go right as possible
Safety management principle	Reactive, respond when something happens or is categorised as an unacceptable risk	Proactive, continuously trying to anticipate developments and events
Role of 'human factors'	Humans are predominantly seen as a liability or hazard	Humans are seen as a resource necessary for system flexibility and resilience
Accident investigation	Accidents are caused by failures and malfunctions. The purpose of an investigation is to identify the causes	Things basically happen in the same way, regardless of the outcome. The purpose of an investigation is to understand how things usually go right as a basis for explaining how things occasionally go wrong
Risk assessment	Accidents are caused by failures and malfunctions. The purpose of an investigation is to identify causes and contributory factors	To understand the conditions where performance variability can become difficult or impossible to monitor and control

Source: Hollnagel, 2014.

FIGURE 10.1 The relationship between Safety I and Safety II.

created the situation in the first place. There are of course still relatively simple failures grounded in individual or organisational function such as an unintended control movement or interpersonal miscommunication. In these cases, there may be straightforward design solutions (e.g. change the control layout, ensure prominent well-designed warnings in place, eliminate gaps or overlaps in responsibilities, etc.). Hollnagel's argument is simply that the conventional ('Safety I') approach has severe limitations when we try to deal with increasingly complex systems such as healthcare and the like where the human is expected to simultaneously be both a faultless rule-follower and a creative problem-solver. Hollnagel suggests that 'Safety II' provides a framework for better understanding the nature of such work and ways to better design and manage systems to maximise success rather than minimise failure.

RESILIENCE IN ACTION

IN-FLIGHT UPSET OFF THE COAST OF WESTERN AUSTRALIA – QANTAS FLIGHT 72

At 09.32 a.m. local time on 7 October 2008, an Airbus A330-303 departed Singapore on a scheduled flight (QF72) to Perth, Western Australia, with 303 passengers and 12 crew. At 12.40 p.m. while cruising at 37,000 ft the autopilot disconnected. At about the same time multiple failure indications appeared on the electronic flight displays. At 12.42 p.m. whilst the crew were evaluating the situation the aircraft abruptly pitched nose down, reaching a maximum pitch-down angle of 8.4 degrees and dropping 650 ft. The crew returned the aircraft to 37,000 ft but at 12.45 p.m., a second abrupt pitch-down occurred dropping 450 ft. The crew contacted air traffic control to urgently request a diversion to the nearest capable airport at Learmouth. Having heard about the serious injuries sustained in the passenger cabin from the cabin crew, the flight crew declared a MAYDAY subsequently landing at Learmouth at 1.50 p.m.[18]

A380 IN-FLIGHT UNCONTAINED ENGINE FAILURE– QANTAS FLIGHT 32

Just before 10 a.m. local time on 4 November 2010, an Airbus A380 operating as Qantas Flight 32 departed Singapore for a scheduled flight to Sydney,

Australia. Nearly 500 people were on board – 440 passengers and 29 crew including five on the flight deck. Shortly after take-off, climbing through 7,000 ft, 2 loud bangs were heard followed by indications of a failure of the number 2 engine on the left wing. The aircraft was put into a holding pattern whilst for the next 50 minutes, the crew attempted to deal with a cascade of failure indications involving the flight controls, hydraulic systems, fuel system, avionics system, brake system, wing slats and autothrust system.

The aircraft touched down safely back at Singapore Changi a little under 2 hours after taking off. Since the number 1 engine continued to run despite being shut off, it was nearly another hour later before passengers and crew were able to evacuate the aircraft.[19]

In addition to the official investigation reports produced by the Australian Transportation Safety Bureau, unusually the captains of both flights subsequently wrote books on the events giving us their detailed first-person observations about both episodes. The bland technical language of the official investigative reports fails to convey the drama and tension of the lived human experience. Having spent most of his career analysing the commonplace 'unsafe acts' of the human operator and some of the structural failings of organisations, Jim Reason later used the term 'heroic recoveries' to describe the upside of human performance in systems. These creative and adaptive capacities, he suggested, were 'potentially more beneficial to the pursuit of improved safety in dangerous systems' than focusing on the downside of the human propensity to error.[20]

In contrast with the well-understood mechanisms involved in human error, as discussed previously (see Chapter 3), Reason discerns little commonality behind the actions taken by the participants what he calls these 'heroic recoveries'. However, the detailed first-person accounts of the QF32 and QF72 events do show marked similarities despite the very different nature of the events themselves. Both captains were faced with previously unencountered and unimagined situations – the former involving a series of internal computer processing failures resulting in corrupt data being fed to the flight control computers resulting in autonomous action by the A330 to 'recover' from a non-existent condition and the latter an avalanche of systems failures originating in the uncontained failure of a fan blade in one of the A380s' engines.

Both captains faced multiple alerts and warnings and multiple checklists to be actioned. Both refer to the negative effects of almost continuous aural and visual alerts, with Captain **Richard de Crespigny**, the commander of QF32, likening the situation to 'being in a military stress experiment'. Both captains drew on their wealth of previous experiences – as a former naval fighter pilot, in the case of Captain **Kevin Sullivan** (QF72) and an avid interest in engineering and software design, in the case of Captain de Crespigny. Both relied on one of the most basic principles of piloting which is to prioritise aviating, over navigating and then over-communicating. Both captains adopted a critical approach to procedure-following rather than an unquestioning obedience and both made the best possible use of the other pilots on the flight deck. By a stroke of good fortune, the flight deck of QF32 contained probably the most A380 experienced crew ever assembled in one place, as in addition to Captain de Crespigny and his first and second officers were two Qantas check captains. Good fortune also played a further part in both events as they occurred in the daytime, in good weather and visibility. Had they occurred at night and in bad weather, the outcomes might well have been different.

The performance of these highly trained aircrews in exceptional circumstances shares something in common with the everyday performance of workers in many areas, which is an ability and willingness to adapt procedures and processes to the immediate demands of the task and situation with the aim of achieving the desired goals. Tasks that require unthinking procedure-following (such as assembly-line work) are increasingly subject to automation. Other tasks from policing to healthcare to aviation all require procedure-following but with the added flexibility to also adapt them to current circumstances. We might refer to this as *suppleness* – a quick-thinking deftness, a capacity to adapt and absorb stresses and shocks whilst remaining intact and functional. This quality shows itself most strongly when a system begins to function in entirely unexpected and unforeseen ways leading to intense pressures on the human operators to move beyond the comfortable familiarity of skill- and rule-based control to the remoter outer-lying regions of knowledge-based control beyond.

In both of these examples, involving the highest levels of computer-driven automation available, disaster was averted by the suppleness and flexibility of human performance in the face of completely unknown situations. Ironically, in the case of QF72, it was faultless rule-following by the automated flight control systems that almost led to catastrophe. In the QF32 case, the automated systems tried to transfer fuel into the damaged wing with potentially disastrous consequences. It was the ability of the crew to problem-solve at a higher level and respond flexibly, rather than rigidly adhere to this procedure, that was essential to the safety of the operation.

SAFETY MANAGEMENT: THE NEW AGENDA

Hollnagel's new agenda for safety management is essentially based on getting rid of the notion that all deviations in human performance from prescribed procedures and processes are necessarily bad. This is replaced with the idea that as systems become increasingly complex, with many hierarchical levels and convoluted networks of people and technology, it is the ability of operators to vary and adjust their performance that ensures successful outcomes. He argues that our traditional methods of safety management, developed from the simple technologies and workplaces of the 1930s, are no longer fit for purpose in the new age of advanced technologies. However, in most parts of the world, and even in large parts of the more prosperous economies, a good deal of work is actually not all that different in organization from that which characterised the world of the 1930s. Safety II, therefore, represents a new perspective on problems that arise in a relatively small proportion of workplaces but these are the ones, such as healthcare and transportation, with which we are all likely to have some direct personal involvement and which pose a significant potential risk to those involved.

Hollnagel has listed what he considers the six main myths that characterise current approaches to safety management in most industrial workplace settings.[21] Hollnagel suggests an alternative to each myth or widespread assumption. The myths and their alternatives are set out in Table 10.2.

Hollnagel argues that these 'myths' or widespread assumptions underlie the traditional (Safety I) approach that has undoubtedly led to many improvements in various industries and activities for the past century or more. However, these underlying assumptions can be questioned and examined for 'fitness for purpose' in regard to the increasingly hierarchical and complex sociotechnical systems that have developed in the 21st century. Challenging these assumptions is the basis for Hollnagel's 'Safety II' agenda which tries to focus on the positive side of human performance as flexible, adaptive and sensitive to current demands and therefore critical in keeping systems operating within safe bounds. When things do go wrong, system-based safety investigation tools (e.g. AcciMap, STAMP and FRAM) are required to provide a sound evidential basis for any subsequent changes. The application of these techniques in accident investigation will be described in the final chapter.

TABLE 10.2
Hollnagel's Six Myths of Current Safety Management.

Myth	Alternative
Human error is the largest single cause of accidents and incidents	More productive to understand why the behaviour that normally leads to successful outcomes occasionally leads to unwanted outcomes
Systems will be safe if people comply with procedures, rules and regulations	Since actual operations usually differ from the ideal, blind compliance may actually undermine safety and efficiency. Safe operations depend on adapting procedures to operating conditions
Safety can be improved by adding layers of barriers and protection	People can adapt to additional protections as described by Risk Homeostasis Theory (RHT) in Chapter 2. Additional layers can also increase complexity making it more difficult to understand what's happening
Root cause analysis can identify why mishaps happen in complex systems	In complex systems 'things go wrong in the same way as things that go right' and require systems-based analysis techniques such as AcciMap, STAMP and FRAM
Accident investigation is the logical and rational identification of causes based on facts	Accident investigations are based on underlying assumptions which can be characterised as the 'what you look for is what you find' principle
Safety always has the highest priority and will never be compromised	In practice, safety is as 'high as affordable'. Affordability and economics always have an influence

SAFETY AS A SYSTEM SCIENCE

The important groundwork for understanding how complex systems function was laid by an influential thinker born at the start of the 20th century in Austria. **Ludwig von Bertalanffy** studied at the University of Vienna at a time when Vienna was the intellectual heart of European thought. His studies involved psychology, philosophy and art history and later biology, particularly theoretical biology. This diverse intellectual background informed his thinking about biology and resulted in an approach which favoured a broad perspective on the interrelationships between organisms and the environment over the standard reductionist paradigm. Much of his post-war career was spent in North America where he shuttled between a number of institutions. He is most well-known for the publication of the book *General System Theory* in 1968, which actually contains material written much earlier, from the 1930s and 1940s onwards.[22]

Bertalanffy was concerned with two key aspects of systems – complexity and interrelatedness. Everything from organic life forms to human societies and technologies demonstrates a tendency towards greater complexity. In modern times, for example, we have moved rapidly from the simplicity of the steam engine to the complexity of autonomous vehicles. At the same time, every kind of system from the ecological to human society displays hierarchical organisation and a high degree of interconnectedness. These properties of systems required a different approach from the reductionist science that had hitherto achieved enormous success in advancing our understanding of physical and chemical processes. Bertalanffy proposed a general system science to provide ‘a broad view which far transcends technological problems and demands’ and demonstrating that lack of self-confidence was not a problem, modestly adding that this ‘heralds a new world view of considerable impact’.

As noted previously in this book, the development of cybernetics, with its emphasis on the importance of feedback in various kinds of systems, along with other advances in techniques for analysing network structures have greatly contributed to the development of general systems theory or systems science. Notably, the kind of explanations for phenomena used in systems science is typically broader than the narrower ‘cause and effect’ chain of events in time that forms the backbone of explanation in conventional science. Many phenomena in systems resist such simple explanations. For example, the process of remembering information can be thought of as one of ‘spreading activation’ across a network of nodes and linkages.[23] The estimated 86 billion (give or take) neurons in the human brain constitute a vast network of interconnections.[24]

Potentiation refers to instances where one aspect of a system acts to increase the potency of another factor in a way that is different from the independent effects of the two factors taken individually. In medicine, one drug (e.g. diazepam) can potentiate the effects of another (e.g. alcohol). Similarly, in a workplace system, poor interface design may potentiate the effects of inadequate training. Neither may, strictly speaking, ‘cause’ an error and the presence of both may be needed before there is a heightened probability of failure. Such notions do not fit well with the simple notions of direct cause and effect that underlie both everyday understanding and most legal concepts of accountability and responsibility.

Bertalanffy recognised that the phenomena manifested in systems could not be explained simply in terms of the properties of their individual components as studied in isolation. General systems theory was conceived as a 'science of wholeness' which recognised the similarities found across systems from the biological to the technological. For Bertalanffy, the idea of emergent phenomena did not mean that phenomena appeared out of nowhere, like a magician pulling a rabbit out of a hat, but that systems behaviour could only really be explained by complex non-linear relationships between components and their interrelationships. Hollnagel's work on 'Safety II' in general, and FRAM in particular, have usefully drawn attention to the multiple varieties of influence that one function can exert on another. As well as providing direct input into another function, for example, a function could simply provide the preconditions for another to take place. In Hollnagel's terms, influences can resonate or propagate through a system using a variety of mechanisms.

'SAFETY II' IN PRACTICE

Hollnagel has developed a technique known as the *Functional Resonance Analysis Method'* or FRAM for analysing the variabilities in performance that underlie work in complex systems. Simple cause-effect models remain adequate enough for many simple situations but struggle to adequately account for how and why outcomes emerge in more complex systems. Complex systems consist of numerous sub-systems with multiple interconnections between them. The human body, for instance, can be thought of in terms of respiratory, nervous, circulatory, endocrine and other systems that connect in often as-yet-unknown ways. It is not uncommon for medical diagnoses to run along the lines of: 'we don't know exactly why A is happening but it is related to processes occurring in X, Y and Z'. Complex technological systems can be represented by organisational charts but these do not represent the multiple underlying pathways and networks that link different individuals and groups in different parts of the system. The best way that we have of representing such connections is in the form of networks.

One of the basic principles of systems science is that systems can be represented as networks of relations.[25] A simple network structure is illustrated in Figure 10.2. The circles or nodes represent anything from physical structures, people, or workgroups to more abstract concepts. The lines represent linkages between them. Each node could also be a network in itself. In any complex system, the number of nodes and links grows exponentially so that tracking simple 'chains' of cause and effect becomes virtually impossible as well as conceptually implausible. Instead, the activity in parts of the network can be thought of as setting the conditions for activity in other parts of the network. Essentially, this is the basis for the technique developed by Hollnagel referred to as FRAM.

A FRAM model is a description of the work process in functional terms. In other words, it represents the various steps and stages involved in carrying out some activity in terms of the functions involved. The technique has its own website providing

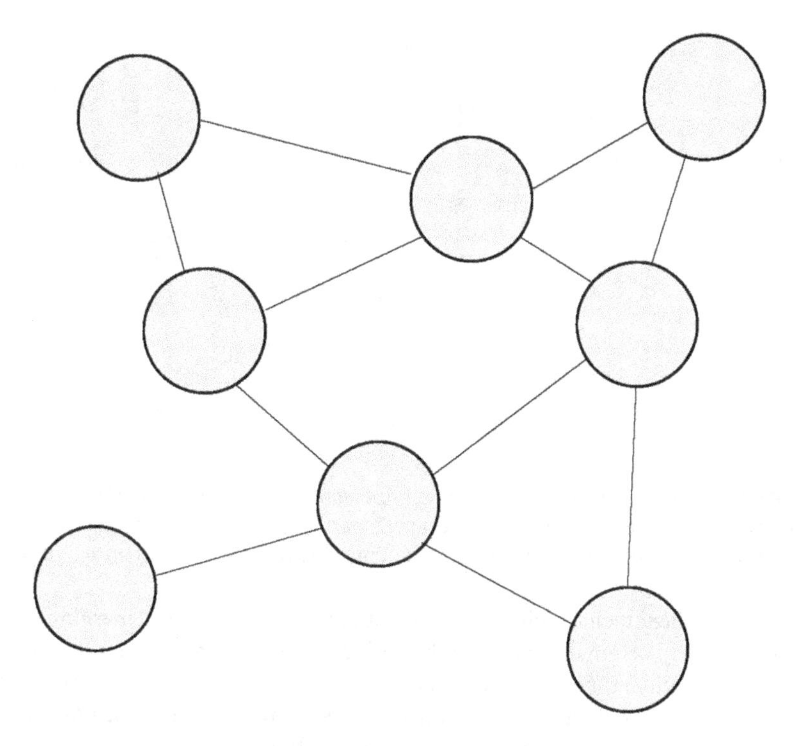

FIGURE 10.2 A representation of a network. Circles represent nodes and lines represent links between the nodes. Note that some nodes have more links than others.

guidance and instructions.[26] FRAM is very much based on Hollnagel's focus on 'work-as-done' rather than how it might be described in official manuals and guide lines. The key step is to identify the functions involved in performing a given task and this can be done through observation of work processes or by careful questioning of the people who actually carry out the work. Functions can be designated as *foreground functions* (i.e. whatever is in focus) or *background functions* (i.e. other necessary but not focal functions). These designations are context-dependent depending on the focus of the analysis. Then determine how each function links to the other functions – does it provide direct input into another function or does it affect it in some other way? In addition to providing basic input or output, FRAM proposes four additional types of input for functions. These are preconditions, resources, control and timing. The ordering and linkages should be consistent with the directional flow of steps within a task. The results can be set out as a network using the 'FRAM model visualiser'. An example is shown in Figure 10.3.

FRAM provides a handy representation of a new way of looking at how outcomes arise in complex systems. In place of causal chains, Hollnagel uses the systems science concept of *emergence* whereby completely new phenomena can arise from existing structures. This might involve dramatic changes at a chemical or molecular level resulting in organic compounds emerging from the inorganic to

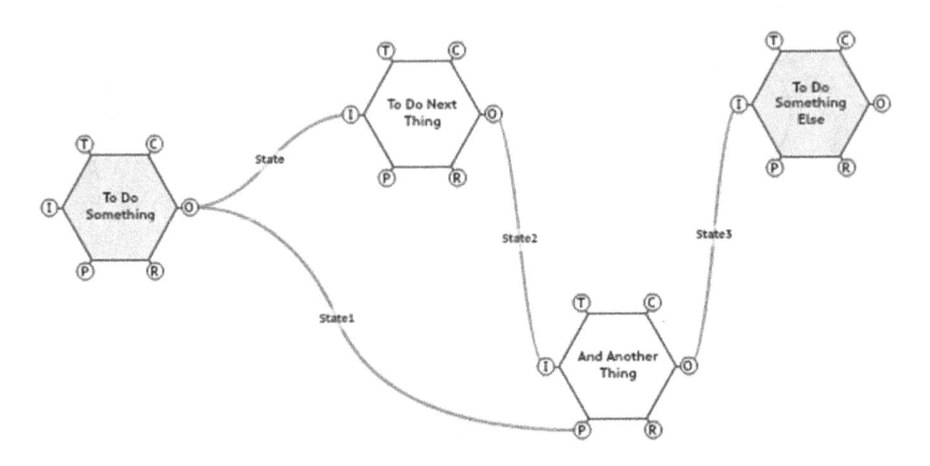

FIGURE 10.3 The elements of a FRAM representation of a simple task. Hexagons represent functions which are normally in the form of an action ('to do something') and connect to other functions in one of six ways (input, output, precondition, resource, control or timing).

new cultural phenomena such as investment banking eventually emerging from a system of money originally based on bartering goods. Hollnagel suggests that the underlying principle that could explain emergent phenomena is that of resonance – a concept taken from physics whereby a small force applied to one part of a system can generate much larger responses elsewhere in the system. A stone dropped in a pool provides a simple analogy – the ripples are going to be affected by static objects in different ways, but the ripples will also affect floating objects in different ways. The ripples, in effect, resonate with and against whatever exists within the same environment.

In sociotechnical systems, these 'ripples' or small forces consist largely of normal variability in performance as people adjust their actions in accordance with the current demands of their working environment. These small, dynamic adjustments can potentially have large implications under certain conditions, yet are entirely 'normal' in themselves. In Hollnagel's view, the usual accident investigation approach, based on a philosophy of 'root causes', involves looking for significant errors and failures amongst the precursors to the event, most often operationalised as variations from 'normal' practice. This approach is likely to fail in complex systems where outcomes can emerge as a result of subtle variations in the interactions between different elements of the system.

A number of studies utilising FRAM have already been reported in the scientific literature. Two literature reviews have also been published very recently.[27] The first involved screening five research databases over a 14-year period for published articles in English resulting in the identification of 108 published studies. The second scanned three of the same databases plus Google Scholar, over an 11-year period and identified 52 published peer-reviewed articles. Both reviews noted a steady increase in the number of FRAM publications over time and both noted an overwhelming preponderance of European authorship with very few articles emanating from North

America. Both reviews found that the domains that most often featured were healthcare and aviation.

At this point, researchers have been using the FRAM framework fairly flexibly with considerable variation in how the method has been applied. Both the reviews noted this as well as made a number of other critiques. Most obviously, the FRAM method relies very heavily on subject matter expertise (SME) to elicit the functions involved in accomplishing a task and thus is a highly time- and resource-intensive procedure. However, many other systems modelling techniques can also be similarly resource-intensive. At this stage, the approaches have been primarily qualitative with little attempt to quantify the variability observed in operations. This makes it difficult to generalise outside the very specific context under analysis to other contexts or to make comparisons between different tasks and activities. One of the reviews pointed to a lack of direct connection between the FRAM analysis and specific safety recommendations.

The first step in applying the FRAM technique is to identify the functions involved in a given task. Typically around 30 such functions have been identified for tasks such as 'drilling offshore oil well'[28] or 'hospital to home transitional care'.[29] Using the available software a visual representation of the functions and the linkages between them can be built fairly easily. The combination of 30 functions each with up to six possible kinds of links to other functions can generate highly complex figures resembling an elaborate 'cat's cradle'. This is useful in more realistically representing the way in which influences propagate or resonate in a network than can be achieved with simplistic 'links in a chain' diagrams.

The second step involves characterising the actual and potential variability associated with each function. Normally this involves variations in timing (e.g. too early or too late) or precision (e.g. accurate or not accurate) but could involve other things such as sequencing, force or duration. In some studies, this characterisation consists of a descriptive narrative of ways in which the function could vary and an assessment of possible impacts of this variability on other functions that occur further downstream.[30] In this way, a picture of the *work-as-done* (WAD) is developed along with potential insights into the impact that natural variations in performance could have elsewhere in the system. This kind of systems-based analysis represents a paradigm shift from the individualistic approaches to risk and error described in Section 1 of the book. It represents an evolution in approach from the organisational perspectives outlined in Section 2, which considered both organisational structure and individual responsiveness. The 'Safety II' approach as instantiated in techniques such as FRAM is essentially systems-based in its underlying description of processes such as resonance leading to emergent phenomena.

CRITIQUE OF 'SAFETY II'

Nancy Leveson (see Chapter 8) has written a very extensive (109 pages) and somewhat scathing appraisal of Hollnagel's 'Safety II' agenda.[31] Her first objection is that for some time the theories and methods utilised in workplace or occupational safety have differed significantly from the approaches taken in product and system safety engineering. Certainly, there are major differences in occupational background and training of practitioners in these two areas. Although never explicitly stated by Hollnagel, the historical background outlined in his books is predominantly that of workplace/occupational safety and it seems clear that this is the area that he is addressing in 'Safety II'.

Leveson argues that the whole 'Safety II' approach utilising ill-defined terms such as 'things that go right' and 'things that go wrong' is antithetical to the quantitative, precise world of engineering. She particularly objects to the characterisation of traditional approaches to safety which Hollnagel refers to as 'Safety I'. In Leveson's view, this characterisation constitutes a 'straw man' – a deliberately exaggerated and distorted picture created solely for the purpose of being knocked down and replaced by the favoured 'Safety II' agenda. In her view, the actual practice of safety engineering bears no resemblance to that described as 'Safety I' at all. However, the situation in workplace safety may well be different as evidenced by the fondness for root cause analysis and simplistic explanations in terms of causal chains.

Leveson seems particularly incensed by Hollnagel's arguments for the need to replace standard cause-effect thinking with a broader perspective based on emergent phenomena. Leveson strongly re-iterates a belief in the deterministic world of Newtonian physics where every phenomenon has a preceding cause and where all complex phenomena can be decomposed into constituent phenomena. Leveson makes no reference to modern physics in the form of quantum mechanics which has long since replaced the strictly deterministic Newtonian view with a 'stochastic' or probabilistic view. In the quirky subatomic world of quantum mechanics, particles such as electrons do not occupy fixed places or positions but can only be said to appear in any given location with a certain probability.[32]

If the subatomic world cannot be characterised deterministically but only in terms of probabilities, then it surely seems more than likely that the complex socially constructed worlds of engineering and technology may also be more profitably viewed from a nondeterministic perspective. It is difficult to see how complex phenomena such as the emergence of language in human evolution can be explained atomistically and reductively. It could be argued that a poorly trained worker using a badly designed interface to carry out a critical task simply has some probability of succeeding and some probability of failing. Neither training, interface design nor worker attention or capacity can be said to uniquely cause any given outcome, but these factors can, and do, influence events with certain probabilities. In Hollnagel's view, it is the variability in each of these aspects that results in different outcomes over time. Leveson can reasonably claim, that this is precisely a matter of the outcomes being caused by the interactions between these system components, which is also consistent with a systems science view as originally understood.

In the end, philosophical disputes about the nature of reality are less important in the context of risk and safety management than the practical utility of viewing systems deterministically or otherwise. Simple chains of causation can be constructed in many workplace contexts and provide perfectly satisfactory accounts of 'workers-falling-into-holes' events. It may also be true, however, that interpreting events in enormously complex distributed systems such as healthcare in nondeterministic terms utilising concepts such as resonance or emergence has greater practical utility. The interest in Hollnagel's approach, particularly in healthcare, suggests that practitioners in that area believe that it might. Time will tell whether the approach advocated by Hollnagel proves sufficiently fruitful in terms of enhancing workplace safety and performance or not.

There does seem to be something of a logical flaw in contrasting traditional ('Safety I') and new ('Safety II') approaches on the basis of whether they focus on 'failure' or 'success'. For example, if I am interested in the safety of stairways, I could analyse records of accidents (failures) and look for common features. I might find several variables showing up related to the environment (e.g. poor lighting), equipment (high-heeled footwear) and user behaviour (e.g. being in a hurry). Alternatively, I could observe users coming up and down stairways and look for variations and adaptations that lead to successful outcomes, finding that good lighting, flat footwear and taking time were examples of successful variations. As long as we categorise failure and success as complementary (i.e. The probability of failure is 'p' and the probability of success is '1–p') then the practical outcome will be the same as in the stairway example regardless of whether we focus on 'things going right' (success) or 'things going wrong' (failure). Leveson makes a similar point.

However, there is an asymmetry in how we traditionally view performance and outcome. When investigating negative outcomes (aka 'accidents', 'crashes', etc.) we invariably ask whether components (human or otherwise) operated as expected. For example, whether pilots followed the procedures. If the answer is 'not' then the investigation may, and frequently does, conclude with a pilot error or component failure as 'the cause'. We do not investigate positive outcomes so we do not know about the instances where operators also failed to follow procedures or componentry also didn't function as intended, but no negative outcome occurred. These situations may actually be quite common, as in the case of healthcare 'workarounds' noted earlier.

As another example, the ill-fated Boeing 737 Max grounded in March 2019 after 2 fatal crashes had flown approximately 500,000 flights since being introduced to commercial service in 2017. Some of these flights undoubtedly experienced problems with the MCAS (Manoeuvering Characteristics Augmentation System) control system (an automated system for flight stability), subsequently implicated as the major factor in the two crashes.[33] Unfortunately, relatively little is known about how these were resolved. More information about how pilots succeeded in overcoming the limitations of the MCAS system (i.e. the 'things that went right' in Hollnagel's terms) would surely be extremely useful in preparation and training for future problems. At the same time, it seems very obvious that correcting 'the things that went wrong' with the MCAS system should be the primary key to increasing the safety of the Boeing 737 Max.

Leveson has little patience for 'Safety II' and what she considers the general wooliness of Hollnagel's arguments and flawed logic. There is no doubt that Hollnagel's case is not helped by the liberal use of statements such as 'The aetiology is the way of explaining the phenomenology in terms of the ontology' and 'This can be seen as exogenic and endogenic performance variability and is in some sense irreducible'.[34] These may sound slightly better to the European ear than to those in North America accounting, in part at least, for the much greater enthusiasm by European scientists and researchers for Hollnagel's approach. The idea of looking at work and safety from a systems perspective which underpins Hollnagel's approach was born out of the broad intellectual developments in early 20th-century Europe, described earlier, that gave rise to systems theory and systems science.

CONCLUSIONS

'Safety II' could be viewed largely as a highly successful exercise in rebranding the much less catchy-sounding 'resilience engineering'. With several clever pairings such as 'Work as done' versus 'Work as imagined' and 'Safety II' versus 'Safety I', Hollnagel's ideas have gained significant traction in certain areas – notably in the domain of healthcare and, particularly, amongst researchers outside North America. At one level, as noted previously, whether the focus is on the vaguely defined 'things that go right' rather than the 'things that go wrong' is partly a matter of semantics. However, it is also true that in complex systems, studying the adaptations and workarounds that people use to generate success might yield somewhat different information than studying the precautionary strategies used to minimise mistakes, but overall the difference between the two approaches may not be as profound or practically significant as hoped. It's also hard to disagree with those who find that it provides a more rounded and realistic view of the challenges workers actually face, nor that its emphasis on 'understanding why things go right' is more humanising than the focus on errors and mistakes. Time will tell whether 'Safety II' turns out to be genuinely better science or simply better marketing and PR.

Nancy Leveson has strongly criticised the scientific and philosophical assumptions of 'Safety II' in no uncertain terms. Some of her criticisms have already been mentioned. The three frameworks described in this third section of the book are all, in their own way, oriented towards advancing the theory and practice of safety science beyond the limitations of simplistic thinking exclusively in terms of human error or organisational failure to a much broader concern with systems as a whole and the extremely complex pattern of interrelationships between every part of these complex systems. To illustrate the extent of the complexity involved, Leveson cites one concrete example where it was calculated that one aircraft system (the TCAS collision avoidance system described in Chapter 6) could in principle exhibit a staggering 10^{40} (that is, 1 with 40 zeros after it!) specifiable different states. This represents an impossible number of functions and pathways to map out even the smallest fraction using something like FRAM. To put this in perspective, it is estimated by the European Space Agency that the number of stars in the visible universe is a mere 10^{22} (1 with 22 zeros or 1 billion trillion!)

TCAS is just one of numerous systems that make up a modern commercial aircraft and these systems are themselves just one part of a complex hierarchical structure encompassing governments, regulators, operators, manufacturers, pilots, etc. There is no doubt that to understand how these vastly complex systems function and to be able to effectively learn from experiences (good and bad) with these systems, we need a much broader conceptualisation of how systems work and how their outcomes are produced. Fortunately, systems science provides a natural framework in which to locate the questions that arise in the context of workplace system safety.

Viewing safety as an emergent phenomenon along with the introduction of broader explanatory concepts based on systems as networks of components with complex interlinkages immediately raises important questions about the process of accident investigation. In the final chapter, we will review the traditional basis of accident and incident investigation and introduce the reader to several of the newer approaches based on systems theory. Whilst engineers may have long been comfortable with exploring non-linear dynamics in complex systems, public understanding and legal frameworks do not fit with explanations of events that do not follow an 'A leads to B leads to C' linear logic. Systems-based approaches to accident investigation often attempt to bypass such limitations through the use of graphical representations that can illustrate complex patterns of non-linear interactions without the need to understand differential equations and the like. The final chapter looks at how the systems science-based perspective can help provide answers to the socially and practically important questions that arise in the aftermath of a major incident or event.

NOTES

1. See for example: Ji, C., Wei, Y., Mei, H., et al. (2016). Large-scale data analysis of power grid resilience across multiple US service regions. *Nature Energy, 1*, 16052. https://doi.org/10.1038/nenergy.2016.52; Rachunok, B., & Nateghi, R. (2020). The sensitivity of electric power infrastructure resilience to the spatial distribution of disaster impacts. *Reliability Engineering and System Safety, 193.* https://doi.org/10.1016/j.ress.2019.106658
2. Hollnagel, E., Woods, D.D., & Leveson, N. (Eds.). (2005). *Resilience Engineering: Concepts and Precepts.* Aldershot: Ashgate.
3. Woods, D.D. (2003). Creating foresight: How resilience engineering can transform NASA's approach to risky decision making. Testimony on '*The Future of NASA*' to the Committee on Commerce, Science and Transportation. Creating-Foresight-How-Resilience-Engineering-Can-Transform-NASAs-Approach-to-Risky-Decision-Making.pdf (researchgate.net)
4. Hollnagel, E. (2018). *Safety-II in Practice: Developing the Resilience Potentials.* Abingdon: Routledge.
5. Hale, A., & Heijer, T. (2004). Is resilience really necessary? The case of railways. In E. Hollnagel, D.D. Woods, & N. Leveson (Eds.), *Resilience Engineering: Concepts and Precepts* (pp. 125–147). Aldershot: Ashgate.
6. This is from another chapter in the same book by the same authors as above (Note 5) entitled 'Defining resilience' (pp. 35–40).
7. Rodriguez, M., Lawson, E., & Butler, D. (2020). A study of the resilience analysis grid method and its applicability to the water sector in England and Wales. *Water and Environment Journal, 34*(4), 623–633.
8. Fleming, M. (2001). *Safety Culture Maturity Model. Offshore Technology Report for the Health and Safety Executive.* Norwich: HMSO. Patrick Hudson at Leiden University has also developed a safety culture maturity model that has been extensively applied in aviation and the oil and gas industry. See for example: Parker, D., Lawrie, M., & Hudson, P. (2006). A framework for understanding the development of organisational safety culture. *Safety Science, 44*(6), 551–562.

9. https://en.wikipedia.org/wiki/Fukushima_Daiichi_nuclear_disaster
10. Hollnagel, E., & Fujita, Y. (2013). The Fukushima disaster – Systemic failures as the lack of resilience. *Nuclear Engineering and Technology, 45*(1), 13–20.
11. Nelson, P.F., Martin-Del-Campo, C., Hallbert, B., & Mosieh, A. (2016). Development of a leading performance indicator from operational experience and resilience in a nuclear power plant. *Nuclear Engineering and Technology, 48,* 114–128.
12. Patriarca, R., Bergstrom, J., Di Gravio, G., & Constantino, F. (2018). Resilience engineering: Current status of the research and future challenges. *Safety Science, 102,* 79–100.
13. Dekker, S.W.A., & Woods, D.D. (2010). The high reliability organization (HRO) perspective. In E. Salas & D. Maurino (Eds.), *Human Factors in Aviation, 2nd Ed* (pp. 123–143). New York: Academic Press Elsevier.
14. Institute of Medicine. (2001). *Crossing the Quality Chasm: A New Health System for the 21st Century.* Washington, DC: The National Academies Press. https://doi.org/10.17226/10027
15. Leape, L., & Berwick, D.M. (2005). Five years after *To err is Human* what have we learned? *Journal of the American Medical Association, 293*(19), 2384–2390.
16. Chassin, M., & Galvin, R.W. (1998). The urgent need to improve health care quality. *Journal of the American Medical Association, 280*(11), 1000–1005.
17. Hollnagel, E. (2014). *Safety-I and Safety-II: The Past and Future of Safety Management.* Boca Raton, FL: CRC Press.
18. Australian Transport Safety Bureau. (2011). *In-Flight Upset 154km West of Learmonth, WA 7 October 2008 VH-QPA Airbus A330-303. Aviation Occurrence Investigation AO-2008-070.* Civic Square, ACT: Australian Transport Safety Bureau.
19. Australian Transport Safety Bureau. (2013). *In-Flight Uncontained Engine Failure Airbus A380-842, VH-OQA overhead Batam Island, Indonesia 4 November 2010. Aviation Occurrence Investigation AO-2010-089.* Civic Square, ACT: Australian Transport Safety Bureau.
20. Reason, J. (2008). *The Human Contribution: Unsafe Acts, Accidents, and Heroic Recoveries.* Farnham: Ashgate.
21. Besnard, D., & Hollnagel, E. (2014). I want to believe: Some myths about the management of industrial safety. *Cognition, Technology and Work, 16,* 13–23.
22. Von Bertalanffy, L. (1969). *General System Theory, Revised Ed.* New York: George Braziller Inc. The book consists mainly of chapters, originally published in German, between 1940 and 1969.
23. Anderson, J.R. (1983). A spreading activation theory of memory. *Journal of Verbal Learning and Verbal Behavior, 22,* 261–295.
24. See: https://www.nature.com/scitable/blog/brain-metrics/are_there_really_as_many/
25. The topic of networks is covered in Chapter Four of: Mobus, G.E., & Kalton, M.C. (2015). *Principles of Systems Science.* New York: Springer Science.
26. The website is: functionalresonance.com. Hollnagel has also published a book describing the RAG and FRAM techniques: Hollnagel, E. (2018). *Safety-II in Practice: Developing the Resilience Potentials.* London: Routledge.
27. Two recent reviews of FRAM have been published: Tian, W., & Caponecchia, C. (2020). Using the functional resonance analysis method (FRAM) in aviation safety: A systematic review. *Journal of Advanced Transportation.* https://doi.org/10.1155/2020/8898903. The second review was: Salehi, V., Veitch, B., & Smith, D. (2021). Modelling complex socio-technical systems using the FRAM: A literature review. *Human Factors in Manufacturing, 31,* 118–142.

28. Franca, J.E.M., Hollnagel, E., Luquetti dos Santos, I.J.A., & Haddad. (2019). FRAM AHP approach to analyse offshore oil well drilling and construction focused on human factors. *Cognition, Technology and Work*. https://doi.org/10.1007/s10111-019-00594-z
29. O'Hara, J.K., Baxter, R., & Hardicre, N. (2020). 'Handing over to the patient': A FRAM analysis of transitional care combining multiple stakeholder perspectives. *Applied Ergonomics, 85*, 1–10.
30. Bridges, K., Corballis, P.M., & Hollnagel, E. (2018). 'Failure-to-Identify' hunting incidents: A resilience engineering approach. *Human Factors, 60*(2), 141–159.
31. Leveson, N. (2020). Safety-III: A systems approach to safety and resilience. *MIT Engineering Systems Lab*. Available for download at: Safety III: A Systems Approach to Safety and Resilience (mit.edu)
32. Carlos Rovelli's wonderful brief introduction to physics has a chapter on quantum mechanics. See Rovelli, C. (2014). *Seven Brief Lessons on Physics*. UK: Allen Lane.
33. Boeing continues to provide updates on the re-designed MCAS. See: https://www.boeing.com/commercial/737max/737-max-software-updates.page An overview of MCAS can be found at: https://en.wikipedia.org/wiki/Maneuvering_Characteristics_Augmentation_System
34. These two statements can be found in Hollnagel's (2014) book – see Note 17 above.

11 Making Sense of Failure *Beyond Accident Investigation (with Karl Bridges)*

History doesn't record the first 'accident' – presumably, a cave-dweller making unintended contact with a rock or a hunter damaging himself or a fellow hunter with a spear rather than the intended prey. We do have records of the first modern technology-inspired accidents involving vehicles such as the motor car and the airplane. The first automobile accident occurred in 1891 in Ohio when a car hit a tree root and spun out of control. Not hard to find the 'root cause' for that one! The first airplane crash happened almost five years after the first powered flight by the Wright Brothers when an aircraft also piloted by Orville Wright crashed in Virginia injuring Wright and killing his passenger. Although major accidents involving aircraft, nuclear power stations, large ships and spacecraft are carefully noted, the same cannot be said for workplace accidents and transportation accidents involving smaller aircraft and ships, and especially automobiles. Accuracy of statistical records also varies markedly from country to country.

The very term 'accident' is freighted with unwanted meanings involving 'chance' and 'unpredictability' – essentially events that are inexplicable. For this reason, the use of the term amongst safety professionals has declined over recent decades. Aircraft accidents are generally referred to as 'occurrences' and motor vehicle accidents are usually referred to as 'crashes'. The US Occupational Safety and Health Administration refers to its procedures for 'fatality and catastrophe' investigation. Many workplace safety investigations are driven by the occurrence of notifiable injury – generally, something that requires medical treatment. Events where the possibility of loss or harm existed, even if it did not actually occur (such as when two airliners pass rather too close to one another), may also be subject to investigation. The term 'adverse event' is also widely used particularly in healthcare. With these caveats in mind, for the sake of simplicity, we will use the term 'accident' as a well-known cover-all for all these other descriptive terms.

ACCIDENT INVESTIGATION

Part science and part craft, there is no one universally adopted method of accident investigation. However, the major workplace safety and transportation investigative agencies tend to follow a roughly similar approach. The International Civil Aviation

DOI: 10.1201/9781003038443-14

Organization (ICAO), for example, defines accident investigation as 'a systematic process whereby all of the possible causes of an adverse event are evaluated and eliminated' involving several stages.[1] The first step is always to gather the available evidence. This may mean closing off a road or airport or temporarily shutting down a workplace to preserve the scene. Any available witnesses should be identified and any recordings (video, voice recorders, data recorders, etc.) identified and secured. Large-scale transportation investigations can involve a staggering array of information from numerous different sources requiring a team of investigators to coordinate. In cases such as an airliner crash, these separate teams may be formed to cover topics such as aircraft structure, power plants, avionics, meteorology, and very often and most importantly, human factors covering the human performance aspects of the event of which there are invariably many.

The second step involves some form of analysis, most commonly beginning with the construction of a timeline of the event. Getting every event in the sequence is an important part of beginning to understand what has happened. Understanding, however, requires some sort of interpretation of the events and whilst investigators may strive for neutrality and objectivity this is impossible to attain. Whether explicitly acknowledged or not, all analyses are based on underlying assumptions. In accident analysis the key assumptions involve causality. Throughout the book, we have examined a variety of notions about accident causality and safety. Major bodies such as ICAO and the UK Health and Safety Executive (HSE), for example, explicitly follow simple linear chain-of-events models based on Heinrich's Dominos and Reason's 'Swiss Cheese' (see Chapter 3).

A comparison of five well-known methods including SCAT based on Heinrich's Domino model and TRIPOD based on Reason's Swiss Cheese model was reported by William Wagenaar.[2] Each method was evaluated against a list of criteria including reliability, validity, practicality and whether it was 'revealing' in the sense of specifying non-obvious underlying causes. TRIPOD was judged to meet the criteria better than the other approaches. More recently, a team of Greek researchers examined the same five models alongside eight other rather less well-known models in terms of a similar list of criteria.[3] Both studies confirm that what can be generated by any accident analysis method is dependent on the underlying assumptions of the method with respect to causal models.

This has been labelled as the 'What-You-Look-For-Is-What-You-Find' phenomenon.[4] In one study, accident investigation manuals were obtained from eight Scandinavian organisations across a variety of fields from patient safety to transportation (marine, air, road and rail) and nuclear safety. Despite the catchy title of the report, the predominant approach of these organisations to investigating adverse events in a diverse set of work domains was, in fact, largely similar. The dominant basis for analysis was the 'Swiss Cheese' model based on a sequence of active (i.e. 'front-line'), latent (i.e. organisational) and defensive (i.e. protective barriers) failures. The researchers point out, however, that following this kind of causal model inevitably limits the scope of the investigation to certain kinds of failures and cannot capture failures that propagate, resonate or spread through complex systems in a non-linear fashion.

In the remainder of this chapter, we will compare four different accident analysis approaches to the same adverse event. The first approach (Human Factors Analysis and Classification System (HFACS)) is based on the widely used Reason 'Swiss-Cheese' model whilst the other three are more recent systems-based approaches based on the work described in the last three chapters – the Causal Analysis based on STAMP (CAST) model derived from Leveson's Systems Theoretic Accident Modelling and Processes (STAMP) approach (see Chapter 8); the AcciMap model based on Rasmussen's approach (see Chapter 9) and finally the Functional Resonance Analysis Method (FRAM) model developed by Hollnagel described in Chapter 10. We first briefly outline the event itself – the unfortunate capsizing of a crowded British cross-channel ferry in 1987. A capsule summary of this event can also be found in Chapter 2.

THE 'HERALD OF FREE ENTERPRISE' FERRY DISASTER

On Friday evening of 6 March 1987, a special report was broadcast on UK television, detailing the 'Zeebrugge Disaster', to be known later as the *Herald of Free Enterprise* accident. It was described by the BBC as the worst ever channel disaster in peacetime. The merchant ship *Herald of Free Enterprise* had capsized suddenly as it left the port of Zeebrugge in Belgium *en route* to Dover in the UK. The incident resulted in the loss of the lives of 193 passengers and crew.

The UK Department of Transport investigated the accident, eventually resulting in the prosecution of the ferry owner's, Townsend Car Ferries. What makes this accident of particular interest, aside from the high death toll at the time, is the multiple failings that were revealed during the investigation and subsequent analysis. As a result, many researchers have subsequently found it useful as an example to illustrate their approach to incident and accident modelling.

The following sections present four different analyses of the *Herald of Free Enterprise* disaster using HFACS, STAMP, AcciMap and FRAM, with some basic explanation of the approach, purpose and outcomes of each approach. Each approach offers a different perspective on the failures and variations that led to this accident occurring. It can be argued that each of the approaches has both shortcomings and benefits. The opportunity to compare four different approaches to analysing and understanding one single event will assist the safety practitioner to better evaluate the value of these widely used approaches and decide for themselves which tool or tools to use for understanding the course of significant events in their own area of interest.

HUMAN FACTORS ANALYSIS AND CLASSIFICATION SYSTEM

As noted in Chapter 3, the *Human Factors Analysis and Classification System* (HFACS) developed by Scott Shappell and Doug Wiegmann focuses on the human – the humans at the point of failure; the humans that supervise, team lead or make tactical decisions; and the humans that provide strategic direction for the company or governing body. The model is largely based on a linear view of safety – that is using the well-known 'Swiss Cheese' model, HFACS seeks to define what those holes in the model actually are. Both the Swiss Cheese and HFACS models have enjoyed considerable popularity over the years due to their relative simplicity of use and ease of understanding.

An HFACS analysis of the *Herald of Free Enterprise* Accident has been conducted by a team of researchers from Taiwan and Liverpool.[5] The researchers first combed through the official inquiry and related material to derive a set of 'causal factors and the causation between them'. These identified causal factors are listed in Table 11.1. Second, these factors were then classified using a slightly modified HFACS system. The results were represented graphically as shown in Figure 11.1.

TABLE 11.1
Identified Causal Factors in the HFACS Analysis of the *Herald* Disaster. Adapted from Chen et al. (2013).

TE	Top event: *Herald* fills with water through the bow doors and lists to port, capsizing on an adjacent sandbank
E1	Large quantity of water entered the car deck
E2	Insufficient upright force to counteract the ingress of water, raising the ship's centre of gravity (C of G) leading to the Free Surface Instability (FSI) effect
E3	Vessel lacks stability
E4	Bow door open
E5	Vessel accelerates to 18 knots
E6	Vessel trimmed by the head to about 0.8 m
E7	Vessel significantly overloaded
A1	Master assumed the ship was ready for sea
A2	Assistant bosun not present at the station to close bow door
A3	The Master did not follow the usual practice to restrict the pitch setting and propeller speed
A4	Master interpreted the absence of any report as a positive
A5	Chief Officer left bow door station unmanned
A6	Master took no account of the additional water ballast used for trimming
P1	Poor Ship's Standing Order (SSO) was given to crews
P2	The harbour loading ramp was unsuitable for all conditions
P3	Insufficient time to empty ballast tanks
P4	Spring tide occurred
P5	No indicators of ship's draught were fitted
P6	No bow-door status warning fitted
P7	No anti 'free-surface instability' devices fitted
P8	No high-capacity ballast pump fitted
S1	Management did not circulate information about other incidents to Masters
S2	Lack of clear orders to crew on duties and responsibilities
S3	Senior Master content with existing SSOs
S4	Harbour station order given before loading completed
S5	Insufficient manning of the vessel
O1	Failures of shore management
O2	Time pressure to sail early and adhere to schedule
O3	Shore management did not consider the fitting of anti-FSI devices necessary
O4	Regulatory authorities did not address the need for anti-FSI devices

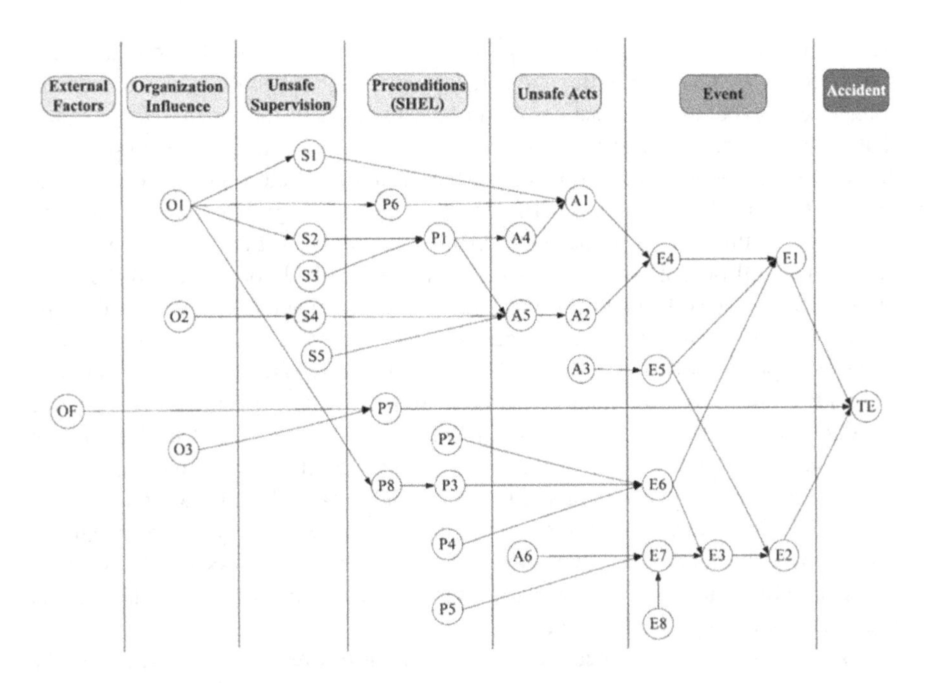

FIGURE 11.1 A graphical representation of an HFACS analysis of the *Herald of Free Enterprise* disaster. The circles represent the identified causal factors as listed in Table 11.1. ©2013. From: 'A human and organisational human factors (HOFs) analysis method for marine casualties using HFACS-Maritime Accidents (HFACS-MA)' by Shih-Tzung Chen et al., published in Safety Science. Reproduced by permission of Elsevier Science and Technology Journals.

The resulting HFACS map shown in Figure 11.1 shows the 'top' event (TE) of capsizing preceded by a range of inter-twined unsafe acts, preconditions, supervision, organisational and external factors. Using this approach, the links between the regulatory authority's failure to address the inherent instability of this design of 'roll-on roll-off' ferry with anti-free surface instability devices and the subsequent events start to become much clearer.

The HFACS approach demonstrates its value in assessing the flow-on effect of organisational problems that lie dormant for many years, akin to old wartime unexploded ordinance, which has the potential to eventually, quite literally, blow up with potentially fatal consequences if subsequent conditions are right. However, the authors of the HFACS model presented in Figure 11.1 pointed out that some of the limitations of HFACS could reside in trying to understand the weight or significance that each human and organisational factor plays in contributing to the overall occurrence of an accident. For example, some of the preconditions and unsafe acts, if they had been addressed before the accident, may not necessarily have prevented it, whereas others, such as A2 (the absence of the assistant bosun), probably would have been instrumental in saving the ship if the bosun had been present at the bow doors on departure.

The most significant benefit of HFACS is that it provides guidance for the practitioner or accident investigator to start thinking about the wider system. That is to go beyond the people or processes that appear to play a role in an accident on the day, to look beyond the apparent failures and to try to understand aspects associated with culture, accepted (but unsafe) practices and organisational decision making, for instance, that have all contributed to this occurrence. With this in mind, the analysts of the *Herald* disaster added an additional layer (i.e. 'External Factors') not in the original HFACS scheme. All too often government legislation and regulatory actions can have significant, if unintended, impacts on operational safety. In the case of the *Herald* and similar roll-on, roll-off (Ro-Ro) ferries, regulators had not addressed the value of anti-free-surface effect devices which might have been able to prevent the capsize.

HFACS has been very widely used. A recent survey found 43 published peer-reviewed reports appearing between 2000 and 2018 where HFACS had been applied to accident analysis.[6] The majority of these studies involved aviation and maritime accidents with some applications to rail, mining, construction and nuclear power. The majority of studies went further than simply classifying accident causes into HFACS categories, by analysing the pathways and relationships between levels and categories with a variety of statistical techniques. HFACS analyses are inevitably limited by the quantity and quality of information available. As this is typically focused on the 'sharp end', HFACS analyses often lack information about events at organisational and governmental levels.

HFACS was clearly grounded in the linear chain-of-event models outlined in the first section of the book, particularly the 'Swiss Cheese Model' of Reason, itself based on the earlier domino model of Heinrich. The techniques discussed in the remainder of this chapter were all developed from the systems-focused approaches described in Section 3. These techniques all assume that non-linear causal pathways are predominant in complex systems and they all provide graphical tools for representing the ways in which such influences can spread in causal networks. In the next section, we look at an '*AcciMap*' analysis of the *Herald* disaster. *AcciMap* was originally developed by Jens Rasmussen primarily as a risk management tool for complex workplace systems but has become increasingly utilised as an accident analysis tool.

ACCIMAP

As discussed in Chapter 9, Jens Rasmussen produced his seminal paper on risk management in a dynamic society in 1997. He argued that accidents generally occur as a result of losing control of physical processes, the result of which results in injury or death, damage of assets, negative media coverage, loss of reputation and a questionable future. At the top, society seeks to control safety with laws and regulations, delivered via government regulators and the guidance of good working practices. This is often delivered through regulators (covering specific industries) and trade unions. However, as is often the challenge when decrypting legislation (the 'what' should be done to make things safe), organisations are often left with applying the laws to their workplaces in many different ways (the 'how' should we do it to make things safe). The decisions made at this level can often make or break safety in the workplace. Indeed, in many accident investigations, the effects of organisational strategic and tactical decision making are readily apparent.

In many respects, one would be forgiven for thinking that the single point of failure most often resides at these higher levels rather than at the level of the human operator, as is more traditionally thought. However, the safety of complex workplace systems is more than the sum of its parts and organisations are often under immense political pressure to deliver, particularly if they are a national flag carrier organisation or state-owned enterprise. Therefore, apportioning blame solely onto the company and managerial decision making is likely to be as short-sighted as solely blaming a single human operator for an accident when it is onerous and questionable industry-wide safety-critical activities that may be significantly contributing. Of course, someone must make sense of the workplace and apply legislation into safe working practices, which will influence the way staff work with each other and with the tools they use to do their jobs. It is important to note the use of the word 'influence' and not control, because as Jens Rasmussen has indicated, the complexity of workplaces rapidly increases as new technology far exceeds human ability to keep up. Indeed, even in highly skilled environments such as an airline cockpit, performance has been fraught with error as a result of overly complex and opaque automation[7].

Human error has often been considered the cause of many accidents. However, as we have seen in the previous chapters, it would be better if the term were restricted to describing the *consequences* of system complexity. Like, many other proponents of safety modelling, Jens Rasmussen demonstrates his theory of accident causation with the presentation of the *AcciMap* methodology and applied it to the *Herald of Free Enterprise* to illustrate how the interaction of several *actors* in the complex workplace system of merchant seafaring ultimately led to the capsize of the vessel.

The complexities of what happened at Zeebrugge can be roughly mapped out as shown in Figure 11.2. This general schematic outline formed the basis of an *AcciMap* representation initially reported by British ergonomist **Barry Kirwan**, based on an analysis of the lower four levels done by Jens Rasmussen for a workshop in 1997[8]. We have adapted this further to include levels 1 and 2 in the model to illustrate the impact of government policy and regulators on the outcome. The developed *AcciMap* representation of the *Herald* disaster is shown in Figure 11.3.

The AcciMap analysis begins by organising contributory factors according to their position in the sociotechnical hierarchy. Relevant factors are generally those without which the outcome would not have occurred. The linkages between factors represent identifiable paths of influence of one factor on another. *AcciMap* clearly shows the significance of actions closely related in time and place to the event on the day – such as the bosun on the *Herald of Free Enterprise* not being present to close the bow doors. However, *AcciMap* also provides the opportunity to clearly see how events, at organisational, regulatory and legislative levels, can have made an equally important contribution to the accident. This provides valuable assistance in quickly developing higher-level safety recommendations. The graphical representation helps make the links between different levels of the system particularly evident. In the present example, the lack of attention to the free-surface instability effect inherent in the Ro-Ro design of the *Herald* and her sister ships by regulators, designers and managers has evident cascading effects throughout the system. Approximately 35

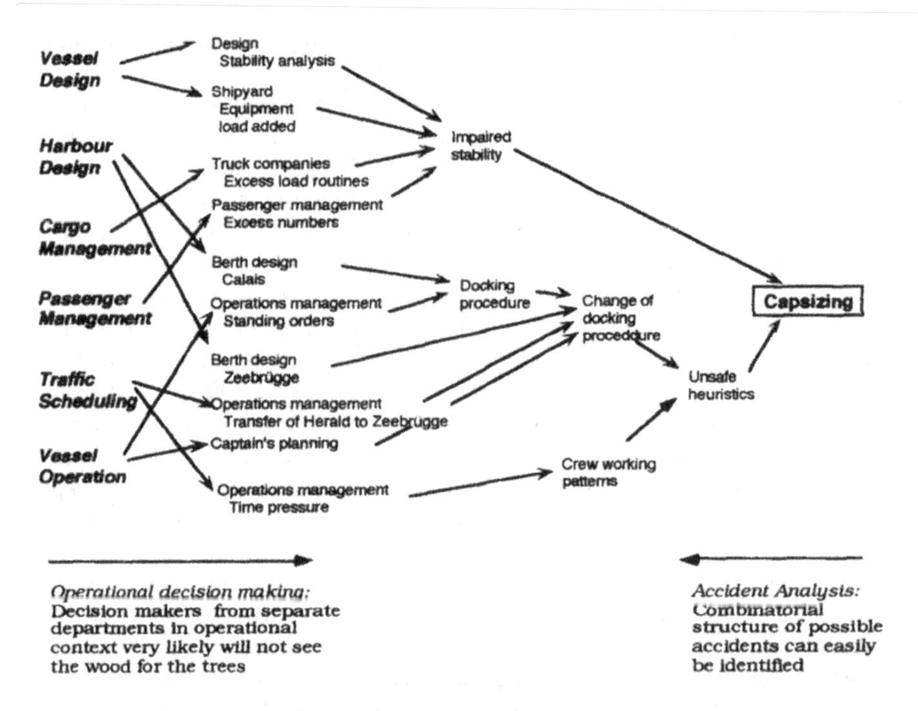

FIGURE 11.2 Initial mapping of the factors involved in the *Herald of Free Enterprise* accident. ©1997. From Risk management in a dynamic society: a modelling problem by Jens Rasmussen published in Safety Science. Reproduced by permission of Elsevier Science and Technology Journals.

Ro-Ro ferries have been involved in major maritime disasters from as early as 1953 through to 2020.

One might see certain similarities with the graphical results of the HFACS approach, in that an arrangement of causal and contributing factors are presented based on specified levels of influence. The difference resides in the way accidents are understood and how they are presented graphically. HFACS takes a more bottom-up approach to accident causation by outlining all the unsafe acts that contributed to the event in question and working back up the presumed chain of causation, detailing preconditions and latent failures Rasmussen's *AcciMap* approach advocates that to build an understanding of a complex workplace system a top-down approach, from the political and strategic decision-making levels, through company management, down to subunits and individuals, is essential. However, to visualise the way accidents, incidents or adverse events unfold, both techniques have their benefits, and it is ultimately the decision of the practitioner as to which approach may best meet their needs. If the practitioner is comfortable detailing the latent conditions that may lead to future accidents, the HFACS model may achieve a more visually pleasing result. For example, using a HFACS model as a visual representation that is easily understood may be more effective in winning over corporate support for financial investment in future incident prevention. Alternatively, if the objective is to detail the

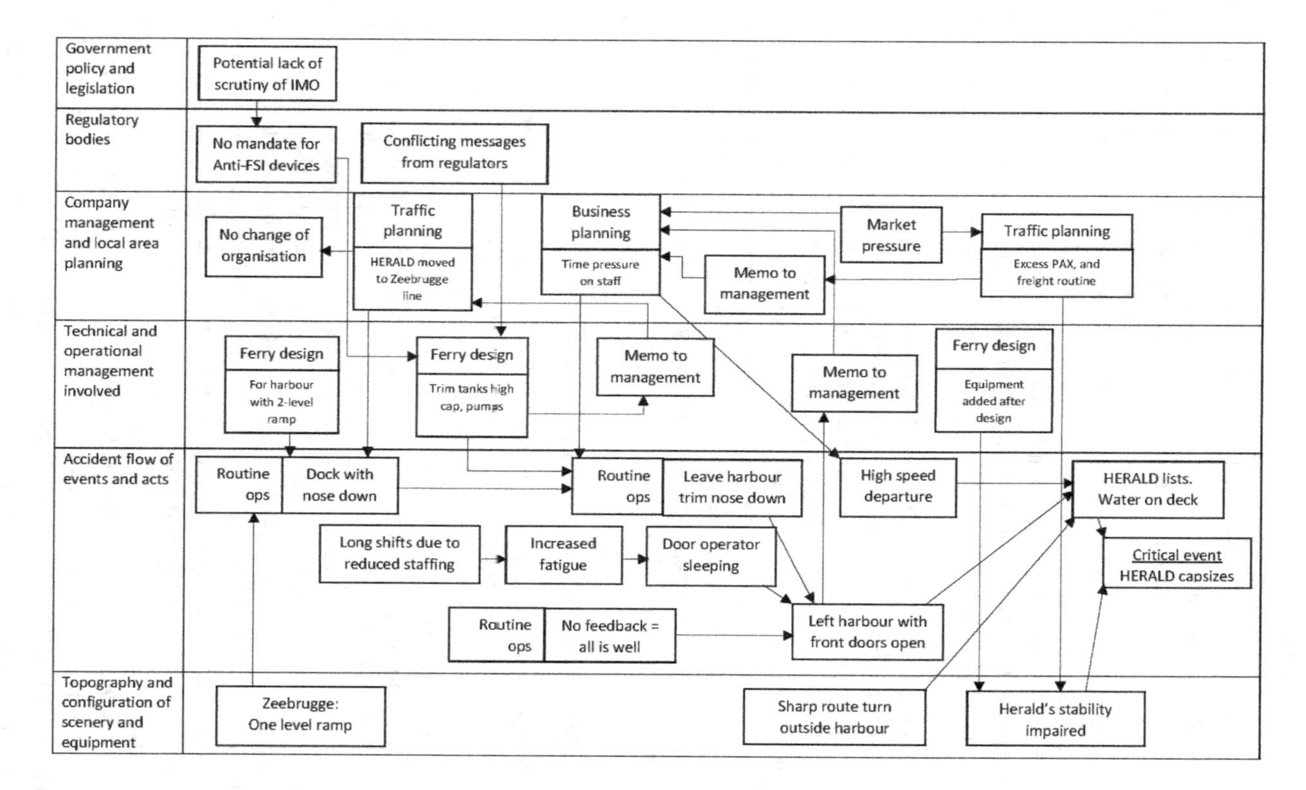

FIGURE 11.3 An AcciMap representation of the links between events at different levels of the system involved in the *Herald of Free Enterprise* capsize.

wider systems implications for risk management, policy and strategy development, and legislative lobbying, then it is likely the *AcciMap* approach may provide a more suitable representation.

AcciMap has been widely applied across a range of topics from a severe road accident in Bangladesh killing 44 people[9], to a major Australian bushfire.[10] A recent review found 23 recently published studies of accident analyses based on AcciMap.[11] The majority (54%) of factors described in these analyses were concentrated at the lower two levels of Rasmussen's six hierarchical system levels, with only 4% relating to the upper two levels. Whilst this primarily reflects the availability of evidence obtained in accident investigations, which tends to be more plentiful in relation to the actions of those closest to the accident, it is also a natural consequence of organising factors hierarchically where a smaller number of factors at a higher level (e.g. lack of resources) would be expected to have multiplying effects the further down the hierarchy you travel.

The researchers classified all the contributory factors identified in the *AcciMap* studies into 79 separate categories and then employed network analysis to explore the relationships between the identified categories of contributory factors. As noted in Chapter 10, networks consist of nodes and linkages between nodes. Network analysis identifies those nodes that have the greatest number of links with other nodes. In this case, these were factors such as 'compliance with procedures', 'violations and unsafe acts' and 'equipment, technology, and resources'. In other words, these nodes (representing these three categories of contributory factors) had the greatest influence on other nodes (factors) in the network.

FUNCTIONAL RESONANCE ANALYSIS METHOD

The Functional Resonance Analysis Method (FRAM) is a tool that embraces more recent philosophical changes to the concept of safety and, in particular, the introduction of 'resilience engineering'. In brief, Hollnagel argues that in order to understand safety and risk it is important to get an accurate view of safety-critical work tasks and then attempt to decipher how everyday variability can cascade to such an extent that an incident or accident results.[12] Many incident investigations result in a plethora of root causes and contributing factors. The FRAM can be used to work out what these factors are likely to be before the incident or accident occurs. The added benefit is that if an accurate map of a specific task is available, then in the aftermath of an accident involving that task, the investigator has a potential baseline or blueprint to draw on as they work through the investigation. Unlike the HFACS and AcciMap approaches the FRAM analysis details how a normal day of ferry operations at Zeebrugge would have occurred – not by presenting standard operating procedures in a different format but by engaging with the key personnel on that day and simply asking 'what is a normal day like for you' concerning the task of interest. A FRAM representation of the *Herald* accident is shown in Figure 11.4.

The map presented is a depiction of 'work as done' (WAD), which presents a task from a realistic point of view, instead of 'work as imagined' (WAI), which encapsulates how work *should* occur. This approach works on the premise that most

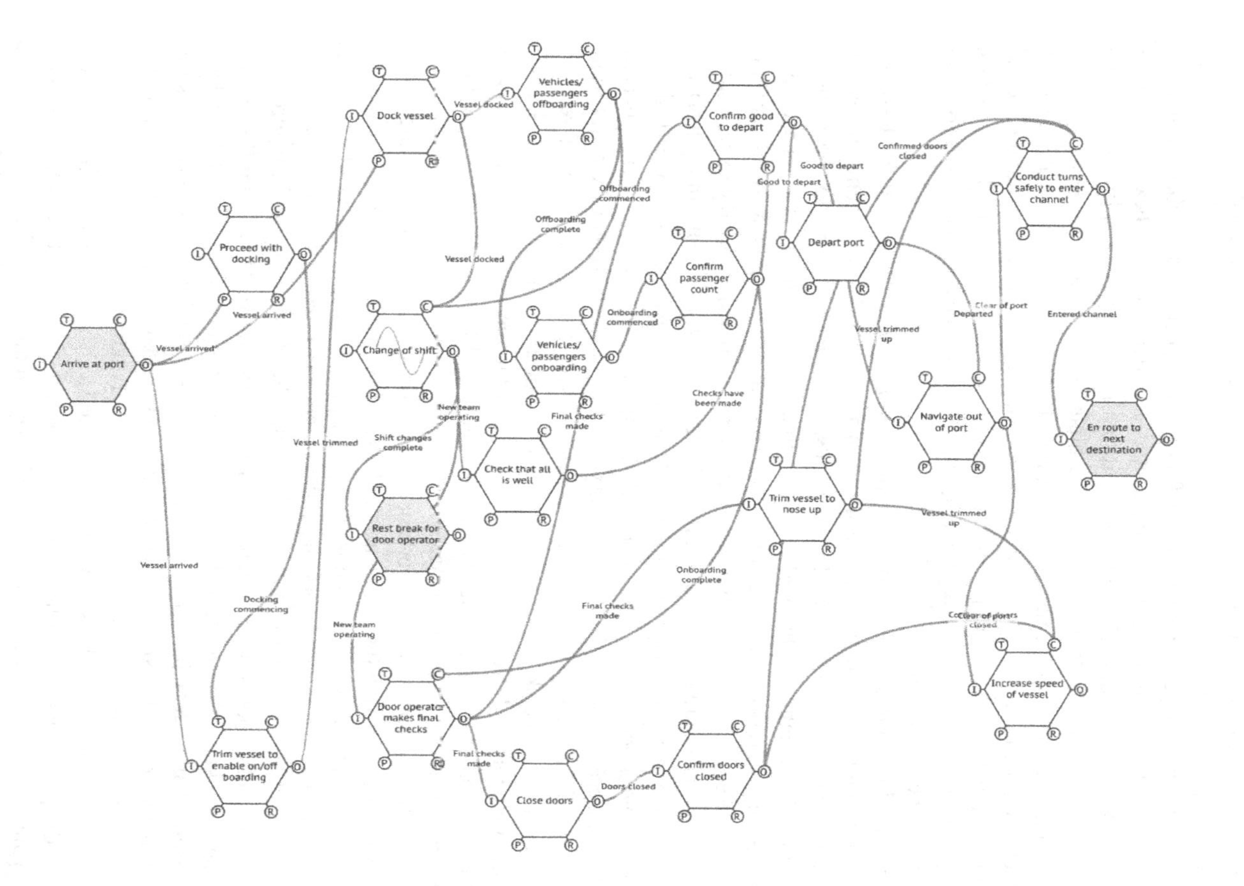

FIGURE 11.4 A FRAM 'map' of the work done in port between the arrival of the inbound ferry and departure of the outbound *Herald of Free Enterprise*.

standard operating procedures (SOP) or rules are quite different to how work tasks are actually completed. It is important to note that a map such as the one presented in Figure 11.4 would normally be created with the help of the workers doing this task, the real subject matter experts.

Each hexagon presents a sub-task (or function as officially described). Each line represents a link between each sub-task, and each letter on the hexagons details the 'nature', of how sub-tasks link up based on the content of each line – officially termed as aspects. The letters stand for (I) Input and (O) Output – which gives an indication of the flow of the task – (C) Control – how one task can have a controlling influence on the next task, (R) Resource – how one task can lead to critical information or product that is consumed in the next task, (P) Precondition – how one task *must* occur before the next task starts and (T) Time – how one task *needs* to occur before the next task starts.[13]

Whilst this approach may appear rather complex and difficult to understand, a properly facilitated process can result in increased worker engagement, insight, learning and anticipation of high-risk events. The first two refer to the value of engaging the insights of the workers who actually carry out the task on a regular basis. Many individuals thrive at the opportunity to discuss their working day, and with a little encouragement, along with assuring them that they are the experts, their engagement is assured. The first draft of the map is likely to result in a surprise at how complex even seemingly simple tasks can be. This realisation is a critical step in the whole process of developing a FRAM diagram. Most things are complicated, and we generally underestimate our ability to adapt to an ever-complex world. With a little bit of feedback from the subject matter experts, at some point, a good realistic representation of a task can be mapped out. The process is an iterative one with successive revisions and tweaks, especially with regard to the way the sub-tasks link up with one another. The outcome is a detailed picture of how tasks occur on an ideal day when there are no errors or issues.

With the map in hand, the researcher or investigator is then afforded the opportunity to work out what the outcomes might be if one of the sub-tasks were to occur differently. Variability can be imposed on the map, and from the details of what tasks follow one another and how they are linked together, a reasonably accurate assessment of the likely outcomes can be made. In the FRAM map shown in Figure 11.4, one of the sub-tasks is marked 'change of shift'. In the *Herald of Free Enterprise* accident, we know that this task did not occur, as the door operator was fatigued and overworked and asleep in his cabin. Noting how the variability in this task feeds forward into several other sub-tasks contributes to the goal of learning how variability in certain parts of the system can affect the overall outcome.

Finally, armed with the knowledge of the ways in which worker fatigue could affect the performance of other sub-tasks as well as the overall outcome, the ferry company could have anticipated what might happen and have taken steps to ensure that worker fatigue was minimised. Although, in this case, the harmful effects of fatigue could also have been anticipated from the extensive scientific research literature on fatigue and performance, the FRAM model provides the investigator with a detailed benchmark of what a normal day looks like, as well as some specific insights into the path that variability in this particular function might take on the rest of the system. From a depiction of what a normal day looks like, the FRAM provides accident investigators with a roadmap, allowing investigators to determine what variability did actually occur that was instrumental in bringing about the outcome. As with AcciMap, the FRAM can in this way become an investigative aid as well as a risk-management tool.

The main limitation of a FRAM analysis is due to the complexity of building a representation of even the simplest work process. While detailed manuals and books are available to provide guidance on creating a FRAM map, many professionals have found the required learning to be more challenging than expected, involving more of a lengthy journey rather than a quick injection of new knowledge. Not only are safety professionals then tasked with learning a complex new method of presenting a task as a FRAM map but they also need to have a good understanding of the underlying philosophy of system safety in general and resilience engineering and 'Safety II', in particular. For many safety professionals constrained by time, this may seem like too tall an order in favour of picking a simpler approach, such as HFACS.

The other limitation of FRAM is that the focus is generally on a single task in great detail. *AcciMap* and HFACS have the capability of clearly and concisely presenting the 'wider system' that may influence accident causation, whereas the FRAM model presents an 'inner system' of complexity, variability and dynamics for a single part of the system. However, aside from all the nuances of each model, they do share one common goal, and that is to describe the incident as an interconnected network of events or behaviours. Regardless of whether the practitioner is focusing on the people, the processes and the wider organisation or the legislative and regulatory framework, there is little insight offered into controls, communication and feedback loops and, most importantly, where they are absent or where they fail. One technique that does provide such insight is based on the STAMP approach developed by Nancy Leveson described in Chapter 8.

SYSTEMS THEORETIC ACCIDENT MODELLING AND PROCESSES APPROACH

The Systems Theoretic Accident Modelling and Processes (STAMP) approach has been outlined in Chapter 8. Suffice to say, however, that of the standard four model-based analysis processes that build on the STAMP theoretical approach during a product lifecycle (STPA, CAST, STPA-Sec and STECA), the *Causal Analysis based on STAMP* (CAST) processes is most relevant in this particular context. The basic view of STAMP is that incidents and accidents are caused by the inadequacies of control because of flawed processes and requirements which often are revealed through the interactions of system components. Systems components can include people, engineering activities, organisational structures and, indeed, physical components.

Again, it must be noted that, much like the other methods and models presented in this chapter, STAMP/CAST is not an investigative technique. STAMP/CAST is an analysis technique which serves to support the investigative process by helping to ensure the right questions are asked and the right information is gathered to create a comprehensive view of what happened. At the risk of causing the authors and followers of both STAMP and FRAM to baulk, there would appear to be a broad similarity between FRAM and STAMP in this respect. FRAM, itself, is not an investigative tool, but creating a map of when everything goes according to plan enables an investigator to ascertain what changed in each sub-task (function) to generate a different outcome. Otherwise, there are clear differences: FRAM focuses on relationships of activities and STAMP on relationships of system components.

The focus of the STAMP paradigm is on safety control and thus the method is used to determine where control has broken down.[14] The CAST analysis achieves this by following the five steps shown in Figure 11.5.

The first step of creating a STAMP/CAST analysis is the assembly of basic information and presentation of the incident, the hazards, the constraints and the control structures in place. By now the reader should be more than familiar with the *Herald of Free Enterprise* accident given it has been discussed using the three previous approaches and already should be able to see how some 'systems' components may be applicable. The basic hazard that is directly relevant to the *Herald of Free Enterprise* accident relates to the ingress of water through the open bow doors and the subsequent flooding of the car deck. As a result, the following constraint can be proposed:

> The vessel must not allow ingress of water as such to hinder control, threaten the public, threaten operational staff, and threaten nearby vessels.

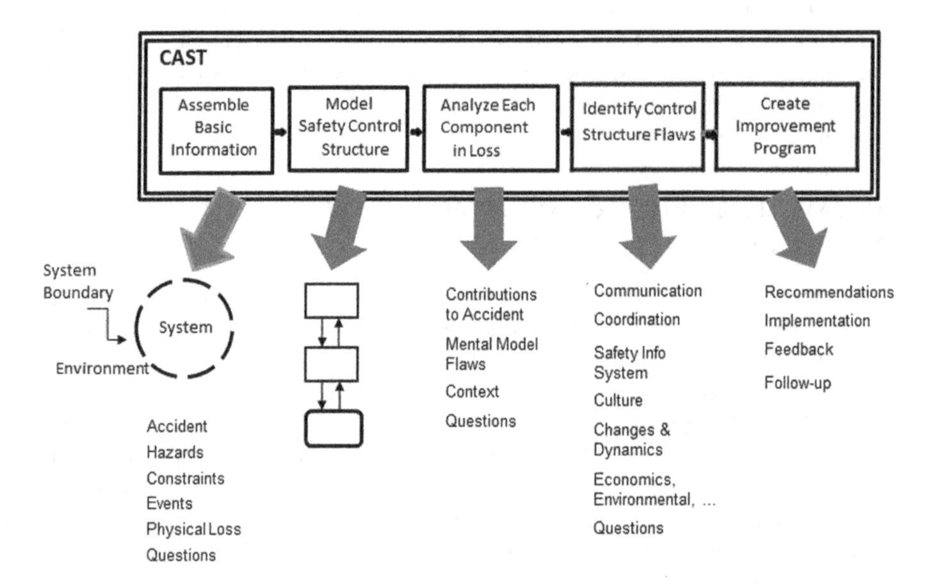

FIGURE 11.5 The five steps of a CAST analysis. ©2019. From CAST Handbook: How to Learn More from Incidents and Accidents by Nancy Leveson. Reproduced by permission of the author.

On the premise of this basic constraint, the entire structure of stakeholders and relevant assets must be sufficient to manage the required controls. The key stakeholders in the *Herald* accident are shown in Table 11.2.

The next step is to model out the relevant safety control structure. Figure 11.6 presents the basic maritime safety control structure surrounding the *Herald* accident. Note that the structure shown is unlikely to be fully comprehensive due to the time and resources normally spent constructing such a presentation but should be sufficient to illustrate the model for the reader and facilitate a comparison between the four different approaches discussed in the present chapter. Another example has been presented by Leveson, showing the structure and communication links between the various agencies and people involved in the Uberlingen mid-air collision described in Chapter 6.

The STAMP/CAST analysis involves examining each of the relationships depicted in the control structure to determine if the relevant responsibilities were fully implemented or not. Normally, significant analysis and discussion would be involved in detailing the multiple circumstances that could have contributed to the accident. In the *Herald* case, these would likely include questions concerning communications between operational staff; communications between regulator and industry; communications with employers; limitations of the vessel and so on. Once this information has been thoroughly analysed, a revised control structure could be presented, which enables the investigator to ask why these breakdowns in control occurred. The modified control structure, reflecting the functioning communication links and information flows at the time of the event is shown in Figure 11.7.

TABLE 11.2
The Stakeholders and Their Responsibilities in the *Herald* Accident.

Stakeholder	Responsibilities
International Maritime Organization (IMO)	• Ensure regulations exist to ensure egress of water does not compromise the safety of the vessel
Port of Zeebrugge	• Perform checks on the port and communicate appropriate berthing and departure instructions • Adhere to IMO regulation
P&O Ferries	• (At the time of the incident) owned European Ferries Group PLC and thus ensure that its subsidiaries were following regulation
Townsend Car Ferries	• Oversee operations to ensure compliance with regulations
Townsend Car Ferry Rostering Management	• Oversee rostering requirements to meet scheduled services
Townsend Car Ferry Operations	• Monitor operations by reporting to external stakeholders (harbour control, customers, etc.) the current and intended state of the vessel at numerous critical moments in the journey such as berthing and departing • Ensure that all internal stakeholders (crew) are cognisant of the current state of the vessel
Schichau Unterweser AG (the ferry builders)	• Ensure regulations are adhered to during construction • Identify and design vessels to meet the needs of the customers

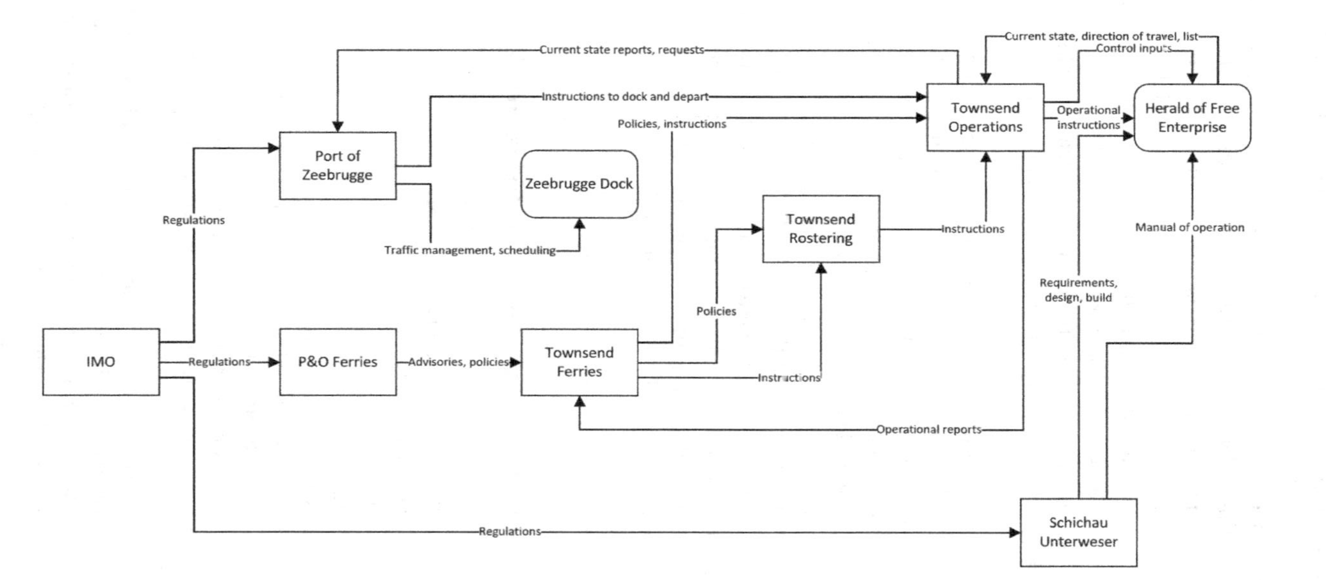

FIGURE 11.6 The Basic Ferry Operations Control Structure. Lines going to the left of a box are control lines, lines from or to the top or bottom represent information or feedback. Rectangles with sharp corners are controllers; rounded corners represent assets.

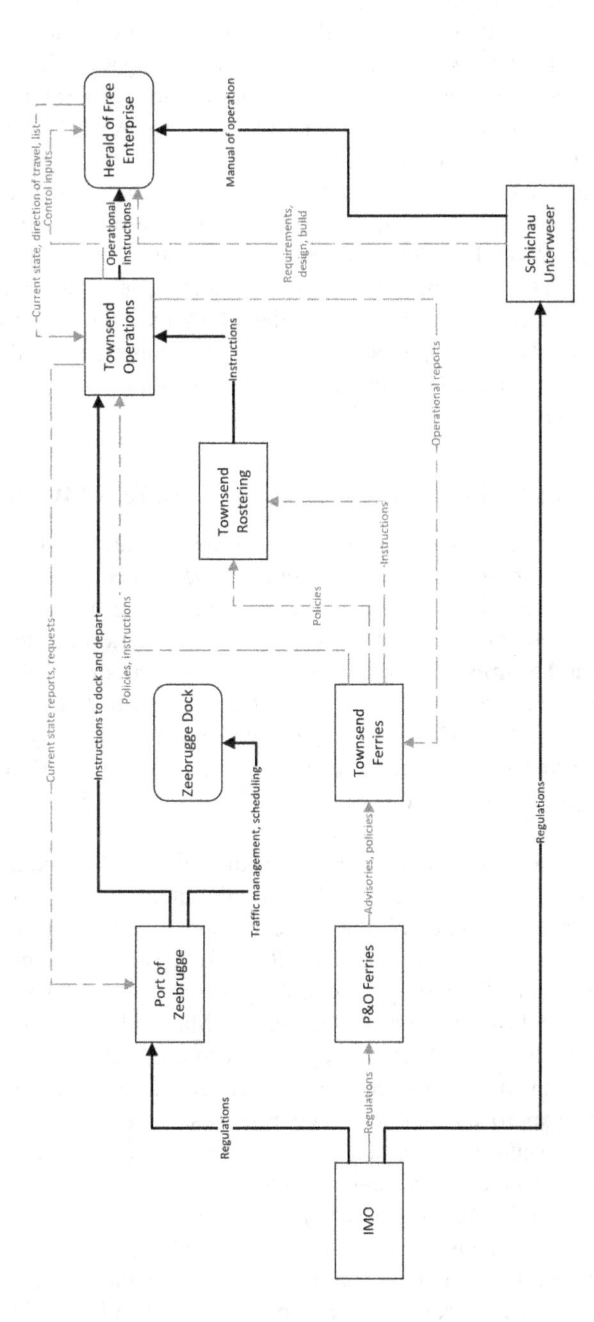

FIGURE 11.7 A revision of the basic control structure at the time of the accident. The greyed-out lines represent where the communication, control and feedback channels were compromised or non-existent at the time of the accident.

The identified flaws in the control structure may lead the investigator to identify issues with the workload, communications, fatigue risk, vessel design and scheduling. Through these flaws the investigator then has a strong foundation as to where to focus their questioning, ensuring a more robust investigation. The follow-up to which leads to the final step of creating an improvement programme ensuring changes are made to avoid repetition of the accident. The STAMP/CAST approach can be applied in more detail to each of the organisations (International Maritime Organization (IMO), P&O Ferries, Port of Zeebrugge, Townsend Thoresen and Schichau Unterweser) involved in the accident. As with the other analytical techniques, a thorough STAMP/CAST analysis can be time-consuming and labour intensive, requiring detailed knowledge of activities at every level of a complex system. Overall, this is a very brief overview of the STAMP/CAST process designed to convey the essentials of the approach and facilitate comparisons with the other approaches. The interested reader is urged to follow up with further reading before embarking on an analysis.[15]

COMPARISONS BETWEEN THE ACCIDENT ANALYSIS MODELS

Each of the four accident analysis models has been individually applied to a wide variety of events. In the past decade, several comparative studies have been reported in the scientific literature where multiple methods have been applied to a single occurrence and the outcomes evaluated against some set of criteria. One of the first was reported by **Paul Salmon** and colleagues, at that time from Monash University in Australia. They analysed an outdoor pursuits tragedy that had occurred in New Zealand in 2008 in the Tongariro national park where a professionally led outdoor activity group exploring a gorge were swept away by a flash flood. AcciMap, HFACS and STAMP were used to analyse this occurrence.[16]

These three methods have been compared in three other studies, along with FRAM in two of the comparisons as well as a handful of other techniques. In addition to the outdoor pursuits case described above, the other cases involved a road accident[17], a rail accident[18] and an oil refinery explosion[19]. A summary of the main conclusions of these four studies is presented in Table 11.3. Two of the studies recommended AcciMap first and foremost, with one study recommending STAMP and one opting for a modified HFACS approach. Overall, AcciMap and STAMP/CAST were generally viewed as the most complete approaches to conducting a systems-level analysis of an accident or occurrence. The consequent recommendations arising from each analytical approach are likely to be broadly similar with STAMP–CAST providing the most insight into communication gaps and overlaps which are prevalent in many workplace and transportation accidents.

Recent extensions to HFACS to cover upper systems levels have arguably made this a somewhat more systems-friendly approach. In addition, some of the more narrowly focused traditional analysis techniques such as Fault Tree Analysis (see Chapter 2) remain perfectly useful for many practical applications. This qualified conclusion encapsulates the current state of the art in accident analysis, which might be expressed as 'there is no one best method for conducting a systems-level analysis

TABLE 11.3
Summary of the Key Advantages and Disadvantages of the AcciMap, HFACS, STAMP and FRAM Accident Analysis Techniques.

	Advantages	Disadvantages
AcciMap	• Encompasses the whole system • Effective visual representation for communicating outcomes • Highlights contributing factors across the system • Ease of use and resource demands	• Subjective, low reliability? • No taxonomic support for classifying contributing factors
HFACS	• Strong taxonomic support • Links between levels can be analysed statistically • Evidence for reliability • Simple to use	• Lack of coverage of government/ regulatory level
STAMP	• Encompasses the whole system • Some taxonomic support • Highly structured process • Highlights contributing factors across the system	• Detailed knowledge of the whole system required • Resource intensive to learn and use • No simple graphical overview • Reliant on subject matter experts
FRAM	• Identifies functional couplings contributing to the accident • Reveals complexity of ordinary work	• Too narrowly focused to provide system-wide recommendations • Complex to use and understand

of complex accidents but it is essential to incorporate systems thinking into any analysis to avoid superficial conclusions and ineffectual remedies'.

It is also possible to incorporate some of the key concepts in systems thinking into occurrence investigation and analysis without adopting any one single method in its entirety. Some of the essential components of systems thinking have been outlined in the previous three chapters. These include the importance of feedback and control; the hierarchical order of complex systems; and the prevalence of non-linear interactions between the numerous elements of a system. At best, traditional methods of accident analysis can only deliver an approximation of the true nature of accident causation in a complex system. Where does that leave the safety manager or engineer looking to incorporate more systems-based thinking in their approach to analysing accidents and other untoward occurrences but unable, or unwilling, to invest significant time and effort in mastering an approach such as STAMP/CAST or FRAM?

SYSTEMS THINKING IN ACCIDENT ANALYSIS

Even the simplest industrial or transportation accident takes place in a social context. This is why we refer to 'sociotechnical' systems where both social (intrapersonal, interpersonal, organisational and sociological aspects) and technological (equipment design and usage) factors are involved in generating the outcomes. In the simplest occurrences (e.g. 'worker falling into hole') straightforward hazard remediation strategies may be relatively obvious and easily applied. These may also be quite sufficient to prevent recurrences at a local level. What may not be immediately apparent is whether similar occurrences have become widespread across a whole sector or activity. In this case, a broader system-based approach may be required. Accidents can have a long trajectory, with influential decisions made many years previously, such as changing the servicing frequency of key components (e.g. Alaska Airlines Flight 261, Chapter 7), gradual changes to organisational culture (e.g. NASA Shuttle Disasters, Chapter 7) or in the design of complex technology (e.g. Uberlingen midair collision, Chapter 6). Long chains of causation do not adequately capture the complexities of such influencing factors that build, interact and develop over time.

As previously discussed, systems are complex. This manifests in two forms. Most systems are composed of numerous parts arranged in a hierarchical order. This is *structural complexity.* In addition, the interactions and interrelationships between the parts can be enormously complex and this can be described as *dynamic complexity.* In the case of a simple industrial accident, the structural complexity can be revealed by charting or mapping the hierarchical arrangement of relevant organisations and agencies. A small bus company, for example, operates within a framework of both local laws and regulations and national laws and regulations, usually under the auspices of a national workplace safety body such as Occupational Safety and Health Administration (OSHA) in the United States, the HSE in the UK or *Worksafe* in New Zealand. Individuals are controlled through the entry and licencing requirements and equipment standards may be influenced by standards associations. The way in which each of these bodies discharged its functions is of potential relevance to the outcome of any analysis.

The first step in applying systems thinking to accident investigation is therefore to draw up a list, or chart, of the various organisations whose activities may be relevant to our workplace accident. These will naturally fall into some kind of hierarchy and may immediately reveal gaps, ambiguities or overlapping responsibilities in requirements, specifications, processes or operations. For example, if there were no legislative or regulatory requirements for a Safety Management Scheme (SMS) then the relevant hazard identification may have been done haphazardly or not at all. There are good reasons why mapping out the system surrounding a given event is an essential component of structured, systems-based approaches such as STAMP–CAST and AcciMap. For any workplace or transportation accident, the first step should involve a consideration of this structural complexity.

As previously noted, simple linear cause-effect models such as those proposed by Heinrich and Reason (see Chapter 3) have been found lacking in their ability to satisfactorily account for the more complex routes to failure in complex systems. If straight-line notions of cause (i.e. A causes B which causes C, etc.) are inadequate, then a more natural candidate to account for dynamic complexity in systems would be circular, in the form of *causal loops*. Nancy Leveson has pointed out that 'causal loops provide a framework for dealing with dynamic complexity where cause and effect are not related in a simple way'.[20] In other words, pretty much everywhere.

The whole subject of *system dynamics*, created at MIT by **Jay Wright Forrester** more than half a century ago, consists of discovering and representing these causal loops in complex systems. Born near Climax, Nebraska, his subsequent career might consequently have been expected to go downhill from there. In fact, Forrester initially (1940s and 1950s) made a significant contribution to the development of magnetic memory storage in computers, before later (1960s) pioneering the systems dynamics approach in the industry. He oversaw applications of systems dynamics to areas as diverse as urban growth and social policy, but it has not, for some reason, been applied in safety science until relatively recently.[21]

Causal loops are broadly similar to the feedback loops discussed in Chapter 9. Reinforcing loops describe a 'virtuous cycle' where a particular behaviour (e.g. smiling at people) leads to people behaving more nicely, which leads to more smiling and so forth. Balancing loops function much like the loop in Risk Homeostasis Theory (RHT) described in Chapter 2, where feedback acts to keep a system in equilibrium by minimising the gap between the existing state and some target state.

Causal loop diagrams can be constructed by hand or with a variety of software packages and can be used to represent complex events such as accidents. The *Herald of Free Enterprise* disaster could be described in a series of causal loops involving vessel stability, port operations and company management, for example. Sketching out the various pathways in a schematic form immediately reveals some of the multiple interacting influences between the different levels of the sociotechnical system. Laying open the inner workings of a system in this way makes it clearly apparent that simplistic 'explanations' of the accident (e.g. 'it was the fault of the bow door operator') simply fail to account for numerous other factors whose influences can be clearly seen arcing through the network of links and relationships.

Causal loop diagrams have been recently used to analyse workplace and transportation accidents. For example, researchers from Curtin University in Perth, Australia, used causal loop diagrams to analyse an industrial accident at a local waste recycling business in Western Australia.[22] A large quantity of flammable solvents stored in around 2,000 drums leaked and ignited. The actions of local and government regulators and their effects on the company were described in the form of a series of four control loops. The first loop described the effects of the regulator's inspections as effectively reinforcing the current management practices. The resulting graphical depiction provides a representation of some of the dynamic forces at work in the complex interactions between external regulators and company managers.

A more complex application of causal loop analysis to China's worst railway disaster was recently reported by researchers from the China University of Geosciences in Beijing.[23] The accident occurred in 2011 when a high-speed passenger train from Beijing to Fuzhou derailed killing 40 and injuring another 172. Three aspects of the disaster were described with causal loops. These were the design of the signal control system equipment, the regulatory approval of the equipment and the response of operating staff to the signal failure. Each of these aspects was represented by a series of causal loop diagrams and then combined to generate one very large and very complex causal loop diagram. Whilst the accident was (inevitably) blamed on the train operators, the analysis again makes plain the complex network of interrelating influences at work here and the necessity for systemic solutions to address the problems revealed in these cases.

CONCLUSIONS

Describing both the structural and dynamic complexity inherent in any accident are important steps in moving beyond a fixation with 'root cause analysis' towards truly systems-based analyses and solutions. Recent tools and techniques such as AcciMap, FRAM and STAMP have started to appear and have been taken up, at least by researchers and academics with an interest in the workings of complex sociotechnical systems. None of the four approaches that have been presented in this chapter constitute an investigative technique *per se*. Instead, they serve to support the process of investigation from slightly differing perspectives on safety and systems.

STAMP represents a natural evolution from earlier ideas about cybernetics and control, as well as the long history of systems dynamics which studies complexity in business, organisational, economic and engineered systems using causal loops as a key tool. AcciMap is based directly on Rasmussen's own influential theory of risk management, whilst FRAM derives from Hollnagel's unique perspective on safety in complex systems. Research comparisons between the techniques have not yielded any clear conclusions as to which method is superior, so whether a practitioner chooses to invest time and resources into one or another is at this stage largely a matter of individual preference based on which model seems most appropriate, least flawed, easiest to apply or which underlying philosophical paradigm of safety one finds most plausible.

Some practitioners may consider that the key to an effective safety modelling approach is associated with how robust the approach is. Other practitioners may focus more on its simplicity or timeliness. However, there is nothing that really dictates the one best method and many practitioners have favoured utilising multiple approaches or combining parts of different approaches, either at different stages of the investigative process or to present findings and recommendations in different ways to meet the needs of different audiences. For example, the general public may find the linear nature of an HFACS presentation more appealing and understandable than the complexity of a STAMP or FRAM model, to get a basic understanding of what happened.

In this chapter, we have compared a small sample of the multitude of ways currently used to understand accidents and safety. Readers are encouraged to find their own pathway to constantly learn, question and think critically both about their own preferences and about the nuances and intricacies of any accident or occurrence they are analysing. We recommend practitioners devote a bit of time and patience to continuous improvement, increasing their underlying knowledge of safety science and helping others to understand that simply blaming those humans most closely associated with an accident (invariably the front-line operators) will never be sufficient to prevent recurrences of the same incidents and accidents in the future.

At the same time, there is also a need to review and re-evaluate the current approaches to risk and hazard assessment as these are also primarily based on simple linear chain-of-effect models of causation.[24] Researchers at the Centre for Human Factors and Sociotechnical Systems in Australia have found at least 35 different hazard assessment techniques cited in the literature, falling into seven distinct groups. One group covered fault tree methods: one HAZOP-based methods, one human-reliability methods and so forth. As discussed earlier, the focus of all these approaches is mainly on chain-of-event models of causation and the actions of the front-line operators and are consequently much less sensitive to risks and hazards elsewhere in the system.

Only three of the 35 techniques could be said to have been derived from systems theory and these included FRAM and another STAMP-related technique (STPA). The third was a little-known extension of quantitative risk assessment methods known as 'SoTeRiA'. There is as yet very limited evidence as to the efficacy and value of these approaches in practical application. Legislative and regulatory requirements largely dictate the methods that are accepted and utilised in practice, and such requirements are very slow to change. There is a clear need for the reliability and validity of systems modelling approaches to be established to provide the evidence base upon which regulatory and organisational changes can be made. Meanwhile, many practitioners of safety will be content to plough the same familiar furrows, but for those that choose to be more reflective and self-critical on current practice, and on the assumptions that underlie the approaches they use, the safety journey that was launched in Chapter 1 has only just begun.

NOTES

1. The International Civil Aviation Organization. (2015). *Manual of Aircraft Accident Investigation, Part 1 Organization and Planning*. Montreal, QC: International Civil Aviation Organization.
2. Wagenaar, W.A., & van der Schrier, J. (1997). Accident analysis: The goal and how to get there. *Safety Science, 26*, 25–33.
3. Katsakiori, P., Sakellaroupoulos, G., & Manatakis, E. (2009). Towards an evaluation of accident investigation methods in terms of their alignment with accident causation models. *Safety Science, 47*, 1007–1015.
4. Lundberg, J., Rollenhagen, C., & Hollnagel, E. (2009). What-You-Look-For-Is-What-You-Find – The consequences of underlying accident models in eight accident investigation manuals. *Safety Science, 47*, 1297–1311.
5. Chen, S.T., Wall, A., Davies, P., Yang, Z., Wang, J., & Chou, Y.H. (2013). A human and organisational factors (HOFs) analysis method for marine casualties using HFACS-maritime accidents (HFACS-MA). *Safety Science, 60*, 105–114. http://dx.doi.org/10.1016/j.ssci.2013.06.009
6. Hulme, A., Stanton, N.A., Walker, G.H., Waterson, P., & Salmon, P.M. (2019). Accident analysis in practice: A review of human factors analysis and classification system (HFACS) applications in the peer-reviewed academic literature. In *Proceedings of the Human Factors and Ergonomics Society 2019 Annual Meeting*, 1849–1853.

7. For example: Federal Aviation Administration. (2016). AV-2016-013: Enhanced FAA oversight could reduce hazards associated with increased use of flight deck automation. *FAA – Office of Inspector General Audit Report*, 7 January 2016.
8. Kirwan, B. (2001). Coping with accelerating socio-technical systems. *Safety Science, 37*(2), 77–107. Kirwan has been both an academic (at Birmingham University) and practitioner, first in the UK nuclear industry, and currently as safety culture programme manager at Eurocontrol.
9. Hamim, O.F., Hoque, S., McIlroy, R.C., Plant, L., & Stanton, N.A. (2019). Applying the AcciMap methodology to investigate the tragic Mirsharai road accident in Bangladesh. *MATEC Web of Conferences, 277*, 1–7. https://doi.org/10.1051/matecconf/201927702019
10. Salmon, P.M., Goode, N., Archer, F., Spencer, C., McArdle, D., & McClure, R.J. (2014). A systems approach to examining disaster response: Using AcciMap to describe the factors influencing bushfire response. *Safety Science, 70*, 114–122.
11. Salmon, P.M., Hulme, A., Walker, G.H., Waterson, P., & Berber, E. (2020). The big picture on accident causation: A review, synthesis and meta-analysis of AcciMap studies. *Safety Science, 126*, 1–15.
12. Hollnagel, E. (2012). *FRAM: The Functional Resonance Analysis Method.* Farnham, UK: Ashgate.
13. Hollnagel, E., Hounsgaard, J., & Colligan, L. (2014). *FRAM – The Functional Resonance Analysis Method – A Handbook for the Practical Use of the Method, 1st Ed.* Retrieved November 2016 from: http://functionalresonance.com/onewebmedia/FRAM_handbook_web-2.pdf
14. Leveson, N. (2019). *CAST Handbook: How to Learn More from Incidents and Accidents.* Downloaded from: http://sunyday.mit.edu/CAST-Handbook.pdf (mit.edu)
15. The CAST analysis of the Uberlingen collision is described in: Leveson, N. (2011). *Engineering a Safer World.* Cambridge, MA: MIT Press. See Note 14 above for a more recent guide to the application of CAST.
16. Salmon, P.M., Cornelissen, M., & Trotter, M.J. (2012). Systems-based accident analysis methods: A comparison of Accimap, HFACS and STAMP. *Safety Science, 50*, 1158–1170.
17. Stanton, N.A., Salmon, P.M., Walker, G.H., & Stanton, M. (2019). Models and methods for collision analysis: A comparison study based on the Uber collision with a pedestrian. *Safety Science, 120*, 117–128.
18. Underwood, P., & Waterson, P. (2014). Systems thinking, the Swiss Cheese Model and accident analysis: A comparative systemic analysis of the Grayrigg train derailment using the ATSB, AcciMap, and STAMP models. *Accident Analysis and Prevention, 68*, 75–94.
19. Yousefi, A., Hernandez, M.R., & Pena, V.L. (2018). Systemic accident analysis models: A comparison study between AcciMap, FRAM, and STAMP. *Process Safety Progress, 38*(2), 1–16.
20. Leveson, N. (2020). Safety-III: A systems approach to safety and resilience. *MIT Engineering Systems Lab.* Available for download at: Safety III: A Systems Approach to Safety and Resilience (mit.edu)
21. Sterman, J.D. (2000). *Business Dynamics: Systems Thinking and Modelling for a Complex World.* Boston, MA: McGraw-Hill.
22. Goh, Y.M., Brown, H., & Spickett, J. (2010). Applying systems thinking concepts in the analysis of major incidents. *Safety Science, 48*, 302–309.

23. Fan, Y., Li, Z., Pei, J., Li, H., & Sun, J. (2015). Applying systems thinking approach to accident analysis in China: Case study of "7.23" Yong-Tai-Wen High-Speed train accident. *Safety Science, 76*, 190–201.
24. Dallat, C., Salmon, P.M., & Goode, N. (2019). Risky systems versus risky people: To what extent do risk assessment methods consider the systems approach to accident causation? A review of the literature. *Safety Science, 119*, 266–279.

Index

J

K

L

M

N

O

P

Q

R

T

U

V

W

Y

Z